Proceedings of the 2015 International Conference on

Food Hygiene, Agriculture and Animal Science

Proceedings of the 2015 International Conference on

Food Hygiene, Agriculture and Animal Science

14–15 November 2015
Wuhan, Hubei, China

Edited by
Hongbiao Ding
Chinese Academy of Agricultural Sciences, China

World Scientific

NEW JERSEY • LONDON • SINGAPORE • BEIJING • SHANGHAI • HONG KONG • TAIPEI • CHENNAI • TOKYO

Published by

World Scientific Publishing Co. Pte. Ltd.
5 Toh Tuck Link, Singapore 596224
USA office: 27 Warren Street, Suite 401-402, Hackensack, NJ 07601
UK office: 57 Shelton Street, Covent Garden, London WC2H 9HE

British Library Cataloguing-in-Publication Data
A catalogue record for this book is available from the British Library.

FOOD HYGIENE, AGRICULTURE AND ANIMAL SCIENCE
Proceedings of the 2015 International Conference on Food Hygiene, Agriculture and Animal Science

ISBN 978-981-3100-36-7

Printed in Singapore

Preface

The 2015 International Conference on Food Hygiene, Agriculture and Animal Science (FHAAS 2015) is held in Wuhan, China from November 14 to November 15, 2015. The conference aims to provide an excellent international academic forum for all the researchers, practitioner, students and teachers in related fields to share their knowledge and results in theory, methodology and application on nutrition and food hygiene, agriculture and animal science. FHAAS2015 features unique mixed topics of nutrition and food hygiene, agriculture and animal science.

FHAAS2015 proceeding tends to collect the most up-to-date, comprehensive, and worldwide state-of-art knowledge on materials and its application. All accepted papers have been submitted to strict peer-review by 2-4 expert referees, and selected based on originality, significance and clarity for the purpose of the conference. The conference program is extremely rich, profound and featuring high-impact presentations of selected papers and additional late-breaking contributions. We sincerely hope that the conference would not only show the participants a broad overview of the latest research results on related fields, but also provide them with a significant platform for academic connection and exchange.

The Technical Program Committee members have been working very hard to meet the deadline of review. The final conference program consists of 38 papers divided into 3 sessions. The proceedings would be published in a volume by World Scientific Publishing Company.

We would like to express our sincere gratitude to all the TPC members and organizers for their hard work, precious time and endeavor preparing for the conference. Our deepest thanks also go to the volunteers and staffs for their long-hours work and generosity they've given to the conference. The last but not least, we would like to thank each and every of the authors, speakers and participants for their great contributions to the success of FHAAS2015.

FHAAS2015 Organizing Committee

Committee

Prof. Zhengchao Ren, Associate professor, Gansu Agricultural University, China

Prof. Zhen ZHANG, Associate professor, Wuhan University of Light industryInstitute of Economics and Management, China

Dr. Hongbiao DING, Professor, The Chinese Academy of Agricultural Sciences Institute of Feed, China

Prof. Libin WEN, Director, Jiangsu Academy of Agricultural Sciences Veterinary Institute, China

Dr. Xiaohua HE, Professor, Foodborne Toxin Detection & Prevention Research Unit,USDA, Agricultural Research Service, America

Prof. Xuefeng YU, Associate professor, Shenzhen Institute of Advanced technology, Chinese Academy of Sciences, China

Prof. Gan ZHAO, Associate professor, College of Life Science, South China Agricultural University, China

Prof. Zhonglin TANG, Associate professor, Beijing Animal husbandry and Veterinary institute, Chinese Academy of Agricultural Sciences/Shenzhen Agricultural Institute for Genomic Research, China

Dr. XU Xilin, doctor/associate professor, South China University of Technology, China

Dr./Prof. Nor Ainy Mahyudin doctor/associate professor, Universiti Putra Malaysia, Malaysia

Associate Professor/Dr. Amiza Mat Amin, doctor/associate professor/Deputy Dean, Universiti Malaysia Terengganu, Malaysia

Dr./Prof. Panuwat Suppakul, Associate Professor/Ph.D., Kasetsart University, Thailand

Prof. Irene DINI, professor, Federico II - Pharmacy Department, Italy

Contents

Session 1: Nutrition and Food Hygiene

Research on construction of the quality and safety of agricultural products traceability based on multisided platform-taking beef quality and safety traceability in Xinjiang as an example

Shihong Liu and Tao Ma
Agricultural Information Institute, Chinese Academy of Agricultural Sciences; Agricultural Information Services Laboratory, Beijing
E-mail: liushihong@caas.cn

China has carried out of a lot of useful exploration in the quality and safety of agricultural products traceability, and developed a number of management platforms, but there are still many problems, for example supply and demand sides of information asymmetry, transaction costs and risks are too high, poor circulation products and other regulatory agencies and other parties can't effectively participate. This article from the perspective of the theory of the multisided platforms (MSPs), in the analysis of the important features of the MSPs on the basis of the theory, combined with the quality and safety of agricultural products traceability features, discussed on quality safety of agricultural products traceable method of constructing a multisided platform, particularly designed of the path multi-stakeholder participation, and analyzed its characteristics. It is of great significance for agricultural traceability information sharing to achieve symmetry, reduce transaction costs and risks, and the realization of smooth flow products.

Keywords: Multisided Platform; Traceability; Quality and Safety; Platform Designing

1. Introduction

MSPs theory developed and extended from the bilateral market theory. This is different from traditional markets, which simply analyze the buyer and the seller. MSP is "platform" as the core, and provides a platform for participants to interact with a variety of customer connection to a profit.

Currently, domestic seen in newspapers MSP articles rare, and they are only in the theoretical stage, related article which is positive barely seen. Meanwhile, most of the agricultural markets are the single market, Multisided market almost nothing, bilateral market is also very rare.

China's agriculture MSP is still in its infancy, there are only a few examples of bilateral platform, such as the Chinese food network, Chinese chili network, and Chinese apple net. However, most of these platforms are for the same merchandise category providing services. In addition many provinces and cities have set up a special website for the local agricultural information, but the functionality of these platforms have significant limitations, Tend to specialize in publishing products [1], and lack of appropriate support services, such as legal advice, professional logistics and distribution, and so on. Most farmers also lack of size, or lack of incentives to participate, and the threshold is too high to registered, they abandon the platform services. Therefore, a successful example of regional units agriculture MSP is rare.

Situation of China's current agricultural multisided market mainly as follows: First, agricultural products accounted for a large share of the market, but almost no platform; Second, there is no third-party platform to ensure quality; Third, the vast majority of suppliers are individuals and small organizations, few large supply organization; Fourth, there is no single market technical standards, management structure, methods of payment, etc. [2].

Foreign MSP theory developed rapidly. The main reason is that the entity MSP has achieved rapid development in recent years, and the scale expanded rapidly. Most of them have taken the lead in their industry. Successful MSPs create tremendous value by reducing search costs or transaction costs, and achieve dominance by economies of scale [3]. Currently MSPs theory research, mostly focus on information products, IT applications, such as the media industry, terms of credit cards and other payment systems.

2. The Theory of MSPs

2.1. *The Theory of MSPs*

Through technology, products or services provide direct interaction for two or more customer groups thus create value, this characteristic mainly displays in "cross-side network effects".

Cross-side network effects (also be known as indirect network effect) is an important characteristic of MSPs, that is, the value which obtained by one of the platform participants expand dramatically along with the surge of other participants. Such as different kinds of electronic commerce. The more buyers they have, the more value sellers will get. However, MSPs has its duality. On the one hand, It is hard to join in the platform when one participant ignore the entrance from the other side. This is the hardest difficulty to overcome. On the other hand, the establishment of such a platform often can obtain high barriers, that is why so many MSPs occupied a great advantage in their respective fields. But the Cross-side network effects cannot serve as a good guarantee to access barriers. The platform, only by charging a certain percentage cost, higher cost to the one who need to transfer can keep high barriers, and resist new competitors. In addition, to serve customer or participant, each marginal transaction cost and the marginal cost will reduce along with the increase of the cost amount. Scale economy is the characteristic of most MSPs, Therefore, The barriers will be heightened obviously by the economies of scale.

2.2. *The Conditions of MSPs*

If an organization wants to develop itself into platform enterprises, they need to have the following four conditions:

Firstly, they should have an open platform for carrying supply operating entity and demand operating entity.

Secondly, the platform operator must have the ability, resources and money to gather both supply side and demand side or many industry chain parties. In most cases, they can rely on the core value of this opening platform, the core resources, or special policy of this enterprise, they must have one advantage as three mentioned above. But no matter how, strong financial and brand strength is a must.

Thirdly, the platform operator must have a set of scientific operation and management mechanism. This is a complex system to some extent, it is a "virtual government" which corresponds to the real society, but has stronger uncertainty and more emerging affairs.

Finally, the enterprise itself must find a sustainable way to make profit so that to ensure sustainable development of the platform.

2.3. *The difference between the MSPs trade framework and traditional trade framework*

Picture 1 MSPs trade framework and traditional trade framework have fundamental differences (Pic.1). First, in a sense, every party in the platform is a customer of this MSP, the information publishing and sharing right is equal. Second, the MSP can provide the "edge" direct connection. The product platform mode was shown in picture1 (middle) missing the first demand, that is the end user is not the object of product platform service, so the product package which provided by product platform still exist errors to meet end customers' actual demands. Agency pattern (as shown in picture1the right pattern) is missing the second requirement: both sides have no direct connection, for the product supplier, the demand of the market reaction speed is low, it caused more difficulties to improve product timely according to the need of the end user.

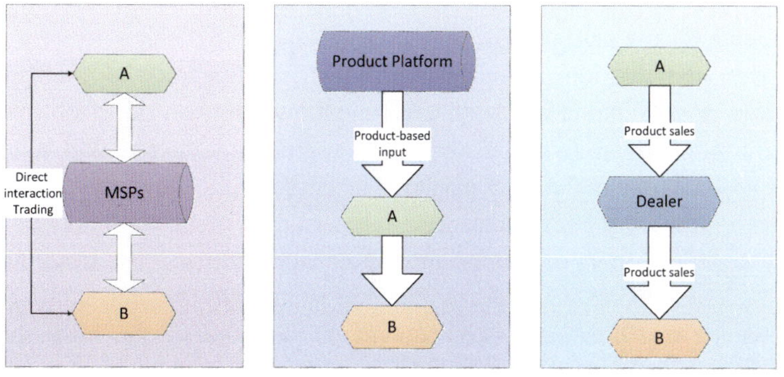

Fig. 1 Multisided trading framework platform compared with traditional trading framework

3. Platform Design

3.1. *Participants Design*

Participants design platform, the platform is the most important issue to establish encountered. In general, participants better. However, increasing the number of participants will also have to coordinate different demands, restrict innovation MSP and other issues. But most of the platform at the beginning will choose as few participants, and then gradually developed, gradually extended to more participants in order for them to provide more services.

In this the "unilateral" to "multisided" process, the most important point is to find a new party (or parties), so that you can, and one that already exists to create a strong indirect network effects. This is primarily in two steps: First, the function first. Identify new participants existing business, we are able to "be served by" providing new services. Second, the benefits first. Benefits in the expansion process of the MSP, the new party must give the existing parties, this can be accepted. For daily trading conditions currently existing "are service groups", the new service participants, whether to reduce costs and increase revenue. And to strengthen the indirect network effects, such as increased effectiveness and efficiency of existing parties.

Quality and safety of agricultural products traceability MSP of information sharing system using a typical platform model, in the form of membership or alliance will produce suppliers, brokers, consumers, government regulators and other contact advertisers to the platform around the quality and safety of agricultural products traceability system, information sharing principle, to establish a database group. On agricultural information, transaction information, quality information, traceability information, updated in real time by the parties and sharing, increase user viscosity by increasing the frequency of specific measures to deal, firmly dominant MSP, to create a strong cross-network effects.

The original main platform middlemen and suppliers, to ensure orderly service, government regulators must be involved in this transaction. Also, because of the participation of regulatory authorities, parties involved in the transaction platform will be more confidence for the platform.

Advertiser participation is recognition of the value to the platform. Platform developed to a certain extent and with a certain flow after, on the platform of the party would benefit more advertisers. Because brokers, product suppliers will have advertising needs. This will make some quality participants come to the fore, and other stakeholders easier to find high-quality trading partners.

Quality certification is an important aspect of deepening the development of the platform. The key is to see whether the function is suitable for "decentralization", that is performed by the platform parties themselves or executed by the MSP. In the current system of our existing quality certification basically using government "focus" certification. Evaluation of the quality and safety of agricultural products quality and safety traceability platform, it is recommended to slowly take "decentralization" process, while reducing costs, while play information platform and other advantages.

3.2. *The design principle of MSPs*

3.2.1. *Management principles*

In essence, MSP is to provide a place, for all third parties to interact, and thus create value. Without restrain the third party's behavior, will affect the ecosystem of the MSPs.

Non-price management principles of the MSPs mainly have the following two aspects, one is platform access rules, the qualification to enter platform has strict limits. The second is to establish platform interaction principles to decide the limit and duty of MSPs participants.

For different MSPs, the degree of management is different. For the MSPs which perform strict management rules, the quality of platform participants is more important than the quantity. Because quantity is too large, and the quality is poor, high quality resources tend to have "crowd out" or “drown” effect. This will generally occurs in the late stage of MSPs development. But in the early stage of the MSPs, the management rules are relatively loose. At this time, it is more important to expand the number of one party's participants. Therefore, the agricultural products quality and safety traces to the MSPs management principles. A relatively loose management principles were carried on in the early stage, then tighten the principles in the middle late stage.

3.2.2. *Pricing principles*

Most MSPs have no charge or offer price subsidies for at least one participant. Taobao provides free paticipation rights for shops, which is a typical case. This helps to attract the "seller" to join in the platform quickly, when the "shops" expand to a huge size, the attraction to customers increased then customers flock to the shop.

Hense, MSPs in charge, and benefits system, tend to follow the following several principles: First, if there is a pricing deal, the recipient is the one who get more value and incomes; Second, if there is no pricing deal, the recipient is the one who obtain more benefits; Third, for every participant, charge lower fees for those who are sensitive for price and charge higher fees for those who are insensitive for price.

4. The quality and safety of agricultural products traceability MSP model

In this paper, the characteristics of Xinjiang as a case of beef cattle, build quality and safety of agricultural products traceability MSP framework model. By implementing the platform can be achieved traceability information sharing, reduce transaction costs and risk, achieve smooth purpose product distribution channels. With the gradual improvement of the future platform, and other products through traceability platform interoperability, open information channels to achieve suppliers, consumers and governments win-win situation.

Quality and safety of agricultural products traceability MSP model is shown in Figure 2.

This MSP provides consumers with quality assessment information, logistics and information services. Suppliers of products: farmers, farm can interact directly with consumers through the platform, and can also produce vendor connectivity platform. The production process can chain flat by introducing brokers, processors, so that the production can be easier to track and trace. The introduction of government regulators to expand the dimensions of functionality and information platform, reducing regulatory costs and make products more credible. User feedback is also timely to reflect consumer

MSP by tracing process. Advertisers can serve as a platform for future profitable expansion interface.

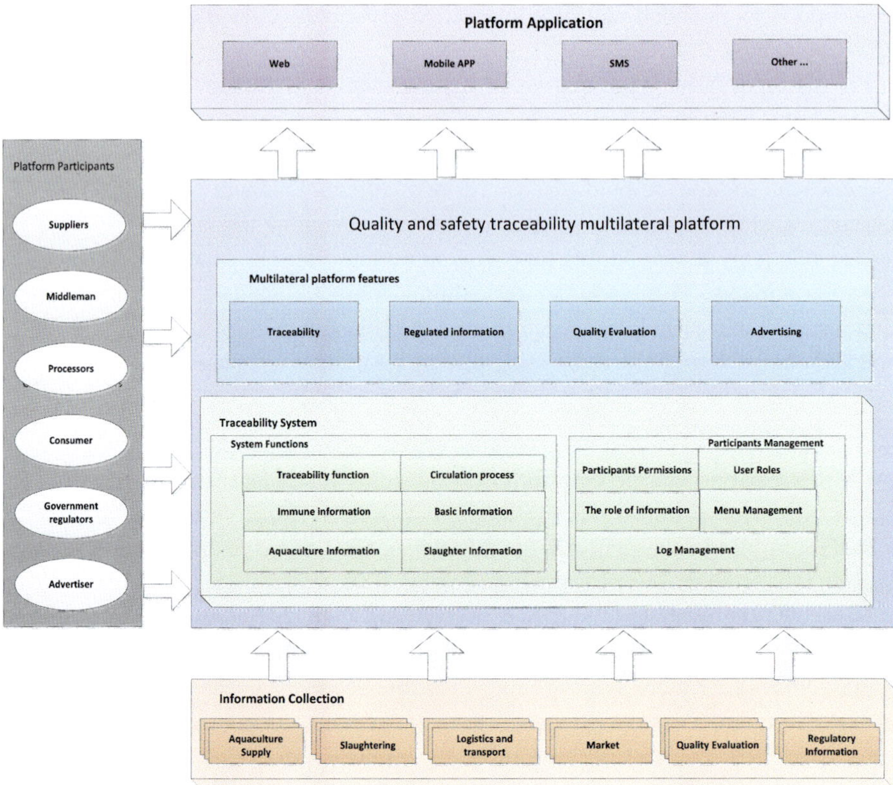

Fig. 2 MSP model quality and safety of agricultural products traceability

5. Questions and Discussion

5.1. *Multisided involved in the design*

Our original traceability platform base suppliers and consumers on both sides included, expanded to have a product supplier, brokers, consumers, government regulators, advertisers and other parties involved in the internet. New participants to join, as a platform for the original participant provides a better income, by building "multisided" to make the platform more deepen operational. This will not only increase the frequency and efficiency of trading participants, but also increase the user bonding of participants.

5.2. *Try "decentralization" policy*

Deepening platform development through quality certification traceability platform, reducing asymmetric information because search costs and brought a crisis of confidence. The key is to achieve quality certification "access" policy; substandard products will lose power on the platform to continue to operate, to ensure healthy and sustainable development of the market. The platform attempts to "centralization" evaluation mechanism, namely the quality of the evaluation provided by the consumer, quality supervision performed by the regulatory authorities.

5.3. *Government regulatory role irreplaceable*

The government regulators platform in the role cannot be ignored. During product testing, information and services, traceability processes, etc. have a strong binding. Government play the role of supervision as a party to the platform to make the platform more credible, beneficial to achieve the combined effect of the MSPs.

5.4. *Benefit expand*

As a MSP for advertisers to reserve a interfaces for advertisers to participate, not only to provide a larger space for the subsequent development platform, but also reduce search costs for consumers, it could be one aspect of the benefits platform expansion.

5.5. *Discussion*

This article is based on the analysis of MSP theory, and put forward some thoughts on the frame, and the practical application of the promotion are still a long way to go. The next lot of theory and technology research, such as how to avoid competition of a party to the MSP is too fierce, leading to unprofitable, weakening its development of quality products or services of incentive to invest; and how in the case of multi-stakeholder, holding platform well run, running and what can be optimized etc.

References

1. Sun Rui, Wen Xiaoqing. Construction and agricultural bilateral market platform of the pricing mechanism [J]. Commercial Times, 2013, (34):31-32.
2. Wang Shuo, Liu Shihong, Zheng Huoguo, Hu Haiyan, Ma Tao, Chen Tao. Xinjiang Beef Traceability System Research and Implementation [J]. Journal of Anhui Agricultural Sciences, 2013, (26):10856-10859, 10861.
3. Andrei Hagiu. Winning multisided platform [J]. Board of Directors, 2014, (2):084-087.
4. ANDREI HAGIU. Strategic Decisions for Multisided Platforms [J]. MIT Sloan management review, 2014, 55(2):71-80.

Construction on food safety traceability chain

Huo-guo Zheng[1], Hai-yan Hu[1], Shi-hong Liu[1]

[1]*Key Laboratory of Agro-Information Service Technology, Ministry of Agriculture; Agricultural Information Institute, Chinese Academy of Agricultural Sciences, Beijing 100081*

E-mail: zhenghuoguo@caas.cn

Abstract: The very important step of food traceability system construction is to figure out what information needs to be traced, and in what way to organize and manage the quality & safety information. The normal way is based on the perspective of information systems development. In food traceability chain view, the quality safety factors from origin to consumption were integrated up management. This is a new way of the traceability system building, which is food traceability chain. In this paper, the food chain was decomposed and refined. The hierarchical model of food safety traceability chain was proposed. Finally, we analyze the granularity of the food safety traceability chain. The establishment of food safety traceability chain is not only the foundation of food hazard analysis, quality control, risk analysis, but also the basis for food safety traceability system construction.

Keywords: Food; Quality and safety; Traceability chain; Traceability system.

1. Introduction

In recently years, some major food safety incidents, such as BSE (bovine spongiform encephalopathy), foot and mouth disease, avian influenza and bacillus coli contaminated food occur frequently around the world. These food safety incidents greatly endanger human life and health, but also serious damage to the economic development of all countries, and consumers are less confidence of food safety but pay more attention to the food origin. In response to food safety issues, food safety traceability system is considered to be an effective solution from farm to table for food quality and safety, which is a tool for food safety information disclosure, and monitor the harmful element of the food chain from production to consumption [1].

Originally, food safety traceability system was established in developed countries mainly for beef and other animal products. In years of 2000, European Union has released (EC) 1760/2000 decree requiring beef corporations to establish traceability systems in the EU and its Member States. Cattle identification system was developed in 2002 in Canada in order to achieve beef traceability [2]. In 2009, the United States launched NAIS (National Animal Identity System) project, to ensure the traceability of beef. Australia also established NALS (National Livestock Identification Scheme, NLIS) for national livestock tracking and tracing system since 2001 [3]. Experience in these countries show that an effective traceability system can enhance their national food safety regulatory level, concurrently can improve requirements on imported food.

Compare with foreign country, the study and application of food safety traceability system in China has a late start, but has also made great achievement. April 2004, the State Food and Drug Administration, Ministry of Agriculture and other eight units jointly launched a comprehensive food and drug safety project, the State Food and Drug Administration and other departments started meat traceability system project not far behind. Up to now, there are several food safety traceability systems established by government [3]. Meanwhile, lots of universities and research institutions have carried out research in food traceability technology and system with the help of the national science and technology plan and the provinces funding. A series of traceability system have been studied, including animal products [4-6], vegetables [7-8], fruits [9], grain and oil products [10], aquatic products [11] and bee products [12].

In this paper, the food chain is analyzed, which is divided into four levels, which are the trace elements, trace elements, trace elements and trace elements.

2. Food chain analysis

In the last few years, the research of food safety traceability system is mainly based on the theory of information system development, which generally includes the following technologies: identification technology, encoding technique, system development technology. Although traceability system is an information processing system, it has its own characteristics. We believe that the core of traceability system construction includes: how to clarify the quality and safety factors from food production to food consumption, and to provide information services for producers, regulators and consumers by means of information technology. Food safety traceability chain collect the participants, the quality and safety factors from the beginning of raw materials planting or breeding to final consumption, which laid a good foundation for the food hazard analysis, quality control, risk analysis, but also for the easy job for food safety traceability system construction.

The food chain is defined as "sequence of the stages and operations involved in the production, processing, distribution, storage and handling of a food and its ingredients, form primary production to consumption" in ISO22000:2005. The definition indicates that the food chain includes the production of feed for food-producing animals and for animals intended for food production. The food chain also includes the production of materials intended to come into contact with food or raw materials.

According to the definition of food chain, we figure out the food production and processing process diagram, including the plant source of primary agricultural products, animal derived primary agricultural products, feed products, as well as multi-component food (shown as Fig. 1).

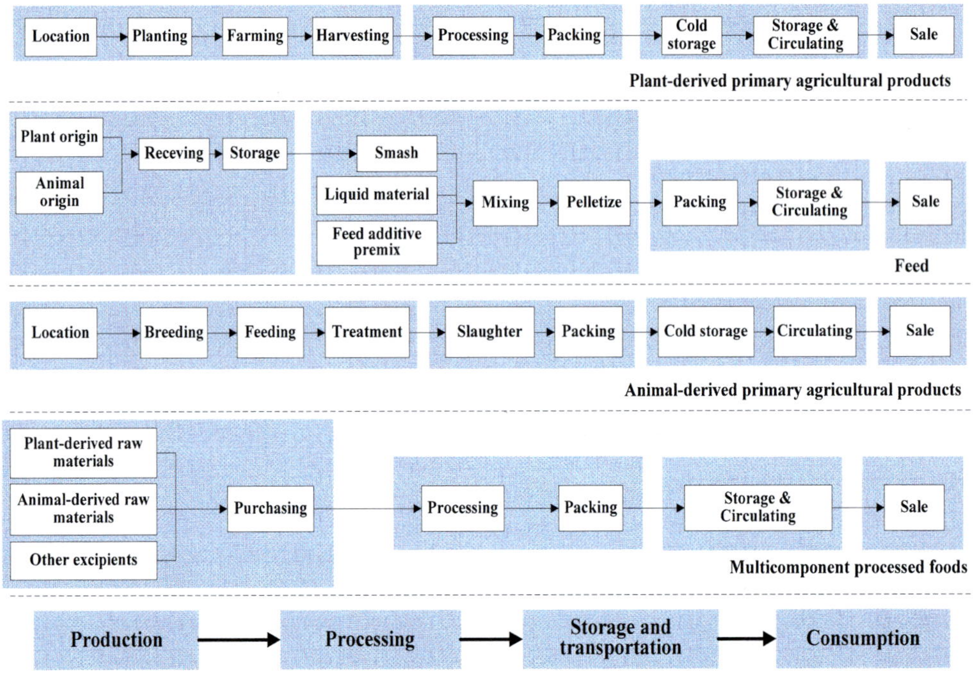

Fig. 1 Flow of food production to consumption

Plant derived primary agricultural products include fresh vegetables, fresh fruits, cereals, oil crops, edible mushrooms, etc. Although there is a long period of time from planting to harvest for agricultural products, but from the perspective of quality and safety, the traceability process involved is relatively small.

However, animal derived primary agricultural products, including livestock and poultry products, aquatic products, etc., compared with the plant derived primary agricultural products; there are more traceability process and more quality and safety factors from the breeding to the consumption.

In terms of multi-component food, from production to consumption generally need to go through food processing, packaging, storage, transportation and sales steps. Because the raw material of the multi-component food may be plant derived primary agricultural products, it may also be animal derived primary agricultural products, and the quality and safety factors involved are more complicated.

3. Components of food safety traceability chain

Generally speaking, the food traceability chain can be divided into four levels: traceability links, traceability units, traceability factors, and traceability indexes. From raw material production to consumption, there are a lot of steps which can be summarized into a number of links, each specific link can be divided into a relatively independent unit, and each unit contains a lot of quality and safety factors, for a specific factor, contains a number of indexes.

3.1 Links in the traceability chain

From raw material production to consumption, food needs to pass many steps. Due to the special nature of the production process, different types of food have the different processing steps. For example, the processing step of primary agricultural products is relatively simple, usually including cleaning, removal of impurities and other operations. The steps of feed products and multi-component processing food are more complex than primary agricultural products because of its raw materials contain the primary agricultural products or animal derived primary agricultural products.

The food chain can be divided into a number of steps or operation sets, which include four parts: production, processing, storage & transportation and consumption according to the operation location or involved organization.

There are two reason we summarize the food safety traceability chain as four core elements: first, the food supply chain involves so many steps, some steps are relatively simple, it is convenient for us to analyze the chain when they are classified and combined; second, it facilitate analysis of quality and safety factors when the steps taken place in the same place.

3.2 Units of the traceability chain

The food safety traceability chain is consist of N (N $\leqslant$ 4) traceability links, and for a specific traceability link, is composed by a number of traceability units. Many steps can be divided into a number of traceability units according to the order of processing time and operation in the same link. For example, link of fresh vegetable production can be separated from four units: vegetable cultivation, the daily operation of the farming, water and soil fertility.

The determination of the unit has important significance for food safety traceability chain. First of all, it is includes several factors in one unit which affect food safety. Secondly, the unit corresponds to the actual operations of the process; it is easy to understand, but also easy to carrying out.

3.3 Factors of the traceability chain

There are biological, chemical and physical hazards in food, which are generally found in food production, processing, transportation and consumption of various links, through the dividing of the trace of the above hazards, the hazards drop to some specific traceability units. It is relatively easy to find out the factors leading to the occurrence of food safety incidents in the retrospective unit.

For a particular kind of agricultural products, the traceability link and unit is relatively fixed, but the traceability factors are changed more frequently. The reason is the food production environment, the processing procedure and transport tools change from tine to tine. And with the help of HACCP (Analysis Critical Control Point Hazard) technology, we can easily pick out the factors in food chain.

3.4 Indexes of the traceability chain

Once we confirm the traceability factor which affect food safety, the indexes of the factor is determined immediately because traceability indexes is attached to the unique factor. Food safety index is a measure of the food safety hazards, and it is also the indicator of the damage to consumers when food safety incidents occur. Such as pesticide residues in the Chinese chives, peanut oil will have a great harm on consumers.

At present, we can convenient find the food safety factor index through the national and industry standards of all kinds of food, which list the limited value of critical factors affecting human health.

4. Food safety traceability chain analyses

4.1 Hierarchical structure of food safety traceability chain

As mentioned above, food safety traceability chain include four levels: traceability links, traceability units, traceability factors and traceability indexed. Traceability links describe how many gates the food goes through from raw material production to sales. From the perspective of comprehensive traceability based on the food chain, each type of food should include production, processing, storage and consumption four links. However, in the process of building the system, the traceability chain may not have four full links because of the existence of incomplete chain, quality and safety information cannot be collected, or the specific link has no quality and safety factors at all. The hierarchical structure of food safety traceability chain is shown as Fig. 2.

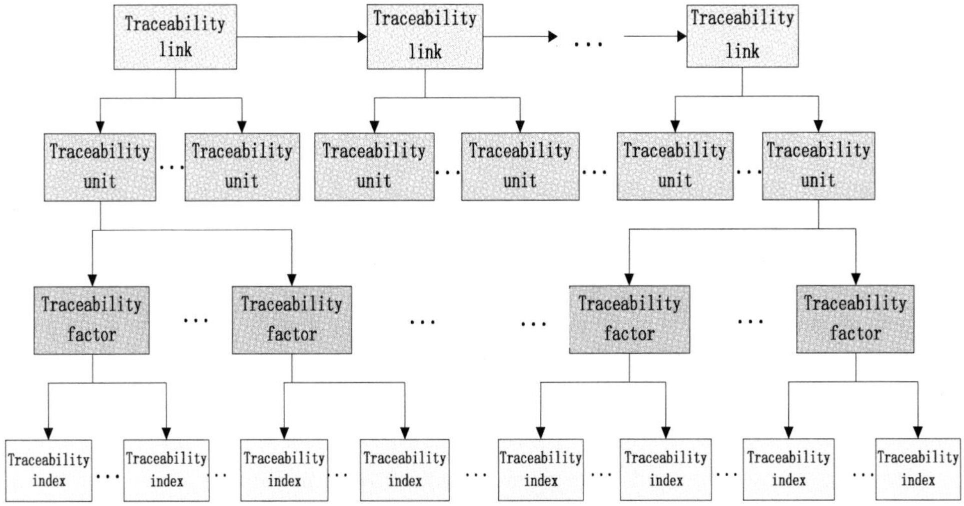

Fig. 2 Diagram of Food Safety Traceability Chain Hierarchy Model

Food safety traceability chain model has the following characteristics:

(1) For a specific food, the traceability chain includes K (K $\leqslant$ 4) traceability links;

(2) For each specific link, there are a number of traceability units;

(3) Each traceability unit contains several traceability factors, which may be the same one in different units;

(4) For a specific factor, it may be have at least one index. They may have the same index in different factors, but the limit value of the index is not necessarily the same.

4.2 Granularity of food safety traceability chain

The granularity of food safety traceability chain is an important index to measure the effectiveness of a traceability system. Through food safety traceability chain granularity, we can clearly distinguish how many important traceability factors or indexes the traceability system achieved. Generally speaking, there are three kinds of granularity, which is the accuracy, the width and the depth of food safety traceability chain.

The accuracy of food traceability chain is determined by the accuracy of analysis unit and the acceptable error. It has two indicators: the accuracy of the traceability unit and the accuracy of traceability index. For the primary agricultural products as example, due to the production scale and the characteristics of the product itself, we generally take a parcel of land as the traceability unit, and take the single cabbage or single cucumber as the minimum traceability unit is not practical obviously.

The width of food safety traceability chain refers to the range of the forward or backward tracing or tracking, but also the width of the traceability system. For a specific food, the width is related to the traceability links. And the width of the range is from 1 to 4.

From the perspective of model, the depth of food safety traceability chain refers to the level of the system can cover, including the four levels of the chain. Generally, the depth of the food traceability chain is determined by the numbers of quality and safety control point.

Overall, if we want a higher requirement for the accuracy, width and depth of food traceability chain, we will encounter the higher the cost of building food safety traceability system correspondingly.

5. Conclusion and discussion

The quality and safety elements involved from food production to consumption can be managed in the food traceability chain. And this is the basis for the design and construction of food safety traceability system. In this paper, based on the food chain, we gradually decompose the chain into traceability links, traceability units, traceability factors and traceability indexes in order to meet the need of construction of food traceability system, according to the analysis from food production to consumption, and their location, participation in the organization and other factors. Eventually, we established a food safety traceability chain model. Once we have the food traceability chain, we can clearly define and draw the traceability factors and indexes from different links and units, also can easily recognize the system participants including the producers of primary agricultural products, the food processors, the storage company, the transporter and the retailers. All these things constitute the whole information set that food safety traceability system want to record, tracing and tracking. By analyzing the food safety traceability chain granularity, we provide a reference for enterprises or regulators to establish the food safety traceability system.

Acknowledgements

This research was financially supported by agricultural network technology innovation team, the innovation project of Chinese Academy of Agricultural Sciences.

References

1. Lu Changhua, Wang Changjiang and Hu Sinong. Animals and Animal Products Identification Technology Traceability Management. Chinese Agricultural Science and Technology Press, 2007, 35-36
2. Wang Lifang, Lu Changhua and Xie Jufang. Review of traceability system for domestic animals and livestock products. Transact. Chin. Soc. Agric. Engin., 2005, 21(7):168-174
3. Xing Wenying. The policy of traceability in quality and safety of agricultural products in USA. World Agriculture, 2006, (04):39-41
4. Xiong Benhai, Fu Runting, Lin Zhaohui et al. A solution on pork quality safety production traceability from farm to dining table --taking Tianjin city as an example. Scientia Agricultura Sinica, 2009, 42(1):230-237
5. Jiang Bin. Research on Safety Evaluation System for Animal by-Products Quality based on Traceability System. Shanghai: Donghua University, 2014:3-6
6. Jia Yiniiang, Su Zhongbin and Shen Weizheng. Design of beef cattle breeding information traceability system. Journal of Agricultural Mechanization Research, 2014, (8):185-188
7. Yang Xinting, Qian Jianping, Sun Chuanheng, et al. Design and application of safe production and quality traceability system for vegetable. Transactions of the Chinese Society of Agricultural Engineering, 2008, 24(3):162-166
8. Fu Xiao, Fu Zetian, Zhang Lingxian, et al. Vegetable traceability system based on web. Computer Engineering and Design, 2009, 30(1):85-87,128
9. Pu Yingyan, Wang Yingkuan, Yue Tianli, et al. Construction of traceability system for quality safety of apple and apple juice. Transactions of the Chinese Society of Agricultural Engineering, 2008, 24(2):289-292
10. Zheng Huoguo, Liu Shi-hong, Meng Hong, et al. Construction of traceability system for quality safety of cereal and oil Products. Scientia Agricultura Sinica. 2009(09):3243-3249
11. Sun Chuanheng, Yang Xinting, Li Wwenyong, et al. Design and realization of distributed traceability system of aquatic products based on supervision mode. Transactions of the Chinese Society of Agricultural Engineering, 2012, 28(8):146-153

12. You Zhaotong, Kong Yaguang, Hu Xiaofei, et al. Research and development of quality traceability system based on intellectual technology for bee products. Journal of Zhejiang University (Agriculture & Life Sciences), 2014, (5):533-540

Mixed RNA duplexes, a new type of nucleic acid inhibit A549 cell line growing

Y. Yun[1] and W. G. Duan[2]

[1]*School of Basic Medicine, Kunming Medical University, Kunming 650500, China*

[2]*School of Pharmaceutical Science, Yunnan University of Traditional Chinese Medicine Kunming 650500, China*

E-mail: deardwg@126.com; yunyu_li@126.com

A new type of nucleic acid (mixed RNA, mixRNA) that shared the activity of siRNA was put forward. The mixRNA duplexes contain RNA or DNA elementary units but contain no RNA or DNA segments. The new type of mixRNA duplexes inhibited cell proliferation in A549 cell line (a human lung cancer cell line) in a concentration-dependent manner. Since the mixRNA was theoretically resistant to DNase and RNase, mixRNA could be a new tool for scientific research and for clinic therapy.

Keywords: Small RNA Interference; mixRNA Duplexes; cell survival; A549 Cell Line.

1. Introduction

Small RNA interference (siRNA) is a wonderful tool to manipulate functional protein expression at RNA level without modifying genome, and could result in cell death or functional deficiency [1]. This technology has been widely applied in biological and even in clinical research [2]. However, siRNA highly depends on RNA duplexes, which are usually about 21 bp with two-nucleotide 3` overhangs [3], or vectors, which are able to transcribe and assemble the RNA duplexes [4]. Therefore, there are mainly two strategies involved in siRNA technology that could be used clinically, namely, RNA duplexes and siRNA vectors [4].

siRNA vectors are almost impossible used clinically as drugs, because of their safety[4]. When a siRNA vector injects into a body, it would be difficult for scientists to control its biological behavior. The vector, like a virus, could be able to duplicate itself in the body, or even recombine into host's chromosome(s), and would result in unexpected functions or genome modification. Eventually, the vectors would cause unexpected adverse drug reactions.

By contrast, it is more likely for small RNA duplexes to be therapeutic drugs [4]. The small RNA duplexes can not duplicate themselves, and impossibly integrate themselves into a chromosome. They are safe at this point, at least safer than siRNA vectors. However, RNA duplexes are not stable, and they are easily degraded by all kinds of RNase and some physical and chemical factors. Therefore, small RNA duplexes are synthesized at high cost (about as expensive as ten times of DNA of the same sequences), and only maintains a short duration in body; which would cause siRNA therapy very expensive. If a new type of nucleic acid overcomes the weakness, it could be much better.

Here, we put forward a new type of small nucleic acid, mixed RNA (mixRNA) duplexes that act on ribonucleotide reductase subunit M2 (RRSM2), which shared the siRNA effect.

2. Materials and Methods

2.1. *Materials*

Single stranded RNA or DNA related to RRSM2 [5, 6] (Table 1) were synthesized by GenScrip Co., Ltd. (Nanjing, China). Single or double sequence(s) were denatured at 80 °C for 5 min and annealed at an atmosphere. A549 cell line (a human lung cancer cell line) was purchased from Beijing Dingguo Changsheng Biotechnology Co., Ltd. (Beijing, China) and identified by checking its morphology according to the description of American Type Culture Collection (ATCC). Transfection kits of Lipofectamine 2000® were manufactured by Invitrogen (Carlsbad, CA, USA). Cell Counting Kit-8 (CCK-8) assay kits were manufactured by Genomeditech (Shanghai, China).

Tab. 1 Single nucleic acid sequences

#	Sequence (5`→3`)	Comment
1	GATTTAGCCAAGAAGTTCAGT	DNA
3	GAUUUAGCCAAGAAGUUCAGU	RNA
5	dG-A-dT-U-dT-A-dG-C-dC-A-dA-G-dA-A-dG-U-dT-C-dA-G-dT	mixRNA
7	TGAACTTCTTGGCTAAATCGC	DNA
8	UGAACUUCUUGGCUAAAUCGC	RNA
9	dT-G-dA-A-dC-U-dT-C-dT-U-dG-G-dC-U-dA-A-dA-U-dC-G-dC	mixRNA

2.2. *Cell culture*

A549 cells were cultured in F12K medium with 2 mM L-glutamine, 1.5 g/L sodium bicarbonate, and 10% fetal bovine serum, at 37 °C, in an atmosphere of 95% air, 5% CO_2. When cells were about to capture 80% area of the flask, these were digested and seeded to 96-well plates (5×10^4 per ml, 180 ml per well) cultured with single sequence or double sequences that packed by lipofectamine 2000 (Invitrogen, USA) for 48 h.

2.3. *Cell survival assay*

The cell viability was assayed by CCK-8 assay kits, which principle was very similar to MTT assay. Briefly, when cells were cultured for 48 h, the culture medium was renewed and 20 µl 0.5% 2-(2-methoxy-4-nitrophenyl)-3-(4-nitrophenyl)-5-(2,4-disulfophenyl)-2H-tetrazolium, monosodium salt (WST-8) was added to every well and incubated for 3 h. The reduced WST-8 was water soluble, and the absorbance (OD) was read by a plate-reader (Tecan Infinity 200®) at 450 nm referred with 800 nm. (Reduced WST-8 has the maximal absorbance at 450 nm scanned by the plate-reader, almost no absorbance at 800 nm).

3. Results and Discussion

RRSM2 is a housekeeping gene. If the gene products are interfered, cell survival will be inhibited [6]. In Figure 1, double sequences of 3# and 8# (38#) were able to be assembled to classic RNA duplexes of siRNA on RRSM2, and caused cell death in a concentration-dependent meaner. Also, double sequences of 5# and 9# (59#, double mixRNA) caused cell death, and the pattern was similar to that caused by sequences of 38#. While single sequences and other double sequences (18# and 19#) showed almost no inhibition on cell survival (Fig. 2). The results suggested that the mixed RNA duplexes shared the effect of siRNA at the same level.

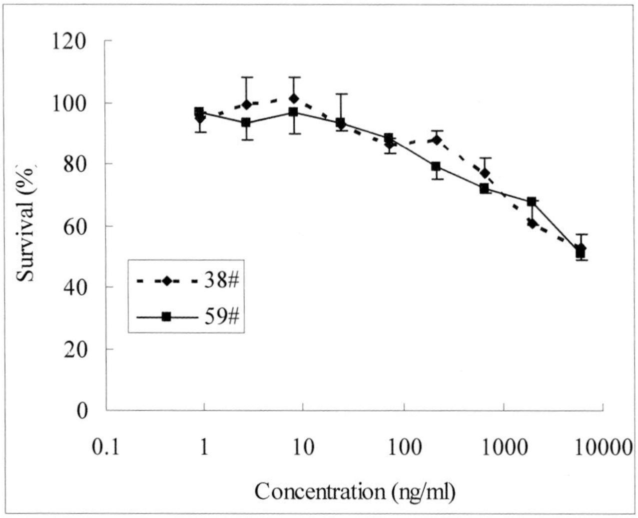

Fig. 1. A549 cell survival was determined by CCK-8 analysis kits (mena ± SD, n = 3). Double sequences of 3# and 8# (38#), which are small RNA duplexes of siRNA on ribonucleotide reductase subunit M2 (RRSM2), caused A549 cell death in a concentration-dependent meaner. Double sequences of 5# and 9# (59#, double mixRNA) caused similar cell death.

According to Watson-Crick law of base pairing, double sequences of 17#, 18# and 38# could exist base pairing formed by hydrogen bonds (Table 2). The elementary unit of mixRNA (5# and 9#) belonged to ribonucleotide or deoxyribonucleotide, and linked one by one, therefore, the mixRNA contained no DNA fragments and RNA fragments. Because base pairing also exists between RNA and DNA, base pairing between the double mixed RNAs highly possibly existed (Table 2). Since the inhibitory activity of double sequence 59# (double mixRNA) was very similar to that of sequences 38# (siRNA), their mechanism in inhibiting A549 cell survival could be also similar.

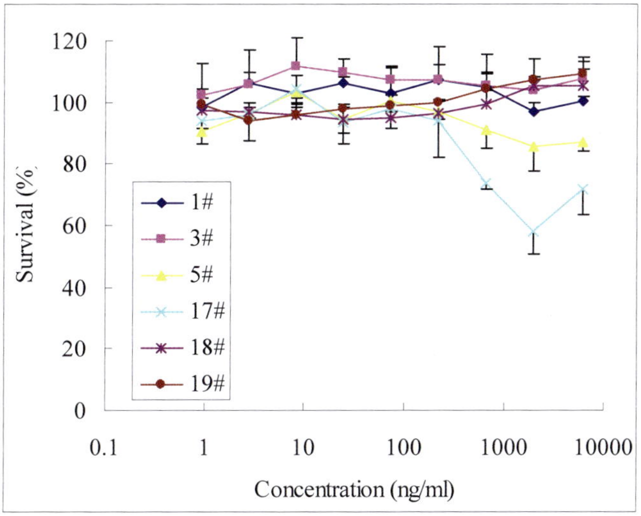

Fig. 2. A549 cell survival was determined by CCK-8 analysis kits (mena ± SD, n = 3). Single sequences and other double sequences (18# and 19#) showed almost no inhibition on cell survival. However, strangely, double sequences of 17# showed cell survival inhibition to an extent at certain concentration(s).

Tab. 2 Hydrogen bonds possibly existed.

<table>
<tr><th>Seq.</th><th>Base pairing</th><th>Note</th></tr>
<tr><td>17#</td><td>5`-GATTTAGCCAAGAAGTTCAGT-3`</td><td>DNA</td></tr>
<tr><td></td><td>||||||||||||||||||</td><td>H. bonds</td></tr>
<tr><td></td><td>3`-CGCTAAATCGGTTCTTCAAGT-5`</td><td>DNA</td></tr>
<tr><td>18#</td><td>5`-GATTTAGCCAAGAAGTTCAGT-3`</td><td>DNA</td></tr>
<tr><td></td><td>||||||||||||||||||</td><td>H. bonds</td></tr>
<tr><td></td><td>3`-CGCUAAAUCGGUUCUUCAAGU-5`</td><td>RNA</td></tr>
<tr><td>19#</td><td>5`-G-A- T-T- T-A- G-C- C-A- A-G- A-A- G-T- T-C- A-G-T-3`</td><td>DNA</td></tr>
<tr><td></td><td>| | | | | | | | | | | | | | | | | | |</td><td>H. bonds</td></tr>
<tr><td></td><td>3`-dC-G-dC-U-dA-A-dA-U-dC-G-dG-U-dT-C-dT-U-dC-A-dA-G-dT-5`</td><td>mixRNA</td></tr>
<tr><td>38#</td><td>5`-GAUUUAGCCAAGAAGUUCAGU-3`</td><td>RNA</td></tr>
<tr><td></td><td>|||||||||||||||||||</td><td>H. bonds</td></tr>
<tr><td></td><td>3`-CGCUAAAUCGGUUCUUCAAGU-5`</td><td>RNA</td></tr>
<tr><td>59#</td><td>5`-dG-A-dT-U-dT-A-dG-C-dC-A-dA-G-dA-A-dG-U-dT-C-dA-G-dT-3`</td><td>MixRNA</td></tr>
<tr><td></td><td>| | | | | | | | | | | | | | | | | | |</td><td>H. bonds</td></tr>
<tr><td></td><td>3`-dC-G-dC-U-dA-A-dA-U-dC-G-dG-U-dT-C-dT-U-dC-A-dA-G-dT-5`</td><td>MixRNA</td></tr>
</table>

As mentioned above, the most advantage of mixRNA duplexes (double mixRNA) is the fact that the cost is much lower and more stable than RNA duplexes; which could make therapeutic siRNA more feasible. mixRNA is neither classical DNA nor classical RNA, but shares the functional base pairing of DNA and RNA. Furthermore, mixRNA can not be degraded by both DNase and RNase theoretically, and its half-life in body could be much longer than that of RNA.

Taken together, we gave a preliminary glimpse to the new type of nucleic acid, double mixRNA, with some advantages, though some mechanisms and hypothesis need further investigation.

Acknowledgments

The work was financial supported by Yunnan Provincial Science and Technology Department (1010CI043).

References

1. Fire, S. Xu, M. K. Montgomery, S.A. Kostas, S.E. Driver and C.C. Mello, Potent and specific genetic interference by double-stranded RNA in Caenorhabditis elegans, Nature 391, 806 (1998).
2. S. Chabot, J. Teissie and M. Golzio, Targeted electro-delivery of oligonucleotides for RNA interference: siRNA and antimiR, Adv Drug Deliv Rev. 81C, 161 (2015).
3. S. M. Elbashir, W. Lendeckel and T. Tuschl, RNA interference is mediated by 21- and 22-nucleotide RNAs, Genes Dev. 15, 188 (2001).
4. S. Subramanya, S. S. Kim, N. Manjunath and P. Shankar, RNA interference-based therapeutics for human immunodeficiency virus HIV-1 treatment: synthetic siRNA or vector-based shRNA? Expert Opin Biol Ther. 10, 201 (2010).
5. J. D. Heidel, Z. Yu, J. Y. Liu, S. M. Rele, Y. Liang, R. K. Zeidan, D. J. Kornbrust and M. E. Davis, Administration in non-human primates of escalating intravenous doses of targeted nanoparticles containing ribonucleotide reductase subunit M2 siRNA, Proc Natl Acad Sci U S A. 104, 5715 (2007).
6. J. D. Heidel, J. Y. Liu, Y. Yen, B. Zhou, B. S. Heale, J. J. Rossi, D. W. Bartlett and M. E. Davis, Potent siRNA inhibitors of ribonucleotide reductase subunit RRM2 reduce cell proliferation in vitro and in vivo, Clin Cancer Res. 13, 2207 (2007).

Investigation of annual optimal time for bromelain production: Approached using statistical analysis

Zhiping Han[1], Yupo Cao[1], Wenhua Zhang[1], Rongqiong Luo[1], Qin Fen[2], Xiaoyi Wei[1*]

[1]*Agricultural Product Processing Research Institute, Chinese Academy of Tropical Agricultural Sciences, China*

[2]*South Subtropical Crops Research Institute, Chinese Academy of Tropical Agricultural Sciences, China*

Corresponding author: Xiaoyi Wei

[*]*E-mail: erping-80@163.com*

Abstract: The paper investigated the activity of bromelain extracted from fresh pineapple juice over three years to figure out the optimal harvest time for bromelain in the Zhanjiang region. The method of statistical analysis had been used for determining the bromelain activity in every month of a year. Meanwhile, the average moisture content and activity of bromelain in were month were linear analysis. The results indicated that November was the best time for bromelain production and processing in a year; the average bromelain activity of November reached 1.089×10^6 U/g. The outcomes demonstrated that both of temperature and humidity played important roles in bromelain production.

Keywords: Bromelain; Moisture content; Enzyme activity; Environmental temperature; Moisture.

1. Introduction

The medicinal qualities of pineapple are recognized in many traditions in China, South America and Southeast Asia [1]. The paper investigated the activity of bromelain extracted from fresh pineapple juice over three years to figure out the optimal harvest time for bromelain in the Zhanjiang region. Bromelain, a proteolytic enzyme from cores, peels, fleshes, stems and leaves of pineapple (Ananas comosus), is a rich source of cysteine proteases [2]. Unlike most enzymes, bromelain has a so wide effective range of activity in both acidic and alkaline conditions that allows it to remain active in a variety of biological environments. It is stable between pH 5.0 and 10.0, thus bromelain is of broad specificity [3]. Unlike to the model plant cysteine protease papain, bromelain is remarkably heat unstable, losing proteolytic activity if temperature above 40 °C where most enzymes are denatured, or destroyed [4, 5]. It has shown potentially beneficial effects due to its analgesic and anti-inflammatory properties. Its several therapeutic properties-including skin debridement, anti-tumor action, enhanced wound healing, and modulation of cytokines and immunity-have been demonstrated in both in vitro and in vivo studies [6]. Currently, bromelain has been used for meat tenderization, dietary supplement, anti-oxidative peptides production, marine product processing, and other health food processing in the food industry [7, 8].

In a number of cases, the effects have been achieved for the successful operation and control of bioprocesses of bromelain activity. The previous work reported that temperature, extremes of pH and the addition of surfactants leaded to the decrease in enzyme activity [9, 10]. However, the studies on the optimal harvest time for bromelain over the year are very limited, as well as the correlation between the local temperature and humidity to and activity of bromelain. Zhanjiang region is the biggest pineapple planting base, where the average temperature over the year is above 20 °C and the mean value of humidity is around 80%. Referring to pineapple could be harvested in every month in this area, the optimal time for bromelain production was investigated in this paper.

To our knowledge, there has no published data concerning the correlation between time over the year to the activity of bromelain. The present work statistically analyzed the bromelain activity over three years, as well as the local temperature and moisture, to figure out the best time in one year to harvest pineapple for bromelain production.

2. Materials and Methods

2.1 Materials

The experiments had been conducted from November 2011 to October 2014 in the Harvest Farm which covers over 6000 km^2 in the rural area of Zhanjiang city. All chemicals used were of analytical grade for the experiments and analyses.

2.2 Bromelain preparation

Pineapples was reaped in the morning and juice was extracted immediately. The juice was kept on ice for 2 hours and only supernatant was used for isolation of enzyme. Bromelain preparation was conducted by Ammonium sulfate [$(NH_4)_2SO_4$] precipitation followed literature [11], and after being centrifuged the pasty bromelain was test activity and moisture.

2.3 Enzyme activity assay

Enzyme activity was determined by using the method which was described by Ota et al. [12]. One unit of bromelain activity was defined as the amount of bromelain necessary to produce 1 μg of tyrosine from casein per minute under assay conditions.

2.4 Data analysis

All experiments were carried out for three times and mean values were showed in the paper.

3. Results and Discussions

3.1 Statistical analysis of the activity and moisture content of bromelain

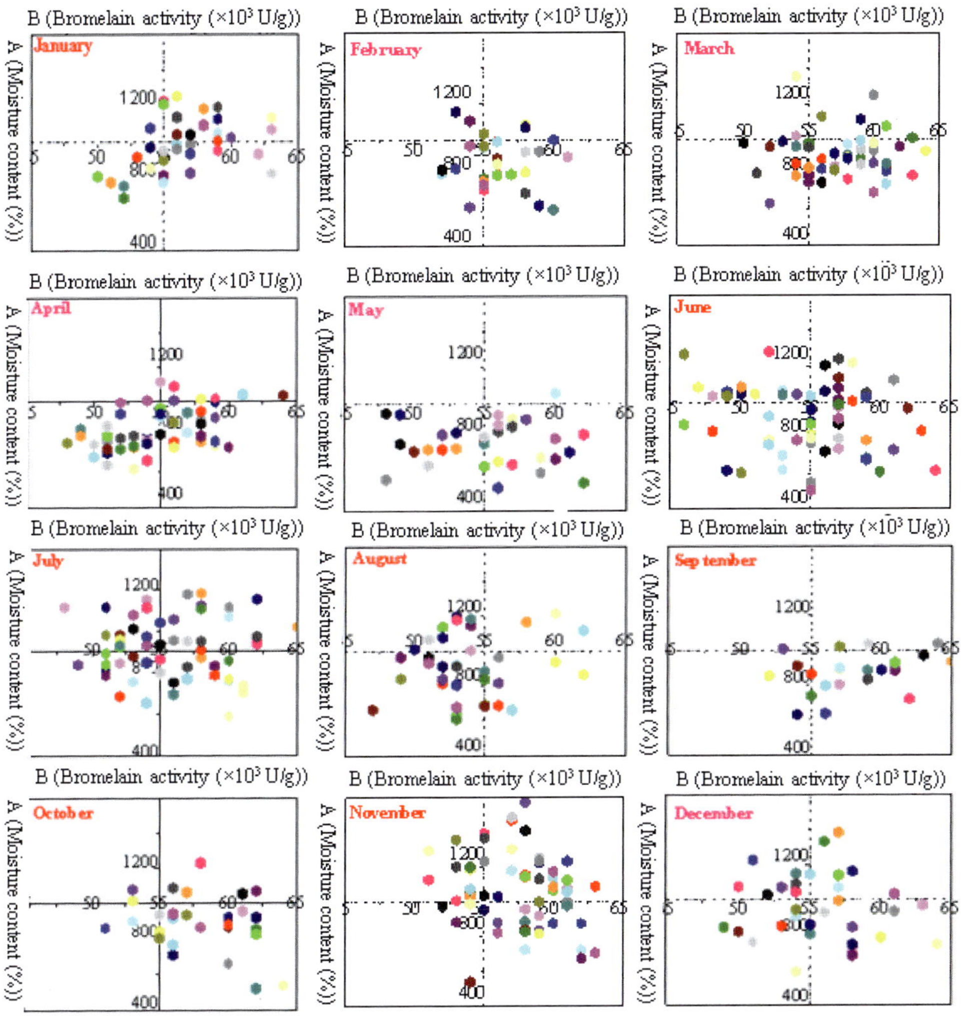

Fig. 1. The statistic of activity and moisture of bromelain over the year. Each point represents the mean value of measurements in three years.

Distributions and regulations of the activity and moisture content for bromelain were showed in Figure 1. These pictures represented the statistical distribution of the activity and moisture content for bromelain in each months over the year. The temperature in summer is above 30 °C which cause decrease in the enzymatic activity. In addition, as the temperature increases, more molecules have enough kinetic energy to undergo the reaction. If the temperature is raised above the optimum point, the kinetic energy of the enzyme and water molecules is so great that the structure of the enzyme molecule starts to be disrupted [13]. Therefore, a decrease in activity was detected. When the temperature drops, the relative humidity goes up on the contrary. Accordingly the moisture content of bromelain becomes high relative. Compared with the moisture content, the bromelain activity had a dropped distribution trend because of lower air temperature.

3.2 Changes in moisture content of bromelain over the years

Figure 2 showed that the average moisture content of bromelain in each month was different It was noticeable that from Sep to Nov the activity of bromelain was relatively higher which was due to the lower temperature and environmental humidity. And the low activity from May to Aug was also related to the high temperature in the summer.

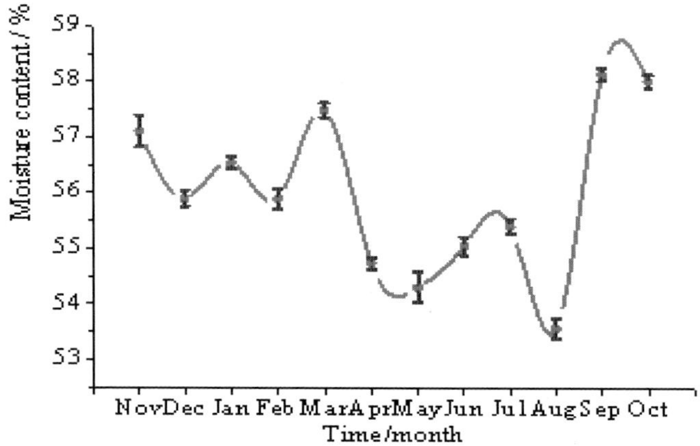

Fig. 2. The average moisture content of bromeain in each month

3.3 Changes in activity of bromelain over the years

The average activity of bromelain in each month was different, as shown in Figure 3. The bromelain activity in every month was complex and was affected by various factors.

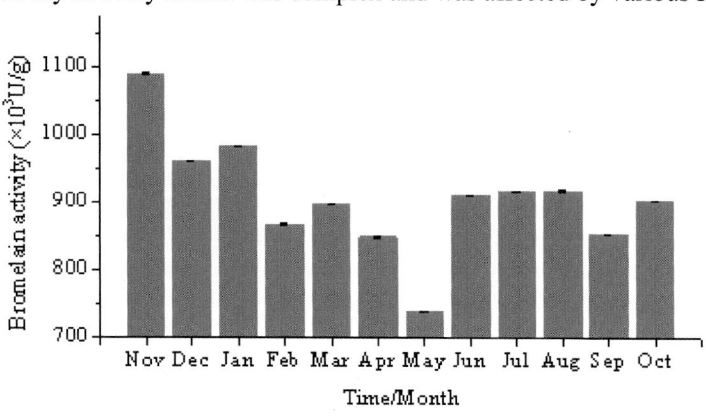

Fig. 3. The average activity of bromelain in each month

To get the optimum of bromelain activity, the effect from other proteases in the matrix should be reduced. From Figure 3, the highest bromelain activity was displayed in May, with activity of around 7×10^5 U/g. The bromelain produced in November had the highest activity of 1.089×10^6 U/g, when the temperature was lower while the humidity was higher.

4. Conclusions

The paper investigated some of the properties of bromelain produced in continuous three years, including activity and moisture content, and the mean values were statistically analyzed. Besides, the effect of local temperature and humidity on the properties of bromelain was discussed as well. The results showed that November was the optimal time for bromelain production in Zhanjiang region, when the average temperature was around 20 °C and the relative humidity was below 80%. Zhanjiang is located on a peninsula which is connecting to Qiongzhou strait and is influenced by tropical marine climate. Generally, from May to September, the local temperature I above 35 °C and bromelain would get denatured at this temperature. And in the months of January, February and March the relative humidity is always above 90%, which may also contribute the lower activity of bromelain. Upon availability of the mass spectrometer and qT-PCR, we plan to study the comparative transcriptomes of bromelain in each month.

Acknowledgements

This work is funded by the Fundamental Scientific Research Funds for Chinese Academy of Tropical Agricultural Sciences(Grant No.1630062015016).

References

1. Chobotova, K., Vernallis, A.B., Majid, F.A.A., Bromelain's activity and potential as an anti-cancer agent: Current evidence and perspectives. Cancer Lett, 290, pp. 148-156, 2010.
2. Umesh, H.H., Sumana, B., KSMS, R., Use of reverse micellar systems for the extraction and purification of bromelain from pineapple wastes. Bioresource Technol, 99, pp. 4896-4902, 2008.
3. Minami, Y., Doi, E., Hata, T., Fractionation, purification, and some properties of proteolytic enzymes from stem bromelain. Agric. Biol. Chem, 35, pp. 1419-1430, 1971.
4. Natalucci, C.L., Priolo, N.S., Buttazzoni, M.S., Caffini, N.O., Proteasas de Bromeliaceae. II. Separación, caracterización y fraccionamiento de una proteasa aislada de frutos de Bromelia hieronymi, Mez. Acta Farm. Bonaerense, 4, pp. 93-98, 1985.
5. Greenberg, D.M., Plant proteolytic enzymes, in: S.P. Colowick, N.O. Kaplan (Eds.), Method Enzymol. Vol. 1. Academic Press Inc., New York, pp. 54-64, 1955.

6. Wu, S.Y., Hu, W., Zhang, B., Liu, S., Wang, J.M., Wang, A.M., Bromelain Ameliorates the Wound Microenvironment and Improves the Healing of Firearm Wounds. J Surg Res, 176, pp. 503-509, 2012.
7. Ha, M., Bekhit, A.E.D.A., Carne, A., Hopkins, D.L., Characterisation of commercial papain, bromelain, actinidin and zingibain protease preparations and their activities toward meat proteins. Food Chem, 134, pp. 95-105, 2012.
8. Siengdee, K., Nganvongpanit, P., Pothacharoen, S., Chomdej, S., OngChai, M.S., Effects of bromelain on cellular characteristics and expression of selected genes in canine in vitro chondrocyte culture. Vet Med-Czech, 55, pp. 551-560, 2010.
9. Kurniawati, S., Nicell, J.A., Characterization of Trametes versicolor laccase for the transformation of aqueous phenol. Bioresour Technol, 99, pp. 7825-7834, 2008.
10. Xue, Y., Wu, C.Y., Christopher, J.B.W., Ning, X., Nie, H.L., Zhu, L.M., Chemical modification of stem bromelain with anhydride groups to enhance its stability and catalytic activity. J Mol Catal B-Enzym., 63, pp. 188-193, 2010.
11. Devakate, R.V., Patil, V.V., Waje, S.S., Thorat, B.N., Purification and drying of bromelain. Sep Purif Technol, 64, pp. 259-264, 2009.
12. Khan, R.H., Rasheedi, S., Haq, S.K., Effect of pH, temperature and alcohols on the stability of glycosylated and deglycosylated stem bromelain. J. Biosci, 28, pp. 709-714, 2003.
13. Ketnawaa, S., Chaiwutb, P., Rawdkuena, S., Pineapple wastes: A potential source for bromelain extraction. Food Bioprod Process, 90, pp. 385-391, 2012.

Effects of hysterectomy on ovarian function in patients with retaining uterine blood vessel

Hongxia Sun[1] and Yufei Cai[2]
[1]*School of Pharmacy, Beihua University, jilin 132013, China*
[2]*Obstetrics and Gynecology Department, Affiliated Hospital of Beihua University, jilin 132013, China*
E-mail: sunhongxiajl@126.com

Objective To evaluate the effect of hysterectomy of reserving the uterine blood supply on uterine artery hemodynamics, blood lipids level and the degree of sexual satisfaction before and after operation. Methods One hundreds of patients with benign lesions of uterus were randomly divided into subtotal hysterectomy of maintaining uterine vessels group, STHMUV (research group, n=50) and traditional subtotal resection of uterine group, SROU (the control group, n=50), comparing the recent impact of two operation on sexual life satisfaction and changes of uterine artery blood flow in two groups,blood tests were taken to check the serum blood lipid levels before and after operation three months, six months, one year and two years at the same time respectively. Results There was no significant difference between the two groups in operation time, bleeding volume, postoperative fever, and residual polyps; There was no significant difference among research group before and after operation in serum lipid level ($P>0.05$), total cholesterol (TC), triglyceride (TG) and low density lipoprotein (LDL) level increased significantly in control group ($P<0.05$), high density lipoprotein (HDL) level decreased ($P<0.05$) in Control group and no significant difference between the two groups in sexual life quality. Comparison between Va and Vb in Research group before and after operation shows obvious difference. Conclusion The effect of uterus hysterectomy for retaining blood supply on ovarian endocrine function is less than that of the traditional total hysterectomy, this operation is very important for preserving ovarian function and delaying ovarian decline.

Keywords: Preservation of Uterine Blood Supply; Subtotal Hysterectomy; Ovarian Function.

1. Introduction

The endocrine function of ovary can be affected by hysterectomy. In traditional hysterectomy, cutting off the ovarian branch of uterine blood vessels could reduce the ovarian blood supply by fifty to seventy percent, leading to ovarian follicle degeneration to reduce hormone production and ovarian function declined four to five years after hysterectomy ,which increased the occurrence of complications such as peri-menopausal symptoms, osteoporosis, high blood lipids and coronary heart disease and so on. Subtotal hysterectomy for maintaining uterine vessels (STHMUV) is a kind of surgery that does not affect ovarian hormone secretion, which can ensure the blood supply to the ovaries after removal of uterus, STHMUV can delay the natural aging of the ovary. So that the ovary is still able to secrete estradiol and other hormones, which delayed the emergence of ovarian physiological decline [1].

2. Materials and methods

2.1 General information

One hundreds cases of patients which had sign of subtotal resection of uterus (SROU), aging between thirty to forty-five years old, was selected from the Department of gynecology in the affiliated hospital of Beihua University during the period from December 2014 to October 2010. and were randomly divided into STHMUV group (research group, n=50, indicating R group) and traditional SROU group (control group, n=50, indicating C group). There was no significant difference in age, types of diseases in the two groups (P>0.05).Cervix was examined and serum lipid levels were detected before operation. All women were admitted to hospital within three to five days after the last menstrual period. There was no significant difference in serum TC, HDI, TG and LDL in the two groups (P>0.05).

2.2 Choice of operation mode

The control group with reference to Li Guangyi, editor in chief of "Practical Gynecology laparoscopic surgery" for the traditional subtotal hysterectomy [10]. The study group adopted the reservation of the uterine vessels subtotal hysterectomy. The study group and the control group were the same, both of which were combined with the spinal epidural anesthesia. The oxytocin and saltwater were mixed to be injected into the uterus, its vasoconstrictive effects lead to the whole uterine tissue became pale due to ischemia, the remaining part of the liquid was dumped into the contralateral body layer during the surgery; cross-cutting and episiotomy surgery could be taken in lower abdomen, each layer was cut to be checked in order by open view according to its anatomical position, the uterus was maintained in a raised position during the operation, both sides of the uterine appendages and uterine blood vessels was clamped with tissue forceps. The position for uterine resection was about 1 cm interior to uterine horn, toward the direction of uterine isthmus, resection of uterine body by the shape of triangle. The lower position was close to the inside of uterus cervix, with about 2 cm thickness of both side of uterine body.The muscle and endometrial layers can be clearly seen at this point, if there was,such as the obvious hysteromyoma, which can be removed, with hemostatic forceps or electric hemostasis. The uterine wound was sutured by U-shaped, and check whether there is staxis or not meanwhile.

2.3 Observation index

Operation time, bleeding during operation, postoperative fever and residual polyps were compared between the two groups. Serum lipid levels were taken to detected three months, six months, one year and two years after operation and compare the serum lipid levels, hemodynamic of uterine artery and survey of sexual satisfaction in different time periods.

2.4 *Statistical methods*

SPSS 17.0 statistical software was used for statistical analysis, all data were expressed as mean±the standard deviation ($\bar{x}$±s), single factor analysis of variance was used for comparison among groups, LSD-t test was used for comparison between two groups, the difference was statistically significant when *P*<0.05.

3. Results

3.1 *Operation effects*

Operation time, intraoperative blood loss, postoperative body temperature and postoperative residual segment polyps were compared between two groups of, the results showed that there was no significant differences (P>0.05), see Table 1.

Tab. 1 Comparison of two groups of patients with operation($\bar{x}$ ±*s*)

Group	*n*	Operation T/min	Blood loss/ml	BT/°C	Residual polyps/%
Research(R)	50	53.2±7.6	90±12.6	37±0.5	0/0
Control(C)	50	55.1±8.3	95±16.7	37.5±0.2	28/56

3.2 *Serum lipid levels*

The changes of serum lipid levels of patients before and after operation.

Tab. 2 Comparison of blood lipid levels in two groups of patients before and after operation

Group	n	TC/mmol·L^{-1}				
		Before-	After-3m	After-6m	After-1y	After-2y
Research(R)	50	4.18±0.52	4.38±0.57	4.45±0.62※	4.44±0.73※	4.36±0.82※#
Control(C)	50	4.23±0.65	4.52±0.63	5.73±0.69△	6.87±1.52△	6.88±1.25
Group	**n**	**TG/mmol·L^{-1}**				
		Before-	After-3m	After-6m	After-1y	After-2y
Research(R)	50	1.25±0.14	1.35±0.42	1.21±0.22※	1.28±0.23※	1.25±0.32※
Control(C)	50	1.32±0.18	1.38±0.37	1.52±0.45	1.96±0.30△	2.76±0.41△
Group	**n**	**HDL/(mmol·L^{-1})**				
		Before-	After-3m	After-6m	After-1y	After-2y
Research(R)	50	1.62±0.34	1.63±0.31	1.68±0.39※	1.78±0.30※	1.96±0.29※
Control(C)	50	1.57±0.31	1.42±0.21	1.38±0.18	1.35±0.34	1.23±0.25△
Group	**n**	**LDL/(mmol·L^{-1})**				
		Before-	After-3m	After-6m	After-1y	After-2y
Research(R)	50	2.16±0.26	2.23±0.24	2.65±0.24※	2.70±0.28※	2.84±0.27※
Control(C)	50	2.35±0.35	2.79±0.38	3.62±0.41△	3.89±0.46△	3.91±0.54

△P<0.05 vs group compared in the previous observation time; ※P<0.05 vs Control Compared in the same observation time.

3.3 Comparison of sexual quality between two groups

The quality of sexual life of the two groups of patients (including low sexual desire, sexual aversion, painful sexual intercourse, vaginal dryness and sexual satisfaction) were compared and there was no significant difference (P>0.05).

3.4 Changes of uterine artery blood flow before and after operation in research group

Tab. 3 Hemodynamic changes in the study group before and after operation

uterine artery(types)	Va (cm/s)	Vb (cm/s)	RI
Left side			
Pre-operation	29.4±6.8▲	9.7±3.5▲	0.82±0.07
Post-operation	18.7±2.7	5.6±3.2	0.79±0.08
Right side			
Pre-operation	28.6±4.6▲	7.5±2.7▲	0.78±0.05
Post-operation	9.2±3.3	4.1±3.2	0.75±0.06

▲P<0.05 vs Compared with the flow velocity of the same side pre- and Post-operation

4. Discussion

The uterus has a close connection with the ovary not only in anatomy and endocrine function but also it is the target organ of ovarian hormones.The uterus and ovary have two sets of blood circulation system, the uterus is innervated by uterine artery and uterine branch of ovarian artery, and the ovary is innervated by ovary artery and ovarian branch of uterine artery. It is reported that the supply of ovarian branches of the uterine arteries is about 50%~70% [2] of the total ovarian blood supply. Traditional uterus resection (TUR) cut off the blood distribution of the uterine artery and vein to the supply of the ovary so that natural recession of ovary emerged rapidly and endocrine function was affected, which appeared a few years after operation [3]. Subtotal hysterectomy for maintaining uterine vessels (STHMUV) does not affect the blood supply of ovarian blood after removal of uterus, blood supply is not affected to maintain the normal state of ovarian secretion function and delayed ovarian physiological failure, the retained part of uterine tissue was still able to secrete hormones. STHMUV can delay the natural ovarian hormone secretion and keep a hormone balance in the maintenance among hypothalamus-pituitary -ovary-uterine [4]. Foreign scholars such as N and Siddle [5] confirmed that the age of ovarian failure in patients with ovarian failure was four years earlier than that of natural menopause, and the degree of performance and severity of menopause were significantly higher than those of normal population.

A large number of studies have demonstrated that estrogen has a protective effect on the cardiovascular system [6], sex hormone levels were associated with atherosclerosis occurrence and cardiovascular events [7]. Estrogen decline can lead to blood lipids disorder [8], and dyslipidemia is an important cause of coronary heart disease in postmenopausal women; similarly, the E2 level decline in pre-menopausal women after removal of uterus can lead to dyslipidemia [9]. Studies have indicated that estrogen levels decline can lead to elevated blood lipids, and high blood lipids will lead to arteriosclerosis. Although traditional subtotal resection of uterine had preserved the ovary, the ovarian blood supply was reduced as a result of cutting off the uterine arteries and veins and the ovarian E2 secretion ability decreased so as to affect blood lipids levels. STHMUV can reduce the effect of operation on ovarian hormone secretion, serum E2 level was not significantly decreased compared with that in pre-operation (data not listed). Our data show that, in the control group of fifty patients with total uterine resection, serum TG, TC, LDL levels were increased, HDL level was decreased,while in research group of fifty patients with STHMUV, which had lower serum lipids compared with that of pre-operation, these results confirmed that ovary branch of uterine artery with STHMUV had little effect on the blood supply of ovary, and protected the blood lipids regulation of ovarian hormone.

Studies have shown that for those women who are not at the menopause stage with uterine benign tumor and need to take subtotal resection of uterine, STHMUV operation style should be chosen as far as possible, which can slow down the natural recession of ovary, reduce hormone levels decline, adjust blood lipids level and reduce the occurrence of arteriosclerosis.

References

1. Xu Xiaofeng, Zhou Huandi, Li Miao. Study on influence of hysterectomy on ovarian function in patients with uterine vessels [J]. Journal of Nanchang University (Medical Edition), 2010, 50(6): 18-22.
2. Wu Fengying, Deng Zhaohong, Tian song. Clinical application of hysterectomy for reserving the uterine vascular ligament [J]. Xinjiang medicine, 2009, 39:86-87.
3. Cicinelli E, Einer-Jensen N, Galantino P. The vascular cast of human uterus: from anatomy to physiology. Ann NY Acad Sci, 2004;1034:19-26.
4. Chen Lu, Li Guangyi, Han Yubin. Clinical control study of laparoscopic total hysterectomy for retaining uterine artery [J]. Chinese Journal of mini-trauma surgery, 2009, 9 (7): 609-612.
5. Siddle N, Sarrel P, Whitehead H. The effect of hystevectomy on the age at ovarian failure: identification of a subgroup of women with premature loss of ovarian function and literature review [J]. Fertil Steril, 1987, 47(1):94-100.
6. Mohamad M J, Karayyem M, Mohammad M A, et al. Serum sex hormones in premenopausal women with coronary heart disease [J]. Am J Phys Anthropol, 2007, 132(1):151-157.
7. Li Kaijun, Wu CongCong, Xiang Yijia, et al. Analysis of estrogen levels and plasma homocysteine, lipids in postmenopausal women with coronary heart disease patients [J]. Prevention and treatment of cardiovascular and cerebrovascular disease 2014, 14 (3): 222-224.
8. Wang Jingjing, He Ruhua, Wu Ling. The relationship between the level of estrogen in postmenopausal women with coronary heart disease and serum lipids [J]. Ningxia medical journal, 2014, 36 (1): 13-15.

9. Sheng Li, Ye Ping. Abnormal blood lipids and cardiovascular disease in women [J]. Chinese Journal of practical Internal Medicine, 2014, 34(1): 15-17.
10. Gimbel H. Total or subtotal hysterectomy for benign uterine diseases? A meta-analysis [J]. Acta Obstet Gynecol Scand, 2007, 86(2):133-144.

Study on optimization of extraction conditions of total alkaloids from *Zizyphi spinosi* semen

Jia-ning Zhang, Ke Ding*, Yan-zhou Hu, Xiang-ning Chen, Tao Han

Department of Food Science and Engineering, Beijing University of Agriculture, Beijing, 102206, China

**E-mail: 20057801@bua.edu.cn*

To obtain the best appropriate method and optimum extraction conditions of total alkaloids from *zizyphi spinosi semen* (*ZSS*), the conditions of reflux extraction were optimized by single factor experiment and orthogonal design respectively. the experiment results were as follows: 20% acetic acid resolved in ethanol as extracting solvent (EtOH:HAc=80:20 (v/v)), ratio of liquid to solid=20:1(ml:g), 4.0 h, 70 °C, and these conditions were optimum conditions after validation of experiment. The total alkaloids of *ZSS* extracted by this method could provide a powerful guarantee for studying ingredients and activities of alkaloids in *ZSS*.

Keywords: Zizyphi Spinosi Semen; Alkaloids; Extract; Reflux.

1. Introduction

Zizyphi spinosi semen (*ZSS*) both belongs to food and medicine as one of a Chinese medicine, is the seed of *Zizyphus jujuba* Mill. Var. *spinosa* (Bunge) Hu ex H. F. Chou(Rhamnaceae). As a Chinese medicine, *ZSS* has a long history, it has been mainly used for curing the insomnia, palpitation and dreaminess, Tixu and hyperhidrosis, impairment of body fluids and thirst, and so on[1]. *ZSS* is produced at the mountain area in the north of China, such as HeBei, ShanXi, ShanDong, ShanXi, Liaoning, HeNan, Inner Mongolia[2]. So far, currnet reports mainly foucs on the function of alkaloids about sedative and hypnotic effect of *ZSS*[3-5]. In addition, Wei Qiao et al. study on the fuction of Anticonvulsive effect[6] and antidepressant effect of *ZSS*[7]. The results of his study show that compared the group of 20 mg/kg cyclopeptide alkaloids, the group of 50, 100 mg/kg total alkaloids of *ZSS* could lead to the death time and seizure time of mice were significantly prolonged. As the conclusion, it means alkaloids of *ZSS* act out the statistical significance that contain anti convulsion effect[6], the total alkaloids of *ZSS* could shorten the immobileity time of mouse tail related to the different kind agent with different dosage, and confront mice body temperature deline due to reserpine effectively, blank experiments in each group between the spontaneous activity of the mice without significant difference also shows that alkaloids of *ZSS*, in a certain degree, it has the effect of anti depression[7].

The mainly methods to extract the total alkaloids are solvent method, infiltration percolation film evaporation continuous extraction, supercritical fluid extraction, ultrasonic assisted extraction, membrane extraction, microwave assisted extraction and so on[8]. The most common one is solvent method, and the efficiency of this method is ralated to the solvent dosage, raw material grinding degree, operating conditions (including temperature, stirring, etc.) closely. Depend on the different character of alkaloids, the solvent extraction could be divided into four types: pure water, acid water-organic solvent (0.5~1.0% acetic acid, hydrochloric acid as solvent), alcohol or alkaline organic solvent. Total alkaloids from securinine, anabasine[9] or nuciferine[10] were extracted by using water as solvent. Alkaloids in Bat Ge crude existing as unstable salt or dissociative alkali were obtained by acid water - organic solvent extraction[11]. Total alkaloids from Wu Zhuyu were extracted by using 70% methanol reflux extraction 24h[12]. Alkalized chloroform is the best for alkaloids from Yan Husuo[13].

Current research shows that, *ZSS* contains two main kinds of alkaloids, namely aporphine alkaloids and cyclopeptide alkaloids[14-20]. Besides, one of cyclopeptide alkaloids-- sanjoinine A has the function as physiological activity[16]. In recent years, with the in-depth study on active components of *ZSS* alkaloid, its function have been achieved more and more attention, so we need to find an effective method of *ZSS* alkaloid extraction that is the basis for the subsequent researches on it. This research foucs on the extraction process of *ZSS* alkaloid in order to get an simple and effciency method.

2. Materials and Methods

2.1 Experimental Materials and Apparatus

ZSS (purchased in Hebei Anguo herbal medicine market); bromocresol green (analytical pure, Beijing XinNuo JiuHeng Technology & Trade Co., Ltd.). Anhydrous ethanol, petroleum ether, chloroform, *potassium* hydroxide, ammonia, sulfuric acid, hydrochloric acid, acetic acid, anhydrous sodium sulfate, adjacent benzene two formic acid hydrogen potassium (analytical pure, Beijing chemical factory).

Ultraviolet - visible spectrophotometry meter (TU-1810 type, Beijing PuXi General Instrument Co., Ltd.); Electronic balance (ALC-210.3 type, Beijing SaiDuoRuiSi instrument Co., Ltd.); Chinese herbal medicine pulverizer (FW135 type, Tianjin TaiSiTe instrument Co., Ltd.); Heat heating thermostat magnetic stirrer (DF-1013, Zhengzhou Greatwall Science Industry & Trade Co., Ltd); Low speed centrifuge (GL-20B, Shanghai Anting Scientific Instrument Factory); Rotary evaporator (RE-2000, the Shanghai YaRongSheng biochemical instrument factory).

2.2 Experimental

2.2.1 The Analysis Method of ZSS Alkaloids

(1) Experiment principle

Alkaloid substances dissolved in aqueous solution with a certain pH value combine with dissociative hydrogen ions to form salt (BH^+), and at the same time, add some acid dyes (such as bromine phenol blue, bromothymol blue, bromocresol green, etc.), then dyes can be dissociated as anions (In^-). The cation BH^+ and anion In^- could be combined to form a colored ion pair (In^-, BH^+) quantitatively, which can be extracted by organic solvent, and enter organic phase by aqueous phase.

$B+H^+ <=> BH^+$ (1); $HIn <=> H^+ + In^-$ (2)

As (1) and (2) showed:

pH↓ (acidity) →BH^+↑, In^-↓→BH+In-↓, HIn is extracted by organic solvent;

pH↑ (alkalinity)→In^-↑, BH^+↓→BH+In-↓, B is extracted by organic solvent.

The staining range of bromocresol green is 3.6~5.2 (Yellow ~ Blue), so the pH 4.5 buffer and pH 4.5 bromocresol green reagent were selected to be dyed.

Because of no standards of *ZSS* alkaloids right, The total alkaloids had been extracted from 50 g defatted *ZSS* powder when using spectrophotometry in each improved experiment to make the results of the experiment could be compared. Then, calculate the mass of extractions from *ZSS* per gram. After that, those extractions from every 2 grams of *ZSS* raw materials were detected by the spectrophotometry to get total amount of alkaloid. At the end, if the light absorption value from aqueous phase of the sample is less and less and from chloroform phase of the sample is more and more that means the content of alkaloids in the extracts was much more. As the result the optimal extraction method can be determined.

(2) Preparation of reagent

(a) bromocresol green solution

0.1 mol/L Potassium hydroxide solution: weigheing 56.000 g potassium hydroxide accurately, adding deionized water until 50 ml

0.2 mol/L Potassium acid phthalate buffer solution (pH4.5): weigheing 40.846 g Potassium acid phthalate accurately, adding deionized water until 1000 ml, adding 0.1 mol/L potassium hydroxide solution to pH4.5; Bromocresol green solution: weigheing 15mg bromocresol green accurately, adding pH4.5 phthalate buffer solution until 500 ml.

(b) preparation of sample solution

Taking an extractive from one extracting experiment as example: first weighing extract total quality, according to the total quality of extract and the use ratio of raw material quality *ZSS* calculated corresponding to every 2 g of *ZSS* extract quality. Firstly measuring the total quality of extractions, according to the ratio of total quality and raw material quality of *ZSS*, calculating the extraction quality from each 2 g *ZSS*; weighting the extractions quality from 2 g raw material of *ZSS*, and dissolved in chloroform. During this process, adding a bit of ammonia to cacelerate the dissolution process. At the end, adding chloroform until 25 ml.

(3) determination of total alkaloids

(a) liquid extraction

Taking the same solution three times separately as 0.50 ml, 1 ml, 2 ml 20 ml to the centrifuge tube, and adding chloroform solution until 5.00 ml, then adding 5 ml bromocresol green solution, shaking well and puting into 30 °C water bath for 5 min, and 1000 rpm speed centrifuge for 5 min.

(b) determination of total alkaloid in aqueous phase

The aqueous phase (upper phase) of each centrifugal tube was transferred into a plug scale test tube by a straw. Additionally, the potassium hydroxide buffer solution was used

as the blank, and the absorbance was measured at 616 nm and recorded the value A.
(c) determination of total alkaloids in chloroform solution

The chloroform phase (under phase) in centrifugal tube was put into another a plug scale test tube with 2 g of anhydrous sodium sulfate, shaking and then keeping staying for 1 hour. Moreover, using chloroform as blank, measuring the absorbance under 416 nm and recording value A.

2.2.2 The Extraction Method of Total Alkaloids of ZSS

(1) process flow chart
Show as Figure 1.

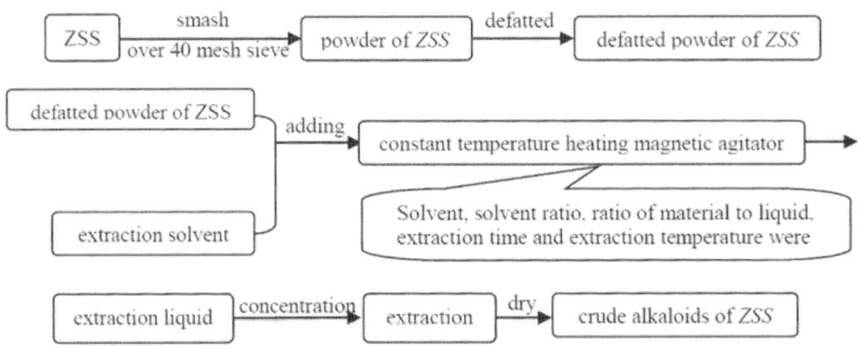

Fig. 1 Processing flowsheet of the reflux exrtraction of alkaloid in *ZSS*

(2) the single factor optimal experiment of total alkaloids extraction conditions from *ZSS*
(a) optimization of extraction solvent

Firstly, putting 50 g skim powder from *ZSS* to 1000 ml three flasks, and adding the 1000 ml different extraction agents to flasks (respectively: 5% glacial acetic acid water solution, 5% glacial acetic acid ethanol solution, 5% ammonia solution) until the liquid to solid ratio was 20:1 (ml:g). Secondly, the three flasks were put in a constant temperature heating magnetic stirrer keeping the same speed at 60 °C for 6 hours. Thridly, extracting solution would be centrifugal under 4000 rpm for 10 min. After that, in order to compared the different effect of different extraction solvent, the centrifugal supernatant was concentrated by a rotary evaporator, and when the liquid volume was less transfer to the 100 ml flask with a plug conical to evaporate until totally dry. Finally, move to the vacuum drying apparatus to dry for 2 or 3 days.
(b) optimization of acid in extraction solvent

Similarity, the process is same as (a), this step focus on adding different acid (Respectively: glacial acetic acid, hydrochloric acid, sulfuric acid) to extraction solvent to adding defatted powder of ZSS constant temperature heating magnetic agitator Solvent, solvent ratio, ratio of material to liquid, extraction time and extraction temperature were optimized extraction solvent concentration dry extraction liquid extraction crude alkaloids of *ZSS* defatted smash over 40 mesh sieve defatted powder of *ZSS* ZSS powder of *ZSS* get the better extraction effect. Besides, the extraction solvent was ethanol as well, and the other conditions were same as (a).

(3) orthogonal test conditions for extracting total alkaloids of *ZSS*

The total alkaloids extraction of ZSS have been effected by 8 factors: extraction method, the granularity of raw materials, solvents, extraction times and solvent concentration, solid-liquid ratio, extraction temperature, extraction time and the eight factors. In this research, extraction method, the size of raw material, solvent type, extraction times, these four factors have been chose. Using granularity of raw material which go through 40-50 mesh and glacial acetic acid ethanol solution as extracting agent, the orthogonal optimization was carried out on four factors: solvent concentration, solidliquid ratio, extraction temperature and extraction time by refluxing extraction. These four factors and level results as shown in Table 1. Additionally, According to the factors and the level of orthogonal table, $L_{16}(4_5)$ was chose to be the orthogonal experiment scheme.

Tab. 1 Factors and levels of the reflux exrtraction of alkaloid in *ZSS*

	Temperature (°C)	Time (h)	Ratio (g:ml)	Ethanol concentration (%)
1	50	2	1:5	95
2	60	4	1:10	90
3	70	6	1:15	85
4	80	8	1:20	80

(4) verification of optimal extraction conditions

According to the results of single factor experiment and orthogonal experiment, the optimal extraction conditions are verified by experiments to make sure this conditions if is the real best extraction way for this research.

3. Results and Discussion

3.1 Optimization Extraction Conditions of Single Factor Experiment

3.1.1 Optimization of Extraction Solvent

Figure 2 is the result of the optimization of the extraction solvent, from the graph (a) can be seen that the ethanol extract in the aqueous phase of the lowest absorbance value, the highest water extract, chloroform extract in the middle.

Figure 2 is the result of the optimization of the extraction solvent. Moreover, from the graph (a) can be seen that the ethanol extract of the aqueous phase is lowest absorbance value, the highest is water extract, chloroform extract is in the middle. From the graph (b), ethanol extract in chloroform phase is the highest absorbance value, the lowest is water extract, chloroform extract is in the middle.

Previous principle of analysis method, the lower the absorbance of the aqueous phase, the higher the value of the content of alkaloids in the extract, which means the higher the content of alkaloids in the extract. Therefore, the content of alkaloids in ethanol extract was the highest, the next was chloroform, and the content of water extract was the lowest.

On the other side, from the results of extraction rate (Figure 2: (c)) showed as, extraction rate from high to low in order is water >ethanol>chloroform. Although, the extract of water is the most, but the content of alkaloid is the lowest, and the impurities are more, and the extraction rate is lowest. Besides, solvents are toxic. Finally, considered abll above factors, ethanol have been chose as the extraction solvent.

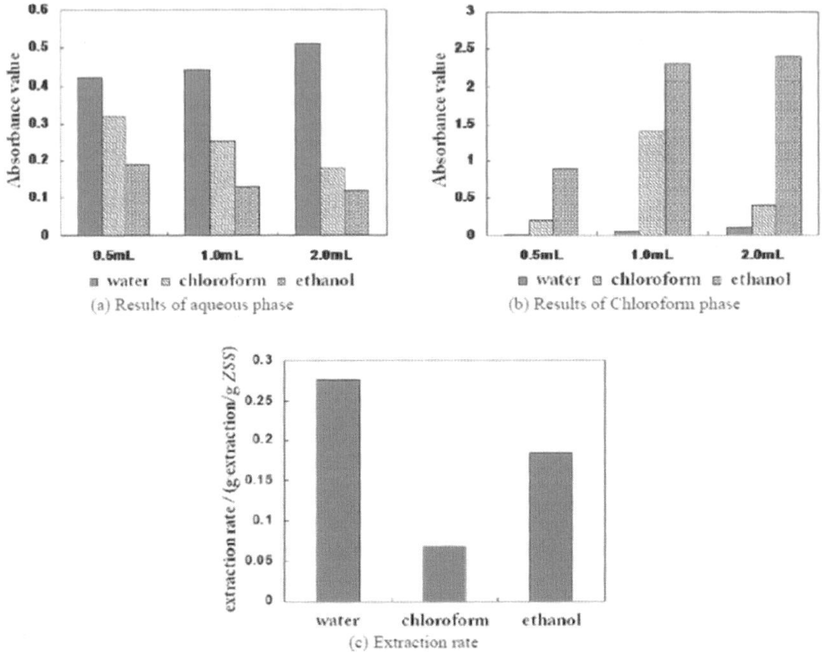

Fig. 2 comparation of different extract solvent

3.1.2 Optimization of acid in solvent extraction

Figure 3 is the result of the optimization of the acid in the solvent. From the graph (a) can be seen that the extract of sulfuric acid in the aqueous phase is the lowest absorbance value, the highest is hydrochloric acid extract, acetic acid extract is in the middle. From the graph (b), acetic acid extractin chloroform phase is the highest absorbance value, the lowest is hydrochloric acid extract, sulfuric acid is in the middle. Therefore, it can be judged that the content of alkaloid in the extract of acetic acid is the highest, and the second is the lowest, and the content of alkaloid in the extract is the lowest. Therefore, the content of alkaloids in adding acetic acid extract was the highest, the next is adding sulfuric acid, and the content of adding hydrochloric acid extract was the lowest. On the other hand, it can be seen from Figure (c), the extraction rate from high to low in order is sulfuric acid >hydrochloric acid> acetic acid. Although, the extract of water is the lowest, but the content of alkaloid is the most, and the impurities are less, and effiency is high. While, the effiency of adding sulfuric acid is really high, but there are impurities as well. So, acetic acid have been chose as the solvent for extraction.

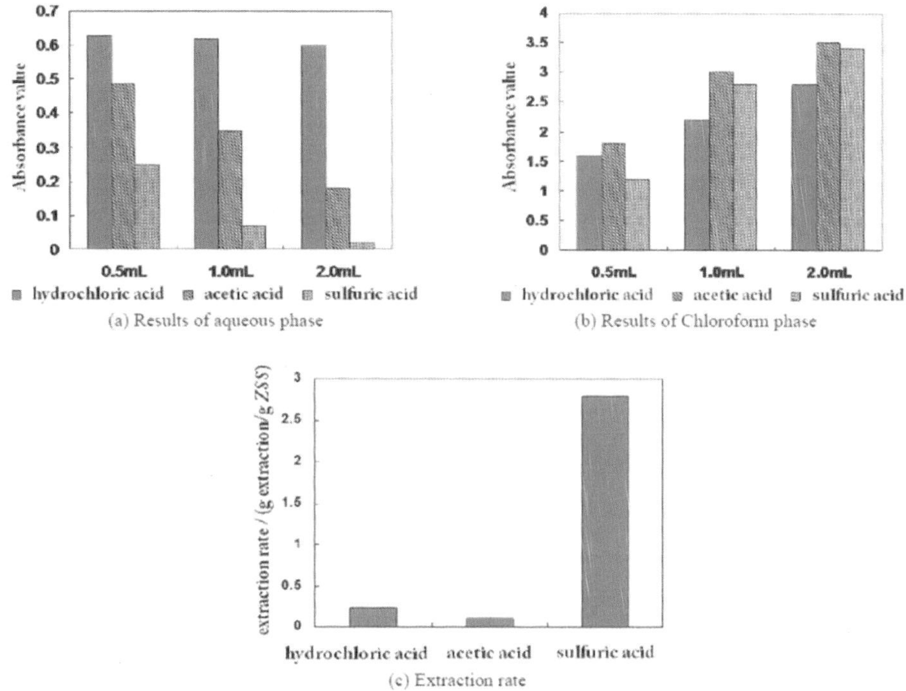

Fig. 3 comparation of different acid in extract solution

3.2 Optimization Extraction Conditions of Orthogonal Experiment from ZSS

3.2.1 Extraction Rate

As Table 2 showed, according to the value of range (R) we can see that the efficacy of *ZSS* alkaloids have been influenced by four factors from high to low in order are solid ratio > extraction temperature > extraction time > ethanol concentration. In summary, the combination of optimal factors are-- 95% ethanol in solid: liquid equal 10:1 solution under 70 °C for 8 h.

Tab. 2 $L_{16}(4_5)$ Orthogonal test and results of the reflux exrtraction of alkaloid in *ZSS*

NO.	Temperature (°C)	Time (h)	Ratio (g:ml)	EtOH conc. (%)	Blank	Extraction rate (mg Extractive/g ZSS)
1	1	1	1	1	1	44.7
2	*1*	*2*	*2*	2	2	87.6
3	*1*	*3*	*3*	3	3	91.1
4	*1*	*4*	*4*	4	4	122.9
5	*2*	*1*	*2*	3	4	98.8
6	*2*	*2*	*1*	4	3	80.2
7	*2*	*3*	*4*	1	2	115.0
8	*2*	*4*	*3*	2	1	96.7
9	*3*	*1*	*3*	4	2	130.8
10	*3*	*2*	*4*	3	1	128.7
11	*3*	*3*	*1*	2	4	57.5

12	*3*	*4*	*2*	1	3	946.3
13	*4*	*1*	*4*	2	3	143.6
14	*4*	*2*	*3*	1	4	98.8
15	*4*	*3*	*2*	4	1	123.7
16	*4*	*4*	*1*	3	2	87.7
Mean Value	86.575	104.475	67.525	301.200	98.450	
Mean Value	97.675	98.825	314.100	96.350	105.275	
Mean Value	315.825	96.825	104.350	101.575	315.300	
Mean Value	113.450	313.400	127.550	114.400	94.500	
Range	229.250	216.575	246.575	204.850	220.800	
Optimal Level	A3	B4	C2	D1		

3.2.2 Determination Content of Extract Alkaloids in Orthogonal Experiment

Tab. 3 $L_{16}(4_5)$ Orthogonal test and results of the reflux exrtraction of alkaloid in ZSS obtaioned from aqueous

NO.	Temperature (°C)	Time (h)	Ratio (g:ml)	EtOH conc. (%)	Blank	Absorbance value		
						0.5ml	1.0ml	2.0ml
1	1	1	1	1	1	0.727	0.651	0.542
2	1	2	2	2	2	0.687	0.536	0.316
3	1	3	3	3	3	0.611	0.592	0.37
4	1	4	4	4	4	0.648	0.497	0.297
5	2	1	2	3	4	0.698	0.605	0.479
6	2	2	1	4	3	0.714	0.538	0.393
7	2	3	4	1	2	0.667	0.559	0.38
8	2	4	3	2	1	0.739	0.537	0.369
9	3	1	3	4	2	0.584	0.469	0.29
10	3	2	4	3	1	0.949	0.38	0.246
11	3	3	1	2	4	0.729	0.65	0.475
12	3	4	2	1	3	0.729	0.612	0.393
13	4	1	4	2	3	0.634	0.467	0.267
14	4	2	3	1	4	0.71	0.537	0.337
15	4	3	2	4	1	0.509	0.491	0.359
16	4	4	1	3	2	0.688	0.683	0.398
Mean Value	0.668	0.66	0.714	0.708	0.731			
Mean Value	0.704	0.76	0.656	0.697	0.657			
Mean Value	0.748	0.62	0.661	0.736	0.672	0.1ml		
Mean Value	0.635	0.70	0.724	0.614	0.696			
Range1	0.113	0.13	0.068	0.122	0.074			
Optimal level 1	A4	B3	C2	D4				
Mean Value	0.569	0.54	0.631	0.59	0.515			
Mean Value	0.56	0.49	0.561	0.547	0.562			
Mean Value	0.528	0.57	0.534	0.565	0.552	1.0ml		
Mean Value	0.544	0.58	0.476	0.499	0.572			
Range 2	0.041	0.08	0.155	0.091	0.057			
Optimal Level 2	A3	B2	C4	D4				
MeanValue	0.381	0.39	0.452	0.413	0.379			
MeanValue	0.405	0.32	0.387	0.357	0.346			
Mean Value	0.351	0.39	0.341	0.373	0.356	2.0ml		
Mean Value	0.34	0.36	0.297	0.335	0.397			
Range 3	0.065	0.07	0.155	0.078	0.051			
Optimal Level 3	A4	B2	C4	D4				

From Table 3 it could be seen that the results of the 3 different sampling volume phase. According to the value of range (R), the efficacy of *ZSS* alkaloids have been influenced by four factors from high to low in order are solid to liquid ratio > ethanol concentration > extraction time>extraction temperature. In summary, the combination of optimal factors are 80% ethanol in solid: liquid equal 20:1 solution under 80 °C for 4 h.

3.2.3 Determination Content of Extract Alkaloids in Chloroform Phase in Orthogonal Experiment

From Table 4 it could be seen that the results of the 3 different sampling volume chloroform phase. According to the value of range (R), the efficacy of ZSS alkaloids have been influenced by four factors from high to low in order are extraction temperature > solid to liquid ratio > ethanol concentration > extraction time. In summary, the combination of optimal factors are 80% ethanol in solid : liquid equal 20:1 solution under 60 °C for 4 h.

Tab. 4 L16(45) Orthogonal tests and results of the reflux exrtraction of alkaloid in *ZSS* obtained from organic Phase

NO.	Temperature (°C)	Time (h)	Ratio (g:ml)	EtOH conc. (%)	Blank	Absorbance value 0.5ml	1.0ml	2.0ml
1	1	1	1	1	1	0.093	0.188	0.712
2	1	2	2	2	2	0.162	0.772	0.688
3	1	3	3	3	3	0.196	0.467	0.628
4	1	4	4	4	4	0.458	1.797	1.302
5	2	1	2	3	4	1.249	1.195	2.302
6	2	2	1	4	3	0.175	0.514	2.315
7	2	3	4	1	2	0.828	1.109	2.495
8	2	4	3	2	1	0.623	1.009	1.436
9	3	1	3	4	2	0.495	0.776	1.065
10	3	2	4	3	1	0.474	0.531	1.179
11	3	3	1	2	4	0.3	0.614	1.107
12	3	4	2	1	3	0.543	0.565	1.403
13	4	1	4	2	3	0.566	0.922	1.347
14	4	2	3	1	4	0.571	0.508	1.836
15	4	3	2	4	1	0.278	0.604	1.139
16	4	4	1	3	2	0.171	0.305	0.534
Mean Value	0.227	0.601	0.185	0.509	0.367	0.5ml		
Mean Value	0.719	0.345	0.558	0.413	0.414			
Mean Value	0.453	0.401	0.471	0.522	0.37			
Mean Value	0.397	0.449	0.582	0.352	0.645			
Range 1	0.492	0.256	0.397	0.17	0.278			
Optimal Level 1	A2	B1	C4	D3				
Mean Value	0.806	0.77	0.405	0.593	0.583	1.0ml		
Mean Value	0.957	0.581	0.784	0.829	0.741			
Mean Value	0.621	0.699	0.69	0.625	0.617			
Mean Value	0.585	0.919	1.09	0.923	1.028			
Range 2	0.372	0.338	0.685	0.33	0.445			
Optimal Level 2	A2	B4	C4	D4				
Mean Value	0.822	1.357	1.167	1.611	1.117	2.0ml		
Mean Value	2.137	1.505	1.383	1.144	1.195			

Mean Value	1.188	1.342	1.241	1.161	1.423
Mean Value	1.214	1.169	1.581	1.455	1.637
Range 3	1.304	0.336	0.414	0.467	0.52
Optimal Level 3	A2	B2	C4	D1	

In conclution, the highese extraction rate of optimum condition is 80% ethanol in solid: liquid equal 10:1 solution under 70 °C for 8 h. From the optimal factors of determination content of extract alkaloids in aqua phase is 80% ethanol in solid : liquid equal 20:1 solution under 80 °C for 4 h. From the optimal factors of determination content of extract alkaloids in chloroform phase is 80% ethanol in solid : liquid equal 20:1 solution under 60 °C for 4 h. Finally, regard the total alkaloids of extraction as the target, the best condition is 80% ethanol in solid : liquid equal 20:1 solution under 60 °C for 4 h.

3.3 Results of Validate Experiment

Followed the best condition from the orthogonal experiment (20% acetic acid ethanol in solid: liquid equal 20:1 solution under 60 °C for 4 h) as verification experiment, determination results of the extracts are shown in Table 5, which can be seen by the result that the extraction of the alkaloid content in the condition is higher than that of the orthogonal experiment. Therefore, the conditions above cloud be determined as the optimal extraction conditions.

Tab. 5 The results of the reflux exrtraction of alkaloid in *ZSS* under optimized conditions

	Volume of Sample solution	Aqueous phase	Chloroform phase
Verification experiment	0.5ml	0.918	1.345
	1.0ml	0.699	1.885
	2.0ml	0.613	2.579

4. Conclusion

The total alkaloids of ZSS was measured by spectrophotometric. Inadditionally, extraction solvent and acid solvent extraction, extraction time, extraction temperature, liquid to solid ratio and solvent concentration were optimized through the single factor experiment and orthogonal experiment. Finally getting the optimal reflux extraction conditions of total alkaloids of ZSS is: Using ethanol as solvent, with acetic acid as assistant extraction, 20% acetic acid ethanol in solid : liquid equal 20:1 solution under 60 °C for 4 h is the best extraction conditions

Acknowledgements

The authors would like to express their gratitude to the Beijing Natural Science Fund (14L00184), the Importation and Development of High-Caliber Talents Project of Beijing Municipal Institutions (CIT&TCD20154045) and The Degree and Graduate Education Reform and Development Project of BUA in 2015 (2015YJS034) for the financial support of this study.

References

1. Chinese Pharmacopoeia Commission (ed.), *Pharmacopoeia of the people's Republic of China*, 254-255, (Beijing: chemical Industry Press, 2005). (in Chinese)
2. Q. R. Shi and Y. Zhou, Research survey on traditional Chinese medicine of the seeds of Zizyphus vulgaris var. spinosus. *Journal of Pharmaceutical Practice* 22, 94-98 (2004). (in Chinese)
3. B. H. Han and M. H. Park, Chemical and pharmacological studies on sedative cyclopeptide alkaloids in some Rhamnaceae plants. *Pure and applied chemistry* 61, 443-448 (1989).
4. B. H. Han and M. H. Park, Alkaloids are the sedative principles of Semen Zizyphi Spinosae. *Archives of pharmacal research* 10, 203-207 (1987).
5. J. W. Fu and W. Qiao, Study on sedative and hyponotic effect of total alkaloid from Semen Zizyphi Spinosae. *Journal of Tianjin Medical University* 11, 52-54 (2005). (in Chinese)
6. L. H. Zhao and W. Qiao, Experimental study on anti-convulsion effect of alkaloids from semen zizyphi spinosae. *Tianjin Pharmay* 19, 4-5 (2007). (in Chinese)
7. T. L. Zhu and Z. S. Gao, Study on the antidepressant effect of active fraction from Semen Zizyphi Spinosae. *Acta Academiae Medicinae CPAF* 18, 420-425 (2009). (in Chinese)
8. X. S. Yao (ed.), *Natural medicinal chemistry*, 459-468, (People's Medical Publishing House, 1994). (in Chinese)
9. Y. Liu and Z. H. Yue, Development of Research on Securinine. *Chinese Pharmaceutical Affairs* 23, 817-818, 828 (2009). (in Chinese)
10. H. G. Chen and Y. G. Yu, Extraction of functional component from lotus leaf. *Food and Machinery* 5, 16-17 (2001). (in Chinese)
11. C. Zhang and H. B. Zhang, Research on extraction process of phenol alkaloids in Menispermum dauricum D.C. *Chinese Traditional and Herbal Drugs* 28, 274-276 (1997). (in Chinese)
12. P. Zhen and F. Z. Yang, Investigation on extraction conditions of total alkaloids in Tetradium ruticarpum. *China Journal of Chinese Materia Medica* 25, 504-505 (2000). (in Chinese)
13. P. Chen and Z. R. Xing, Comaparison of extraction and analysis methods of total alkaloids in Corydalis yanhusuo. *Chinese Journal of Modern Applied Pharmacy* 8, 15-16 (1993). (in Chinese)
14. S. Z. Yin and H. K. Jin, Studies on alkaloids in the seeds of Zizyphus jujuba Mill.. *China journal of chinese materia medica* 22, 296-297 (1997). (in Chinese)
15. H. H. Byung and H. P. Myung, Cyclic peptide and peptide alkaloids from seeds of Zizyphus vulgaris. *Phytochemistry* 29, 3315-3319 (1990).
16. B. H. Han and M. H. Park, Chemical and pharmacological studies on sedative cyclopeptide alkaloids in some Rhamnaceae plants. *Pure and applied chemistry* 61, 443-448 (1989).

17. B. H. Han and M. H. Park, Aporphine and tetrahydrobenzylisoquinoline alkaloids from the seeds of Zizyphus vulgaris var. spinosus. *Archives of pharmacal research* 12, 263-268 (1989).
18. B. H. Han and M. H. Park, Alkaloids are the sedative principles of the seeds of Zizyphus vulgaris var. spinosus. *Archives of pharmacal research* 10, 203-207 (1987).
19. B. H. Han and M. H. Park, Sedative activity and its active components of Zizyphi fructus. *Archives of pharmacal research* 12, 263-268 (1989).
20. M. Tripathi and M. B. Pandey, Cyclopeptide alkaloids from Zizyphus jujube. *Fitoterapia* 72, 507-510 (2001).

Detection of bisphenol A in canned milk by chemiluminescence enzyme immunoassay

K. H. Li and R. R. Liu
School of Life Science, Jiangxi Science & Technology Normal University, Nanchang, Jiangxi, 330013, China
E-mail: likaihong08@163.com and lilirenrong@hotmail.com

A sensitive and rapid chemiluminescence enzyme immunoassay (CLEIA) was established for detection of bisphenol A in canned milk. Five points include antigen coated concentration, the dilution proportion, PH value, ionic strength of the buffer and the concentration of different organic solvents were tested for optimization. A proper condition for CLEIA is that antigen coated concentration was 1ug/ml, the dilution proportion of horseradish peroxidase (HRP)-conjugated anti-antibody was 1:3000, PH value was 8.0, ionic strength of the buffer was 0.8 mol/L, concentration of different organic solvents was 0%. And for spiked samples, the all recoveries ranging from 94.98% to 112.91% were acceptable. The CLEIA is an appropriate method for detection of BPA in canned milk.

Keywords: Chemiluminescence Enzyme Immunoassay; Bisphenol A; Canned Milk.

1. Introduction

Bisphenol A [BPA; bis (4-hydroxyphenyl) propane] is an endocrine disrupting environmental contaminant used in a wide variety of macromolecular compounds [1], such as plastic products and coating products. And those products usually are used as food packaging materials. For example, epoxy resins used to line cans, dental composites and sealants [2]. In order to prevent corrosion, cans of food are usually coated by epoxy resins [3]. Thus, there exists a laggardly and different degree of migration of BPA between coating and food content. Referring to canned milk, BPA will migrate from coating to milk.

The methods for BPA detecting are varied. The chromatographic method and immunological method are commonly used. Chromatographic method such as high performance liquid chromatography (HPLC) with fluorescence detection (FLD) [4, 5] or ultraviolet detector (UV), but FLD is more sensitive than UV. Immunological method such as enzyme linked immunosorbent assay (ELISA) is a normal and sensitive method for BPA detection. [6]. Compared with chromatographic techniques, immunoassays are rapid, specific and sensitive analytical methods, which are cost and time effectiveness. The CLEIA is a combination of a highly sensitive chemiluminescence (CL) detection and specific immunosorbent assay, which can display simplicity, sensitivity, low detection limits, wide linear dynamic ranges, non-radioactive pollution and speed of response with low cost and portable instrumentation [7].

In this paper, a CLEIA was developed based on monoclonal antibodies for the detection of BPA in canned milk. Through optimization of several conditions, an excellent condition for calibration curve and a lower 50% inhibiting concentration (IC_{50}) were obtained. All spiked samples showed good recoveries.

2. Materials and methods

2.1. *Reagents and instrument*

Mouse anti-zearalenone monoclonal antibody; Horseradish peroxidase labelled goat anti-mouse anti-antibody immunoglobulins and hapten were purchased from Sigma; Methanol (HPLC grade) and acetonitrile were obtained from Sigma; Phosphate-buffered saline (PBS); UltraSensitive Substrake Kit for CLEIA was purchased from Heliosense Biotechnologies. Luminoskan ascent (Thermo scientific, USA); Wellwash versa (Thermo scientific, USA); 96-well white polystyrene plates (Costar, USA); 5804R High-speed Refrigerated Centrifuge (Eppendorf, Germany).

2.2. *Experimental details for indirect CLEIA*

The whole experiments were performed in 96-well white polystyrene high-binding microplates. The coating condition was BPA-BSA antigen (120 μl per well) incubated 2h at 37°C lucifuge moisturize. After the wash step, 5% Skimmed milk powder (330 μl per well) was used for blocking the uncoated sites at 37°C lucifuge moisturize incubating 3h. The third step of adding antibody (100 μl per well) and the fourth step of adding horseradish peroxidase (HRP)-conjugated anti-antibody (100 μl per well) take the same incubated condition as the former except for time (40min). Finally, adding the intermixture (100 μl per well) of chemiluminescent substrate A and B (A: B=1:1). The luminescence value was detected by luminoskan ascent and expressed as Relative Luminescence Units (RLU). Especially, there must be a wash step after incubation every time.

2.3. *Experimental details for competitive CLEIA*

The step of coating and blocking were the same as the step of indirect CLEIA. But for third step, a series of 50 μl BPA standards were added per well first, then 50 μl BPA antibody were added per well immediately and gently patted for blending. The remaining steps were the same as the indirect CLEIA.

2.4. *Conditions optimization for CLEIA*

The optimal concentrations of coated-antigen and antibody for the chemiluminescent assay were first carried by checkerboard titration using the twice antigen coated concentration range between 0.25 μg/ml~ 4 μg/ml and doubling dilution for antibody concentration range 1:4000~1:2560000. The diluents of BPA were different concentrations of methanol-water (from 0% to 30%, v/v). The concentrations of horseradish peroxidase (HRP)-conjugated anti-antibody are 1:2000, 1:3000, 1:4000, 1:6000 and 1:8000, respectively. PH value and ionic strength of the buffer range from 5.5 to 9.0 and 0 mol/L to 2 mol/L, respectively.

2.5. *Calibration curves*

Under the optimum conditions, a calibration curve was performed in the 0.1-200ng/ml range. The percent binding (% B/B_0) were normalized between 0% and 100% according to the expression Eq. (1):

$$\%B / B_0 = \frac{E_x}{E_0} \times 100. \tag{1}$$

Where Ex was the chemiluminescent emission value when the concentration of BPA-hapten was x ng/ml. E0 was the emission value of when the concentration of BPA-hapten was 0 ng/ml.

2.6. *Analysis of spiked samples*

The spiked samples were performed based on milk. 100 ng, 500 ng and 1000 ng BPA standard were added into 2 ml milk respectively. The BPA was extracted by 3 ml acetonitrile. After vortex shocking and centrifugation, acetonitrile was separated. Then nitrogen was used to blow-dry the acetonitrile. Finally, the leavings were dissolved by ultrapure water. In this experiment, dilution process was used to eliminate the interference from matrix.

3. Result and discussion

3.1. *Conditions optimization for CLEIA*

The optimal concentration of the coating antigen was 1μg/ml. On the basis of the lowest IC_{50}, ultrapure water was chosen as the diluents of BPA and the influence of organic solvents showed in Figure 1 (a). As the dilution proportion of horseradish peroxidase (HRP)-conjugated anti-antibody increases, the luminous value reduced and the IC_{50} essentially unchanged showed in Figure 1 (b), so we chose an appropriate proportion 1:3000 as the proportion of anti-antibody. On the basis of the higher $RLUmax/IC_{50}$, we chose 8.0 and 0.8 mol/L Na^+ as the PH and ionic strength of dilution buffer of BPA antibody, respectively. And the influence showed in Figure 1 (c) and Figure 1 (d).

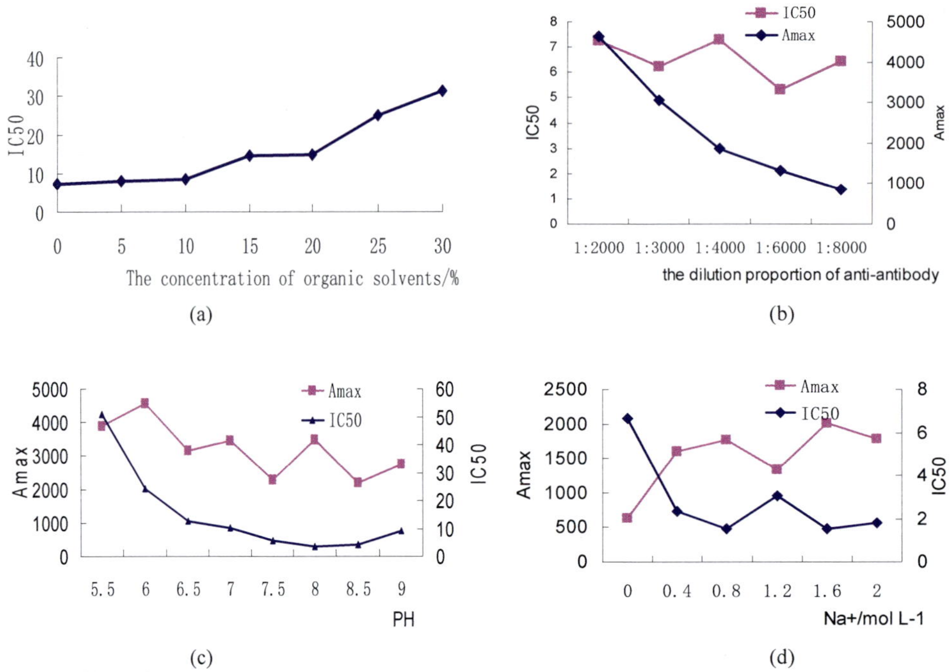

Fig. 1. Effect of organic solvents concentration (a), horseradish peroxidase(HRP)-conjugated anti-antibody dilution proportion (b), PH value (c) and ionic strength (d) on BPA immunoassay based on CLEIA

3.2. *Calibration curves*

The calibration curve is shown in Figure 2. It shows a good linearity correlation (R=0.995) in range of 1 ng/ml to 20ng/ml and was represented by the Eq. (2). According to `the calibration curve, the IC_{50} value is 3.88 ng/ml.

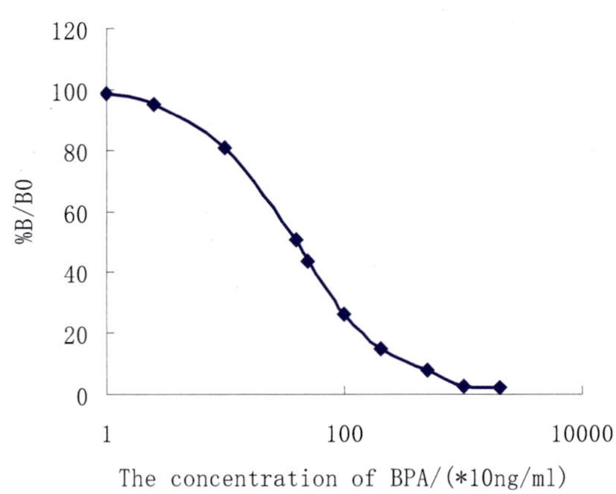

Fig. 2. The calibration curve of BPA based on chemiluminescence enzyme immunoassay (CLEIA)

$$y = -22.566\ln(x) + 132.58. \tag{2}$$

3.3. *Analysis of spiked samples*

With adding standard to three levels, the recovery ratio and RSD value are shown in Table 1. From Table 1 we can find that three levels of the spiked samples, they all obtain good recovery ratios and RSD values. It proves that the CLEIA has a good accuracy and reproducibility.

Tab. 1 The recovery ratio and RSD value of milk

Adding standard [ng]	Recovery ratio [%]	RSD value [%]
100	500	1000
112.91	104.57	94.98
0.93	3.42	3.96

4. Conclusions

In the past decade endocrine disruptors have become a central topic in the international discussion in environmental and food chemistry [8]. The U.S. Environmental Protection Agency (EPA) has set the oral reference dose (RfD) of BPA at 50 μg kg•bw^{-1} day^{-1} (bw means body weight) [9]. Through condition optimization, this method showed a good sensitivity and repeatability. And the method is time and instruments saving. It is not only an appropriate method for detection of trance BPA, but also a basic for monitoring the migration of BPA.

Acknowledgments

This work was financially supported by grants from National Natural Science Foundation of China (No. 81360429) and Natural Science Foundation of Jiangxi Province (No. 20122BAB214006, 20132BAB214004, KJLD14067).

References

1. Z. L. Mei, Y. Deng and H. Q. Chu, et al, Immunochromatographic lateral flow strip for on-site detection of bisphenol A, *Microchimica Acta* **180**, 279-285 (2013).
2. A. M. Hormann, F. S. Vom Saal and S. C. Nagel, et al, Holding thermal receipt paper and eating food after using hand sanitizer results in high serum bioactive and urine total levels of bisphenol A (BPA), *PLOS One* **9**, 1-12 (2014).
3. E. M. Munguia-Lopez, E. Peralta and A. Gonzalez-Leon, et al, Migration of bisphenol A (BPA) from epoxy can coatings to Jalapeno peppers and an acid food stimulant, *J. Agric. Food Chem* **50**, 7299-7302 (2002).
4. T. Watanabe, H. Yamamoto and K. Inoue, et al, Development of sensitive high-performance liquid chromatography with fluorescence detection using 4-(4, 5-diphenyl-1H-imidazol-2-yl)-benzoly chloride as a labeling reagent for determination of bisphenol A in plasma samples, *Journal of Chromatography B Biomedical Sciences & Applications* **762**, 1-7 (2001).

5. X. H. Cao and S. J. Ji, Migration Study of Bisphenol A from Tea Bottle of Polycarbonate Plastic to Tea Water, *Advanced Materials Research* **821**, 929-932 (2013).
6. J. Zhou, S. Zhao and J. Zhang, et al, An indirect competitive enzyme-linked immunosorbent assay for bisphenol-A based on the synthesis of a poly-L-lysine–hapten conjugate as a coating antigen, *Analytical Methods* **5**, 1570-1576 (2013).
7. Y. Zhang, J. Y. Yang and H. T. Lei, et al, Development of chemiluminescent enzyme immunoassay for the determination of malachite green in seafood, *Food & Agricultural Immunology* **26**, 204-217 (2014).
8. R. Braunrath, D. Podlipna and S. Padlesak, et al, Determination of bisphenol A in canned foods by immunoaffinity chromatography, HPLC, and fluorescence detection, *Journal of Agricultural & Food Chemistry* **53**, 8911-8917 (2005).
9. Y. Xia and M. Rubino, Kinetic study of bisphenol A migration from low-density polyethylene films into food simulants, *Industrial & Engineering Chemistry Research* **54**, 3711-3716 (2015).

Thin nutrition pie manufacturing by using radiation effect at stagnation point heat transfer with micropolar flow and multimedia feature

I-Hua Lin and Kai-Long Hsiao*

Department of hospitality management, Taiwan Shoufu University, Tainan city, Taiwan

Department of Digital Entertainment and Game Design, Taiwan Shoufu University, Tainan city, Taiwan

*Corresponding author (Email: hsiao.kailong@msa.hinet.net)

The energy conversion conjugate conduction-convection Ohmic mixed heat transfer of an incompressible viscoelastic fluid on thermal forming thin nutrition pie has been studied. The present study has been applied similarity transformation method to change the partial differential equations transform to a set of nonlinear ordinary differential equations, and it has also been used an implicit finite-difference method to solve the system's equations. The numerical calculating results for the conjugate heat transfer energy conversion problem which have been carried out as functions of different parameters. The parameters of Pr, and k0 are important energy conversion factors in this study, it should be produced greater heat transfer energy conversion effect with a larger values of k0, or Pr, but the parameter Ec will be reduced the heat transfer energy conversion effect. Because of present study has been studied about many parameters physical features, so that it is also belong to a multimedia physical feature study work.

Keywords: Thin nutrition pie, Conjugate heat transfer, Extrusion processing, Stagnation, Multimedia.

1. Introduction

In this study, thin film nutrition pie processing has been extruded through a slit die, stretched and cooled in micropolar fluid with radiation effect. The energy conversion heat transfer problem for radiation and electromagnetic effects micropolar fluids past through a thin nutrition pie have been studied. Micropolar fluid is a fluid containing micrometer-sized particles, called microparticles. These fluids are engineered colloidal suspensions of microparticles in a base fluid.

Recently, the energy conversion for radiative nanofluid and micropolar fluid flow are become importance. Hsiao [1] studied about energy conversion conjugate conduction-convection and radiation over non-linearly extrusion stretching sheet with physical multimedia effects. Hsiao [2] investigated nanofluid flow with multimedia physical features for conjugate mixed convection and radiation. Hsiao [3] studied heat and mass transfer for micropolar flow with radiation effect past a nonlinearly stretching sheet.problem.

The heat transfer problem of for flow with radiation effect micropolar fluids past through a thin nutrition pie has been studied. Flow, heat, and mass transport phenomena over a continuous, moving flat surface are important in a number of technological processes. The analysis of such transport phenomena finds applications in different areas such as the aerodynamic extrusion of plastic sheets, the continuous casting, rolling, and extrusion in manufacturing processes, and the boundary layer along a liquid film in condensation processes. A new stage in the evaluation of fluid dynamic theory is in progress because of its increasing importance in processing industries and elsewhere of materials whose behavior in shear cannot be characterized by Newtonian relationships. The theory of micropolar fluids proposed by Eringen [4] deals with a class of fluids which exhibit certain microscopic effects arising from local structure micromotions of the fluid elements. Physical micropolar fluids may present the non-Newtonian fluid models which can be used to analyze the behavior of exotic lubricants by Khonsari [5], colloidal suspensions or polymeric fluids by Hudimoto and Tokuokab [6], liquid crystals by Lee and Eringen [7], Lockwood et al. [8]. The mass and momentum transport in laminar boundary layer on moving, stationary and linearly stretching surface has considerable practical applications in electrochemistry by Chin [9]. MHD mixed convection related problem by Hsiao [10]. However, there are fluids which react chemically with some other ingredients present in them. Das et al. [11] have studied the effect of a chemical reaction on the flow past an impulsively started infinite vertical plate with uniform heat flux. Anderson et al. [12] had studied the diffusion of a chemically reactive species from a linearly stretching sheet. Anjalidevi and Kandasamy [13] had examined the effect of a chemical reaction on the flow along a semiinfinite horizontal plate in the presence of heat transfer. McLeod and Rajagopal [14] had studied the uniqueness of the flow of a Navier Stokes fluid due to a linear stretching boundary.

In the present study, we have investigated the effect of heat transfer of radiation effect viscous fluid on the flow toward a nonlinearly semi-infinite thin nutrition pie stagnation point. It is an important novel study in this specific heat transfer and energy conversion system. The method arrange a sideway flow toward the thin nutrition pie, then become a stagnation point flow field sheet cooling system, and from the results we find the heat transfer effect is very well for stagnation point arrangement. So that, in this study is not only provide a new ways to solve the extrusion sheet cooling problem, but also have a very good effect for this thermal energy conversion system. The paper has been used many multimedia skills, such as figures, tables to show its results.

The physical model for thin nutrition pie extrusion manufacturing process steady forced convection flow with radiative effect for an incompressible micropolar fluid flow has been investigated. In which the micropolar fluid is toward a stagnation point along the thin nutrition pie, the energy radiation is also toward the thin nutrition pie, so that the sheet extrusion manufacturing processing can be obtain a good energy conversion effect.

2. Theory and Analysis

2.1 General Nutrition of the thin nutrition pie

A 130-gram slice of commercially prepared nutrition pie has about 320 calories, 45 grams of carbohydrates, 12 grams of fat, 2.5 grams of saturated fat, 5 grams of protein,

2.3 grams of fiber, 430 milligrams of sodium and 33 milligrams of cholesterol, according to the U.S. Department of Agriculture's National Agricultural Library. There are 20 percent of the recommended daily intake of fat, 19 percent of sodium, 15 percent of carbohydrates, 13 percent of saturated fat, 12 percent of cholesterol, 11 percent of protein and 10 percent of fiber Vitamins. A slice of pie is a significant source of water-soluble B vitamins, offering 20 percent of the recommended daily intake for thiamin and vitamin B-12, 13 percent for riboflavin, 12 percent for folate and 9 percent for niacin. One slice of pie supplies 9 percent of daily recommendation for phosphorous and 7 percent for calcium and iron. Benefits of these minerals include bone, tooth, neurological, cellular and metabolic health. They also aid in muscle contraction, blood clotting and blood oxygenation. Pie contains nominal amounts of other minerals.

2.2 Flow field analysis

In this study, consider the flow of an incompressible viscous fluid past a thin nutrition pie coinciding with the plane $y = 0$, the flow being confined to $y > 0$. Two equal and opposite forces are applied along the x-axis so that the wall is stretched keeping the origin fixed. The steady two-dimensional boundary layer equations for this fluid, in the usual notation, are

$$\frac{\partial u}{\partial x}+\frac{\partial v}{\partial y}=0, \tag{1}$$

$$u\frac{\partial u}{\partial x}+v\frac{\partial u}{\partial y}=U\frac{\partial U}{\partial x}+(\upsilon+\frac{k}{\rho})\frac{\partial^2 u}{\partial y^2}+\frac{k}{\rho}\frac{\partial N}{\partial y} \tag{2}$$

$$u\frac{\partial N}{\partial x}+v\frac{\partial N}{\partial y}=\frac{\gamma}{\rho j}\frac{\partial^2 N}{\partial y^2}-\frac{k}{\rho j}(2N+\frac{\partial u}{\partial y}) \tag{3}$$

Where (x, y) denotes the Cartesian coordinates along the sheet and normal to it, u and v are the velocity components of the fluid in the x and y directions, respectively. U is the free stream velocity, ρ is the fluid density and N is the micro-rotation or angular velocity. By Nazar et al. (2004) $\gamma=(\mu+\frac{k}{2})j$ and j, γand k are the micro-inertial per unit mass, spin gradient viscosity and vortex viscosity, respectively. υ is the kinematic viscosity, and $j=\upsilon/cx^{1-n}$. The boundary conditions to the problem are

$u_w(x)=Bx^n$, $v=0$, $N=0$ at $y=0$,

$u\to 0$, $N\to 0$ as $y\to\infty$,

Where B and n are parameters related to the surface stretching speed. Defining new similarity variables as

$$\eta=y\sqrt{\frac{B(n+1)}{2\upsilon}}x^{\frac{n-1}{2}},\ u=Bx^n f'(\eta), \tag{4}$$

$$v=-\sqrt{\frac{B\upsilon(n+1)}{2}}x^{\frac{n-1}{2}}\left[f(\eta)+\frac{n-1}{n+1}\eta f'(\eta)\right]$$

$$N = Bx^n \sqrt{\frac{B(n+1)}{2\upsilon}} x^{\frac{n-1}{2}} g(\eta) \quad , \quad U = Ax^n$$

and substituting into Eqs. (1), (2) and (3) give

$$(f')^2 \left(\frac{2n}{n+1} \right) - ff'' - (1+K)f''' - Kg - \frac{2n}{n+1}\beta^2 = 0 \tag{5}$$

$$(1+\frac{K}{2})g'' + fg' - \frac{3n-1}{n+1}f'g - \frac{2K}{n+1}(2g+f'') = 0 \tag{6}$$

Where a prime denotes differentiation with respect to the independent similarity variable η. $K=k/\mu(>0)$ is the vortex viscosity or the material parameter, β is the stagnation parameter. The boundary conditions (3) becomes

$f = 0, f' = 1$, $g=0$ at $\eta \to 0$

$f' \to 0$, $g \to 0$ as $\eta \to \infty$

For the linearly stretching boundary problem (i.e., n = 1) the exact solution for the velocity field f is

$$f(\eta) = 1 - \exp(-\eta) \tag{7}$$

and this exact solution is unique, while for the nonlinearly stretching boundary problem (i.e., $n \neq 1$) there is no exact solution. The shear stress at the stretched surface is defined as

$$\tau_w = \mu \left(\frac{\partial u}{\partial y} \right)_w \tag{8}$$

and we obtain form (4) and (8)

$$\tau_W = B\mu \sqrt{\frac{B(n+1)}{2\upsilon}} x^{\frac{3n-1}{2}} f''(0) \tag{9}$$

Where μ is the viscosity of the fluid.

2.3 Heat transfer analyses

By using usual boundary layer approximations, the equation of the energy for temperature T in the presence of radiation and viscous dissipation, is given by:

$$u\frac{\partial T}{\partial x} + v\frac{\partial T}{\partial y} = \frac{k}{\rho c_p}\frac{\partial^2 T}{\partial y^2} + \frac{\upsilon}{c_p}\left(\frac{\partial u}{\partial y}\right)^2 - \frac{1}{\rho c_p}\frac{\partial q_r}{\partial y} \tag{10}$$

where k is the thermal conductivity, ρ is the density, c_p is the specific heat of a fluid at constant pressure and q_r is the radiative heat flux. Using the Rosseland approximation for radiation by Brewster (1972), the radiative heat flux is simplified as

$$q_r = -\frac{4\sigma^*}{3k}\frac{\partial T^4}{\partial y} \tag{11}$$

Where σ^* and k^* are the Stefan–Boltzmann constant and the mean absorption coefficient, respectively. We assume that the temperature differences within the flow such as that the term T^4 may be expressed as a linear function of temperature. Hence, expanding T^4 in a Taylor series about T∞ and neglecting higher-order terms we get

$$T^4 \cong 4T_\infty^3 T - 3T_\infty^4 \tag{12}$$

in view to Eqs. (11) and (12), Eq. (10) reduces to

$$u\frac{\partial T}{\partial x} + v\frac{\partial T}{\partial y} = \frac{\alpha}{k_0}\frac{\partial^2 T}{\partial y^2} + \frac{\upsilon}{cp}\left(\frac{\partial u}{\partial y}\right)^2 \tag{13}$$

where $\alpha = \frac{k}{\rho cp}$ is the thermal diffusivity; $k0 = \frac{3N_R}{3N_R + 4}$ and $N_R = \frac{k_T k^*}{4\sigma^* T_\infty^3}$ is the radiation parameter. Similarity solutions of Eq. (13) can be found by choosing appropriate boundary conditions. It is of a certain interest to consider separately the characteristics of the following two cases of main practical interest. For prescribed surface temperature (PST case), the boundary conditions are

$$T = T_W \left(= T_\infty + Ax^k\right) \text{ as } y=0; \tag{14}$$

$T \to T_\infty$ as $y \to \infty$

Where k is the surface temperature parameter. Defining the non-dimensional temperature $\theta(\eta) = \frac{T - T_\infty}{T_W - T_\infty}$ and using Eqs. (4) and (14) into Eq. (13), we get.

$$\theta'' + \sigma k0 f\theta' - \left(\frac{2k}{n+1}\right)\sigma k0 f'\theta = -\sigma k0 E_c x^{2n-k}(f'')^2 \tag{15}$$

$$\theta(0) = 1,\ \theta(\infty) \to 0, \tag{16}$$

Where $E_C = \frac{B^2}{Ac_p}$, and $\sigma = \frac{\upsilon}{\alpha}$ is the Prandtl number.

If k=2n, we find form (18)

$$\theta'' + \sigma k0 f\theta' - \left(\frac{4n}{n+1}\right)\sigma k0 f'\theta = -\sigma k0 E_C (f'')^2 \tag{17}$$

It is clearly that Eq. (17) becomes a similar type. On the other hand, for an arbitrary value of k and neglecting heat dissipation, we find from Eq. (15)

$$\theta'' + \sigma k0 f\theta' - \left(\frac{2k}{n+1}\right)\sigma k0 f'\theta = 0 \tag{18}$$

The local surface heat flux can be expressed as

$$q_w = -k_T\left(\frac{\partial T}{\partial y}\right)_w + (q_r)_w \tag{19}$$

$$= -\frac{k}{k_0} A x^{\frac{2k+n-1}{2}} \theta'(0)\sqrt{\frac{B(n+1)}{2\upsilon}}$$

3. Numerical Technique

In the present problem, the set of similar equations (5), (6), (16) and (17) are solved by a finite difference method. These ordinary differential equations have discretized by an accurate central difference method, and a computer program has been developed to solve these equations.

4. Results and Discussion

The effects of dimensionless parameters including the nonlinearly number (n), the material parameter (K), the Prandtl number (Pr), the stagnation parameter (ß), the viscous dissipation number (Ec) and the radiation parameters (k0) are mainly interesting for this study. Flow and temperature fields of the thin nutrition pie flow are analyzed by utilizing the boundary layer concept to obtain a set of coupled momentum equation, energy equation and mass equation.

From Figure 1 reveals that the increase of nonlinearly number n results in the decrease of velocity distribution at a particular point of the flow region. This is because there would be a decrease of the fluid boundary layer thickness with the increase of values of nonlinearly number n.

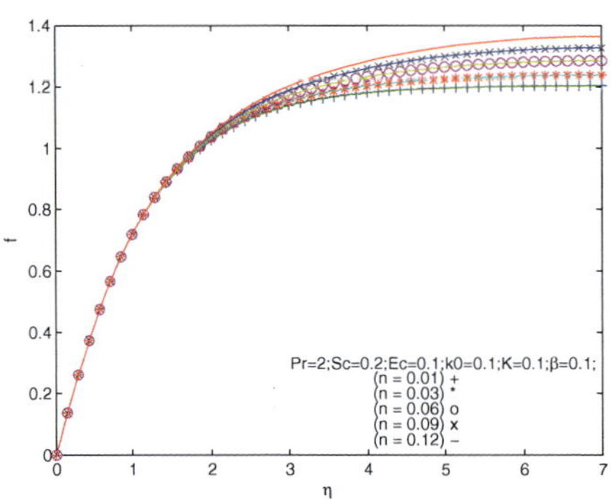

Fig. 1 depicts dimensionless velocity profiles f vs. η as Pr=2.0, Ec=0.1, k0=0.1, ß=0.1, K=0.1 and n=0.01, 0.03, 0.06, 0.09, 0.12.

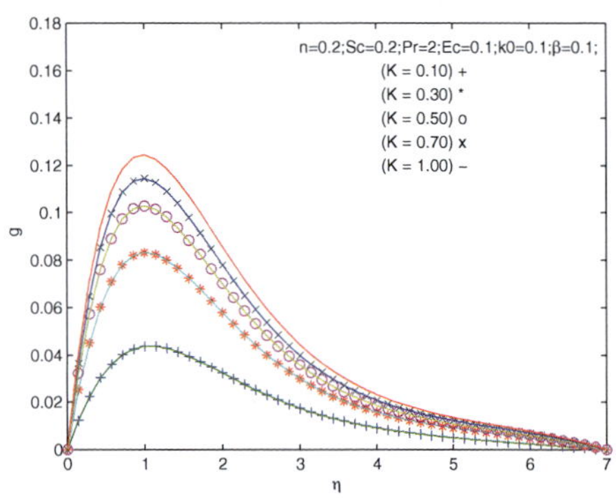

Fig. 2 depicts dimensionless angular velocity profiles g vs. η as n=0.2, Pr=2.0, Ec=0.1, ß=0.1, k0=0.1 and K=0.1, 0.3, 0.6, 0.7, 1.0.

From Figure 2 reveals that the increase of material parameter K results in the increase of angular velocity distribution at a particular point of the flow region. This is because there would be an increase of the fluid boundary layer thickness with the increase of values of material parameter K. It is found that when material parameter K increased, the fluid angular velocity increased.

The Figure 3 reveals that the increase of stagnation parameter ß results in the decrease of dimensionless temperature distribution at a particular point of the flow region. This is because there would be an increase of the thermal effect with the increase of values of stagnation parameter for $-\theta'(0)$ increasing. It is seemed that the increase of stagnation parameter ß resulting in the decrease of temperature distribution at a particular point of the flow region. The Figure 4 reveals that the increase of Prandtl number Pr results in the decrease of dimensionless temperature distribution at a particular point of the flow region. Figure 5 reveals that the increase the parameter k0 results in the decrease of dimensionless temperature distribution at a particular point of the flow region. This is because there would be an increase of the thermal effect with the increase of value of k0. On the contrary, Figure 6 reveals that the increase Ec will increase of dimensionless temperature distribution and the physical phenomena are contrary to the Figure 3 and Figure 4. This is because there would be a decrease of the thermal effect with the increase of values of Eckert number Ec.

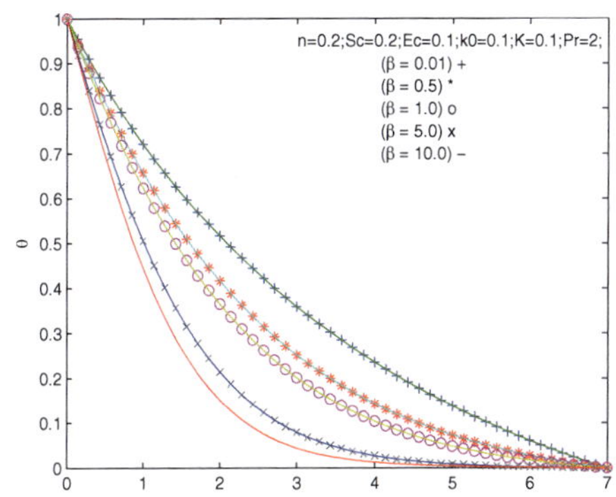

Fig. 3 depicts dimensionless temperature profiles θ vs. η as n=0.2, Ec=0.1, k0=0.1, K=0.1, Pr=2.0 and ß=0.01, 0.5, 1.0, 5.0, 10.0.

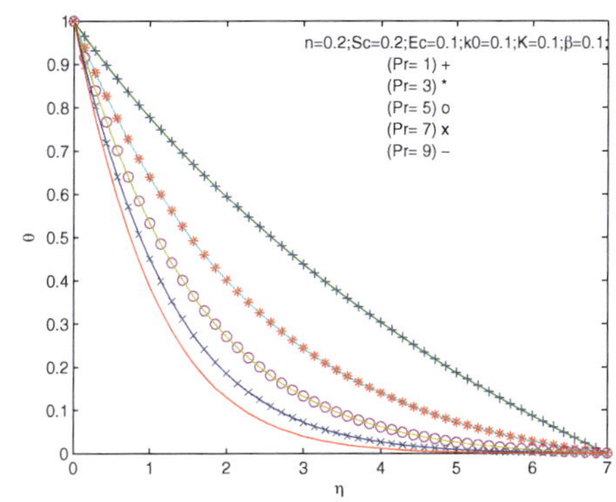

Fig. 4 depicts dimensionless temperature profiles θ vs. η as n=0.2, Ec=0.1, k0=0.1, K=0.1, ß=0.1and Pr=1, 3, 5, 7, 9

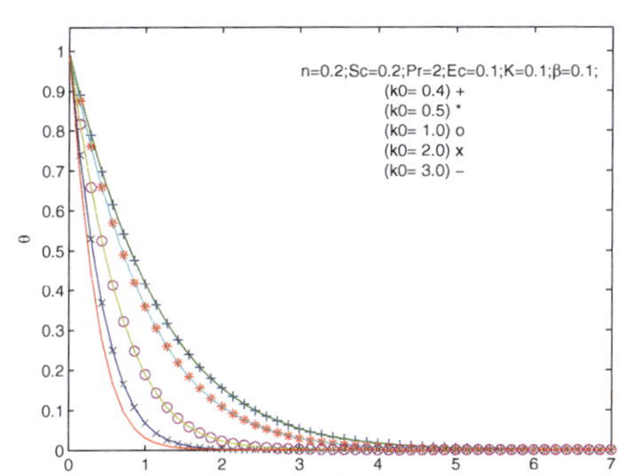

Fig. 5 depicts dimensionless temperature profiles θ vs. η as n=0.2, Pr=2, Ec=0.1, K=0.1, ß=0.1 and k0=0.4, 0.5, 1.0, 2.0, 3.0.

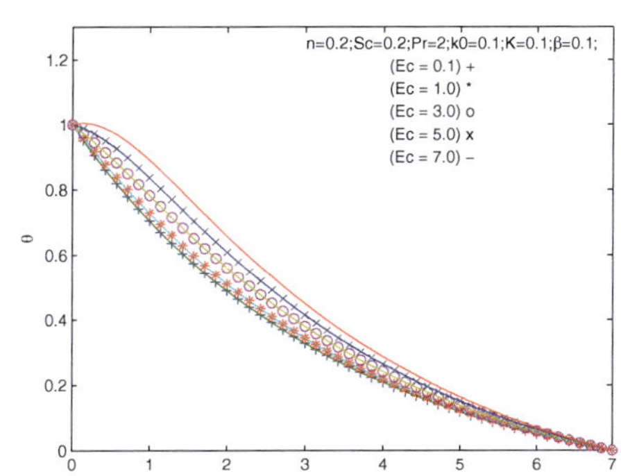

Fig. 6 depicts dimensionless temperature profiles θ vs. η as n=0.2, Pr=2, k0=0.1, K=0.1, ß=0.1 and Ec=0.1, 1.0, 3.0, 5.0, 7.0.

Figures 3 - 6 are the dimensionless temperature distribution for different parameters ß, Pr, k0 and Ec along the thermal boundary layer. These figures indicate that decrease when Pr, k0 increased. The parameter Ec will produce a contrary result from to stagnation parameter ß, Prandtl number Pr or radiation parameter k0.

5. Conclusions

Present studied problems had been solved by using theoretical analysis and numerical solution methods. There are some important conclusions as:

(1) It is found that when nonlinear number n increased, the fluid velocity decreased and it was found that when material parameter K increased, the fluid angular velocity increased.

(2) It is seemed that the increase of stagnation parameter ß, Prandtl number Pr or radiation parameter k0 resulting in the decrease of dimensionless temperature distribution at a particular point of the flow region.

(3) It is become that the increase of Ec results in the increase of dimensionless temperature distribution at a particular point of the flow region.

(4) It is also observed that increase in stagnation parameter ß, Prandtl number Pr or radiation parameter k0 producing a significant decrease in the thickness of the thermal boundary layer of the fluid and so as the increase the heat transfer effect with increased values of stagnation parameter ß, Pr or k0.

(5) All of the study works are depicted by letters, figures, tables, and formulas by different software, and has been published at website, so that it is a multimedia work also.

Acknowledgments

The authors would like to thank Ministry of Science and Technology, R.O.C for the financial support through Grant MOST 104-2221-E-434-001-.

References

1. Kai-Long Hsiao, Energy conversion conjugate conduction-convection and radiation over non-linearly extrusion stretching sheet with physical multimedia effects, Energy, Volume 59, 15 September 2013, Pages 494-502
2. Kai-Long Hsiao, Nanofluid Flow with Multimedia Physical Features for Conjugate Mixed Convection and Radiation, doi: 10.1016/j.compfluid.2014.08.001, Computers and Fluids, Volume 104, 20 November 2014, Pages 1–8
3. Kai-Long Hsiao, Heat and Mass Transfer for Micropolar Flow with Radiation effect past a Nonlinearly Stretching Sheet, Heat and Mass Transfer, Volume 46, April 2010, P413-419
4. Erikson L.E., L.T. Fan, V.G. Fox, "Heat and mass transfer on a moving continuous flat plate with suction or injection", Ind. Eng. Chem. Fundam.5 (1966) 19.
5. Khonsari M.M., "On the self-excited whirl orbits of a journal in a sleeve lubricated with micropolar fluids", Acta Mech. 81 (1990)235–244.
6. Hudimoto B., T. Tokuoka, "Two-dimensional shear flows of linear micropolar fluids", Int. J. Eng. Sci. 7 (1969)515–522.
7. Eringen A.C., "Theory of micropolar fluids", J. Math. Anal. Appl. 38 (1972) 469–480.

8. Lockwood F., M. Benchaita, S. Friberg, "Study of lyotropic liquid crystals in viscometric flow and elastohydrodynamic contact", ASLE Tribol. Trans. 30 (1987) 539–548.
9. Chin D.T., "Mass transfer to a continuous moving sheet electrode", J. Electrochem. Soc. 122 (1975) 643.
10. Kai-Long Hsiao, MHD mixed convection for viscoelastic fluid past a porous wedge, International Journal of Non-Linear Mechanics, January 2011, 46 (2011) 1–8
11. Das U.N., R.K. Deka, V.M. Soundalgekar, "Effects of mass transfer on flow past an impulsively started infinite vertical plate with constant heat flux and chemical reaction", Forsch. Ingenieurwes. 60 (1994) 284.
12. Andersson H.I., O.R. Hansen, B. Holmedal, "Diffusion of a chemically reactive species from a stretching sheet", Int. J. Heat Mass Transfer 37 (1994) 659.
13. Anjalidevi S.P., R. Kandasamy, Effects of chemical reaction heat and mass transfer on laminar flow along a semi infinite horizontal plate, Heat Mass Transfer 35 (1999) 465.
14. McLeod J.B., K.R. Rajagopal, "On the uniqueness of flow of a Navier–Stokes fluid due to a stretching boundary, Arch. Ration". Mech. Anal. 98 (1987) 386.

Study on correlations between interleukin 23 and bronchial asthma

Xinhui Li[1], Hongxia Sun[2,*]
Department of Respiratory Medicine, Affiliated Hospital of Beihua University,
School of Pharmacy, Beihua University
Jilin, 132011, China
[*]E-mail: sunhongxiajl@126.com

Bronchial asthma is an inflammatory response of common chronic respiratory diseases. A variety of cells, such as mast cells, neutrophils, eosinophils and other cells are involved, disorders of cytokine networks mediated the attack of bronchial asthma. Interleukin 23 (IL-23) cytokines is heterodimers which was found to have many different biological effects ten years ago, many scholars have discovered that IL-23 was involved in the allergic diseases. This article will give a introduction about the relationships between IL-23 and asthma.

Keywords: Interleukin-23; Bronchial Asthma; Correlation; Study.

1. Introduction

Asthma is recognized as one of the four major chronic diseases in the world. It is listed as the most among the ten leading causes of death by the medical field, according to medical sources, there were about 1.5 to 2 billion people suffering from asthma in the world, 25 million people with asthma in China, 600 thousands people died of asthma every year. The pathogenesis of asthma is very complicated, which is caused by respiratory tract infection most, induced by climate, emotional change and associated with genetic factors. It can also be caused by certain substances such as allergy. Cytokines such as IL-23 was found to participate in the attack of bronchial asthma during the present study on the pathogenesis of asthma[1].

2. Discovery and source of IL-23

IL-23 was discovered in 2000 year, it was a cytokine of heterodimers produced by the activation of dendritic cells (DC) and macrophages (MC) secretions. IL-23 was composed of P19,a new protein and P40 subunits in IL-12[2]. IL-23 had many biological effects, it could enhance the memory T cell proliferation, improved relevant cytokines secretion, such as interferon-γ(IFN-gamma), IL-17, IL-10, etc by T cell, it could affect monocytes and dendritic cells function still and affect the neutrophil regulating cell(NRC) to adjust IL-17, affecting granulocyte (GC) generation ultimately [2-5].

3. Biological functions of IL-23

It has been confirmed that IL-23 functioned in a variety of diseases. Effects such as anti-tumor, anti-tumor metastasis, promotion of attacks of autoimmune disease, resistance to a variety of pathogenic microorganisms infectious diseases. It was proved that IL-23 receptor consisted of IL-12R beta 1 and IL-23R subunits, which were expressed in memory T cells, NK cells, DCs cells, MCs, microglial cells and was closely related to cytokines, inflammatory cytokines and chemokine production[6]. IL-23 could directly activate MC and DC to produce IL-23R, promoting TNF- alpha and IL-1 inflammatory mediators production, which further promoted the chronic inflammation and cellular immune responses dependent and non-dependent on IL-23 pathways[7]. IL-23 could affect the process of chronic inflammatory disease through matrix metalloproteinase 2 (MMP-2) and IL-22. Studies found that the inflammatory reaction was reduced, the symptoms were alleviated in mice with MMP-2 and IL-22 gene knockout[8].

4. The association between IL-23 and Th17 cells

Activated Th17 cells could express the chronic inflammatory cytokines such as IL-17, IL-17F, IL-6, TNF-alpha and chemokines that promoted the inflammatory response under the effect of IL-23[9]. Recent study found that IL-23 had played an important role in promoting the differentiation of Th17 cells and maintaining cellular stability, IL-23/Th17 cellular pathway could promote the allergic airway inflammation participated by either eosinophils induced by antigen and mediated by Th2 cells or Ag-induced and Th17 cell mediated neutrophils[10,11].

5. Effects of IL-23/Th17 pathway in asthma

IL-23 could also enhance cytokines secretion of Th2 cell induced by antigen and promoted the aggregation of eosinophils in IL-17 knockout mice, indicating that IL-23/Th17 pathway played an important role in asthma[12]. The symptom of EAE was alleviated and the development of encephalomyelitis(EAE) was effectively inhibited in IL-23p19 gene defects mice and the application of anti-IL-23p19mRNA or anti-IL-12/IL-23p40 mRNA antibody after treatment of experimental autoimmune encephalomyelitis. The above results indicated that IL-23/Th17 pathway performed an key role in the process of autoimmune disease and chronic inflammatory diseases[7]. IL-17 was a necessary component of allergic asthma, the level of IL-17 expression in lung tissues and peripheral blood was elevated in asthmatic mice, and the level was proportional to the severity of asthma[13], which showed that Th17 cells and it secreting cytokines played an major part in IL-23-mediated asthma, blockade of IL-23 or IL-17 expression might bring new hope for asthma treatment.

The mechanism of allergic asthma has been considered as the Th1/Th2 cell immune imbalance, representing as relative inhibition of Thl function and relatively high of Th2 function, accompanied by some increased cytokines such as IL-4, IL-5, IL-13 that were secreted by Th2-typed cells, while by decreased IFN-γ cytokine secreted fromThl typed cells, which led to the IgE mediated rapid onset of asthma reaction and the occurrence of the late onset of asthmaic chronic airway inflammatory reaction characterized by eosinophilic infiltration. However, the enhanced Thl cell function did not reduce or reverse the inflammatory response mediated by Th2 cells, which suggested that the classic theory of "Th2 dominance" could not explain all the pathogenesis of allergic asthma. The proposal of IL-23/Th-17 axis[14], also known as the DC-IL-23-IL-17 pathway, make up the insufficiency of allergic asthma pathogenesis induced by antigen, mediated by Th2 cell. It may activate IL-23 cells to proliferate and secrete a large number of IL-17A, which leads to the disorder of IL-23-Th17-IL-17A axis, and causing severe attack of asthma.

The IL-23/Th17 axis is involved in antigen-induced, Th17 cells-mediated airway inflammation of neutrophils accumulation[15], and enhanced Th2 cells-mediated eosinophils accumulation.

Wakashin H et al[16] found that the increased expression of IL-23 in airway can increase antigen induced IL-17A production in transgenic mice(CC10 IL-23) after inhalation of ovalbumin (OVA), thereby recruiting the neutrophils aggregation in airway. but the expression of IL-23 in the airway decreased the antigen induced IL-17A production and neutrophils accumulation in CC10 IL-23 mice with IL-17A deficiency after inhalation of OVA. Song Liyan et al [17] simulated Th17 cell's function via basic immune simulator(BIS) and found that DCs secrete IL-23 during bacterial infection. IL-23 induced Th17 cell to produce IL-17A and IL-17F via activation of transformation factor of STAT3 and RORt, which act on airway epithelial cell to produce CXCL1(chemokine-1) and CXCL8(chemokine-8). Moreover, IL-17A and IL-17F can induce vascular endothelial cells and fibroblasts to produce other cytokines and chemokines, such as IL-8, IL-17F, granulocyte, macrophage colony stimulating factor (GM-CSF) and CXCL10. These cytokines and chemokines are also involved in Th17 cells mediated neutrophil accumulation in airway. Wakashin H et al[16] found that the forced expression of IL-23 in lung tissue or transfer of antigen specific Th17 cells enhanced the antigen mediated airway neutrophil infiltration[16]. These results indicate that IL-23 promotes airway inflammation by activated Th17 cells. These findings suggest that the IL-23/Th17 axis plays an important role in the airway inflammation in patients with airway inflammation, especially in allergic asthma. The specific mechanism for IL-23 / Th-17 axis involved in antigen induced Th17 cells mediated neutrophilic airway inflammation needs further study, but for severe allergic asthma, the molecular target for IL-23 / Th17 axis treatment may become a new treatment method.

6. Study on Si RNA IL-23 and asthma

It was reported that IL-23 protein expression was increased significantly in asthmatic group, the IL-23 protein expression was inhibited in lung tissues of asthmatic mice when siRNA IL-23 was transfected into the epithelial cells of asthma airway of murine and there was significant difference compared with those in asthma group and empty plasmid group ($P<0.01$); siRNA interference could improve the level of interferon-gamma expression, decreased IL-5, ICAM-1 and TNF- levels, compared with asthma group and empty plasmid group ($P<0.01$); siRNA silencing IL-23 inhibited the airway inflammation in asthmatic mice, compared with the control group, the difference was significant ($P<0.01$). All these demonstrated that IL-23 performed an important function in the pathogenesis of asthma, it was the direct factor which caused asthmatic symptoms, blocking its expression could improve asthma incidence, the experiment proved that it could improve asthmatic airway inflammation and hyperresponsiveness effectively through some key genes silencing by siRNA interference technology in the pathogenesis of asthma, which brought new hope for the gene therapy of asthma in the future, IL-23 gene silencing has become a new target for future asthmatic treatment[18].

7. Summary

The study on relationships between IL-23 and asthma is thriving, although the detailed mechanism through which IL-23 mediated bronchial asthma is not yet understood, but it is certain that the important role of IL-23 in asthma, the application of IL-23 antibody and silencing the gene expression of IL-23 etc have pointed out the direction of the research for the future treatment of asthma, we look forward to this research will be applied to clinical patients in the near future[19-20].

Acknowledgements

The research work was supported by the research plan of important technical problem of technological bureau in Jilin city in 2009 year under Grant No. 2009145 for development of science and technology.

References

1. Zhaoli Na, Liuli, Lu Jirong, Li & Chun Yan, Study on the correlation between interleukin and bronchial asthma in children. Chi J Immunol, **28(2),** pp. 165-167, 2012. (in Chinese)
2. Oppmann B, Lesley R & Blom B, et al, Novel p19 protein engages IL-12p40 to form a cytokine, IL-23, with biological activities similar as well as distinct from IL-12. Immunity, **13(5),** pp. 715-725, 2002. (in Chinese)
3. Vanden Eijnden S, Goriely S & De Wit D et al, IL-23 up-regulates IL-10 and induces IL-17 synthesis by polyclonally activated naive T cells in human. Eur J Immunol, **35(2)**, pp. 469-475, 2005.
4. Bastos K R, Marinho C R & Barboza R et al, What kind of message does IL-12 /IL-23 bring to macrophages and dendritic cells? Microbes and Infection, **6(6),** pp. 630-636, 2004.
5. Stark M A, Huo Y & Burcin T, et al, Phagocytosis of apoptotic neutro-phils regulates granulopoiesis via IL-23 and IL-17.Immunity, **22(3),** pp. 285-294, 2005.
6. Belladonna ML, Renauld JC & Bianchi R, et al, IL-23 and IL-12 have overlapping, but distinct, effects on murine dendritic cells. J Immunol, **168(11)**, pp. 5448-54, 2002.
7. Iwakura Y & Ishigame H, The IL-23/IL-17 axis in inflammation. J Clin Invest, **116(5)**, pp. 1218-22, 2006.
8. Munoz M, Heimesaat MM & Danker K, et al, Interleukin (IL)-23 mediates Toxoplasma gondii-induced immunopathology in the gut via

matrixmetalloproteinase-2 and IL-22 but independent of IL-17. J Exp Med, **206(13),** pp. 3047-59, 2009.

9. Romagnani S, Human Th17 cells. Arthritis Res Ther, **10(2)**, pp. 206,2008.
10. Guttman-Yassky E, Lowes MA & Fuentes-Duculan J, et al. Low expression of the IL-23/Th17 pathway in atopic dermatitis compared to psoriasis. J Immunol, **181(10)**, pp. 7420-7, 2008.
11. Wakashin H, Hirose K & Maezawa Y, et al., IL-23 and Th17 cells enhance Th2-cell-mediated eosinophilic airway inflammation in mice. Am J Respir Crit Care Med, **178(10),** pp.1023-32, 2008.
12. Peng J, Yang XO & Chang SH, et al, IL-23 signaling enhances Th2 polarization and regulates allergic airway inflammation. Cell Res, **20(1),** pp. 62-67, 2010. (in Chinese)
13. Schnyder-Candrian S, Togbe D & Couillin I, et al, Interleukin-17 is a negative regulator of established allergic asthma. J Exp Med, **203(12)**, pp, 2715-25, 2006.
14. Iwakura Y, Ishigame H. The IL-23/IL-17 axis in inflammation[J]. J Clin Invest, **116(5)**, pp. 1218-22, 2006.
15. Nakjima H, Hirose K. Role of IL-23 and Th17 Cells in Airway Inflammation in Asthma[J]. Immune Network, **10(1),** pp. 1-4, 2010.
16. Wakashin H, Hirose K, Maezawa Y, et al. IL-23 and Th17 cells enhance Th2-cell-mediated eosinophilic airway inflamma-tion in mice[J]. Am J Respir Crit Care Med, **178(10)**, pp. 1023-1032, 2008.
17. Li Yan Song, Yunbo Guo, Qing Kai Deng et al. TH17 functional study in severe asthma using agent based model[J]. Journal of Theoretical Biology, **30**, pp. 29-33, 2012.
18. Xinhui LI & Fei TE, Effect of Interleukin 23 gene silencing on bronchial asthma in mice. J Beihua Uni(Natural Science), **15(3),** pp. 334-337, 2014. (in Chinese)
19. Lajoie S, Lewkowich IP, Suzuki Y, et al. complement-mediated regulation of the IL-17A axis is a central genetic determinant of the severity of experimental allergic Asthma[J]. Nat Immunol, **11(10)**, pp. 928-935, 2010.
20. Juan Peng, Xuexian O. Yang, Seon Hee Chang et al. IL-23 signaling enhances Th2 Polarization and regulates allergic airway inflammation[J]. NIH Publi Access Author Manuscript, **10 (1)**, pp. 62-71, 2010. (in Chinese)

Influence of environmental factors on freeze-thaw stability of glycosylated soy protein isolate emulsion

Mu-jun Liu, Chao-ran Dou, Guo-ping Yu*, Yu-xiu Yao and Pei-pei Guo

College of food science, Northeast Agricultural University,
Harbin 150030, China

*liumujun0@163.com, 444109958@qq.com, *yuguopingneau@hotmail.com, 1006333472@qq.com, 986325454@qq.com*

The objective of this research was to investigate the effects of various environmental factors, such as pH, NaCl concentrations and the freeze-thaw cycles on the freeze-thaw stability of glycosylated soy protein isolate emulsions. The soy protein isolate-maltodextrin conjugate (SPI-MD) was achieved by Maillard wet heating modification. The freeze-thaw stability was characterized by measurements of creaming index and oiling off after isothermal storage at -20°C for 24h and further thawing. Compared with those stabilized by soy protein isolate (SPI) alone, O/W emulsions stabilized by SPI-MD showed a higher stability against freeze-thaw treatment.

Keywords: Soy Protein Isolate; SPI-maltodextrin Conjugates; Freeze-thaw Stability; Environmental Factors.

1. Introduction

Many proteins are surface active molecules, therefore can be used as an emulsifier to stabilize the emulsion system. In food emulsion systems, soy proteins are often used because of their excellent surface active properties; they absorb as an excess layer at the oil–water interface and lower the interfacial tension between the phases. The protein adsorbed on the surface of oil droplets can form an effective steric hindrance to prevent aggregation of the oil droplets [1]. The ability of protein repulsion force between the oil droplets and interfacial film at the surface of the droplets, which can result in the formation of a stable emulsion [2, 3, 4].

Many food emulsions are frozen to provide long-term stability of the product or as a necessary part of the food itself (e.g., ice cream) [5]. Nevertheless, most of O/W emulsions are highly unstable when they are frozen and rapidly breakdown after thawing. When an O/W emulsion is frozen, a variety of physicochemical processes can occur including fat crystallization, ice formation, freeze-concentration, interfacial phase transitions and biopolymer conformational changes [6].

At present, there are few systematic studies of freeze-thaw stability of O/W emulsions prepared with soy protein isolates as the sole emulsifying agent [7]. Therefore, the objective of our study was to better understand the influence of different environmental factors including pH, NaCl concentrations and the freeze-thaw cycles on the freeze-thaw stability of oil-in-water emulsions.

2. Materials and methods

2.1. *Materials*

Soy protein isolates (HAREBIN High Group Co. Ltd). Soybean oil (JIUSAN Oil Industry Group Co. Ltd). Maltodextrin (Changchun Dihao Foodstuff Development Co. Ltd). All other chemicals were of analytical grade.

2.2. *Preparation of SPI-Mds conjugates*

Based on preliminary experiments, the SPI-MD was prepared by Maillard wet heating modification. The parameters were 1:2 weight ratio of SPI to maltodextrin, 10mg/mL protein content at 90°C for 3h. The samples were taken after heating for 3h. The heated samples were cooled immediately in iced water. The samples obtained were kept at 4°C.

2.3. *Preparation of model oil-in-water emulsions*

Oil-in-water emulsions were prepared by mixing soy oil (oil volume fraction, Φ=0.2) with 1.6% w/v protein dispersions in a high-speed blender at 20,000 rpm for 60s.

2.4. *Freeze-thaw protocol*

Emulsion samples were transferred into thin-walled glass tubes. They were held for 24 h at -20°C, then thawed at 37 °C for 2 h and stored at 25°C for analysis.

2.5. *Evaluation of O/W emulsion freeze-thaw stability*

2.5.1. *Determination of particle size and distribution*

The emulsion particle size and its distribution were determined by laser scattering using a Micro Particle Analyzer. The d4, 3 expressed as emulsion particle size.

2.5.2. *Creaming index*

The prepared emulsion (10 g) was transferred into a test tube that was tightly sealed with aplastic cap. After freeze-thaw cycle, emulsions separated into an optically opaque 'cream' layer at the top and a transparent (or turbid) 'serum' layer at the bottom. The total height of the emulsions (HT), the height of the serum layer (HSL), and the height of the cream layer (HCL) were measured. The extent of creaming was characterized by creaming indices, which were defined as

$$CI(\%)=HSL/HT\times 100\% \quad (1)$$

Where HSL was the height of the serum layer, and HT was the total height of the emulsions.

2.5.3. *Transmission electron microscopy (TEM)*

Transmission electron microscopy (TEM) was carried out on the emulsions before and after freeze-thaw treatment with Japan JME-100CXII TEM. Briefly, samples diluted 15-fold were collected on copper grids and subsequently stained using saturated uranyl acetate solution for 1 min.

2.6. *Statistical analysis*

The experiment was replicated twice with at least triplicate analyses. Data were analyzed by using the General Linear Models procedure of Statistic 8.1 software package (Analytical Software, St. Paul, MN) for microcomputer. Analysis of variance (ANOVA) was done to determine the significance of the main effects. Significant differences ($P < 0.05$) among means were identified using Turkey procedures.

3. Results and discussion

3.1. *Influence of pH on freeze-thaw stability of emulsion*

This experiment examines the modified protein emulsion particle size change at different pH values. In the preparation of the emulsion pH 8, followed by adjusting the pH to 3.0-10.0.

Tab. 1. Particle size and creaming index of the emulsions at different pH values

pH	SPI		SPI-MD	
	$D_{4,3}$(μm)	Creaming index(%)	$D_{4,3}$(μm)	Creaming index(%)
3.0	6.33 ± 0.1^{b}	23.1 ± 0.1^{c}	2.91 ± 0.1^{a}	4.9 ± 0.1^{bc}
4.0	7.45 ± 0.1^{a}	24.8 ± 0.1^{b}	2.80 ± 0.1^{b}	5.1 ± 0.1^{b}
5.0	7.76 ± 0.2^{a}	25.4 ± 0.3^{a}	2.92 ± 0.3^{a}	5.7 ± 0.3^{ab}
6.0	6.63 ± 0.1^{b}	19.7 ± 0.2^{d}	2.72 ± 0.2^{bc}	6.1 ± 0.1^{a}
7.0	5.44 ± 0.3^{c}	18.7 ± 0.5^{e}	2.68 ± 0.1^{c}	5.2 ± 0.7^{b}
8.0	4.29 ± 0.2^{d}	18.0 ± 1.0^{e}	2.60 ± 0.1^{c}	5.0 ± 1.0^{b}
9.0	3.98 ± 0.1^{e}	17.2 ± 1.1^{f}	1.78 ± 0.1^{d}	4.7 ± 1.0^{c}
10.0	3.76 ± 0.1^{e}	17.8 ± 0.9^{ef}	1.57 ± 0.1^{e}	4.2 ± 0.8^{d}

Note: Values are means of three replicate values. Different superscript letters represent a significant difference ($P < 0.05$).

The freeze-thaw stability of SPI and SPI-MD emulsion at different pH is showed in Table 1. SPI emulsions exhibit a pH-sensitive behavior, compared with the SPI-MD emulsion, when the pH is adjusted to 4.0-6.0, the particle size of the emulsion is increased, creaming phenomenon obviously, this is due to the pH of the emulsion near the isoelectric point of the protein, the low solubility of proteins, aggregation occurs between the droplets. For SPI-MD emulsion, within pH3.0-10.0 range, the emulsion can remain stable, even the solution pH close to the protein isoelectric point, the emulsion can be maintained good freeze-thaw stability. This is due to the sugar molecules grafted on soy protein isolate molecule can shield the charge on the protein molecule.

3.2. *Influence of NaCl concentration on freeze-thaw stability of emulsion*

The particle size of SPI and SPI-MD emulsion in various concentrations of NaCl (100, 200, 300, 400, 500mM) was presented in Fig. 1. The particle size of SPI and SPI-MD emulsion was increased gradually with NaCl concentration increasing. When the NaCl concentration less than 200mM, the particle size change of SPI and SPI-MD emulsion was not significant, when the NaCl concentration higher than 200mM, SPI emulsion particle diameter significantly increased, while SPI-MD emulsion particle size was not significant change in a NaCl concentration higher than 300mM, and in the range of 0 - 500mM NaCl concentration, the particle size of SPI-MD are less than that of SPI emulsion.

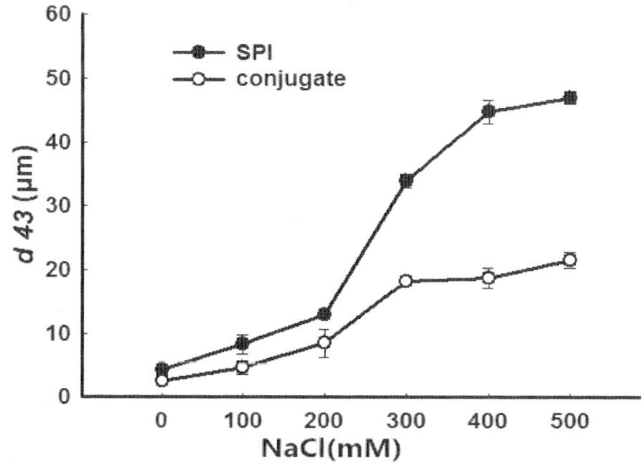

Fig. 1. Effects of potassium chloride on particle size of emulsions

The TEM images of SPI and SPI-MD emulsion with 300mM NaCl after one freeze-thaw cycles were presented in Fig. 2.

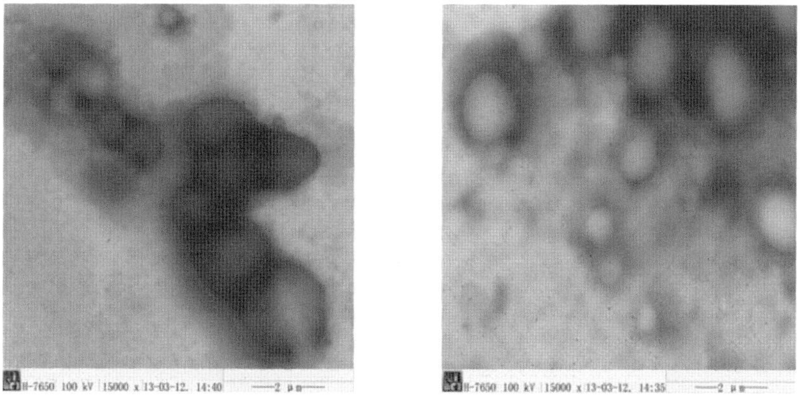

a) SPI emulsion(15000×) b) SPI-MD emulsion(15000×)

Fig. 2. TEM analysis of SPI and SPI-MD emulsions at 300mM NaCl

As is showed in the picture, at high salt ion environment, the SPI and SPI-MD emulsion droplets occurred varying degrees of flocculation and coalescence, but SPI emulsion flocculation is more serious. For SPI-MD emulsion, some black areas exist between the droplets, which indicate that between the droplets occurred "bridging flocculation", but still visible droplets with smaller particle size did not happen coalescence phenomenon.

3.3. *Influence of freeze-thaw cycles on freeze-thaw stability of emulsion*

Effect of the freeze-thaw cycles on the particle size and distribution of SPI emulsion is presented in Fig. 3. The figure shows, the fresh SPI emulsion average particle size is 2.02μm, particle size is unimodal distribution, the SPI emulsion were freeze-thaw process, SPI emulsion particle size increased significantly, and with the number of cycles increased, the particle size tends to increase more significantly. After three freeze-thaw cycles, the emulsion particle size distribution was multimodal, particle size mainly distributed in the range of 10-100μm and appeared large diameter (>100μm), freeze-thaw cycle destroyed the protein-protein interactions of the emulsion, leading to droplet aggregation and flocculation.

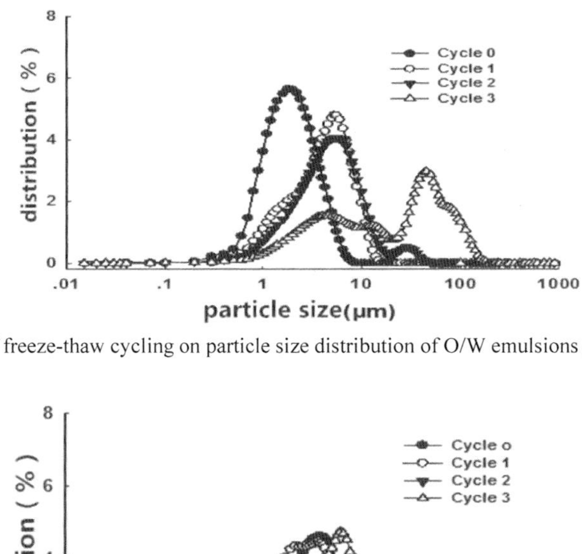

Fig. 3. Impact of freeze-thaw cycling on particle size distribution of O/W emulsions prepared with SPI

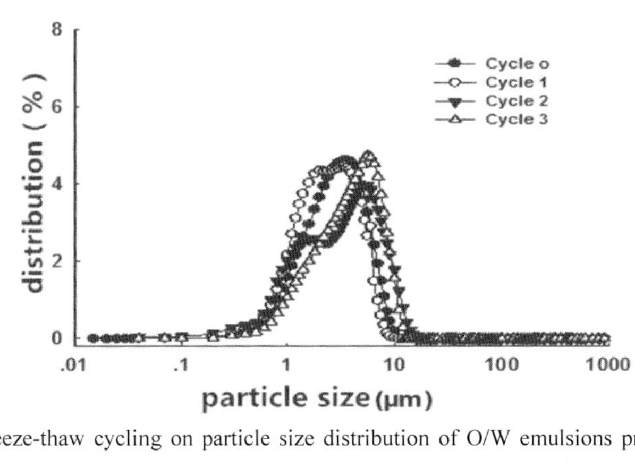

Fig. 4. Impact of freeze-thaw cycling on particle size distribution of O/W emulsions prepared with SPI-MD conjugate

Effect of the freeze-thaw cycles on the particle size and distribution of SPI-MD emulsion is presented in Fig. 4. Compared with the SPI emulsions, in the process of three freeze-thaw cycles, the particle size and distribution of SPI-MD emulsion did not change significantly, unimodal distribution, and concentrated in the range of 1-10μm, illustrated that by SPI-MD stable emulsion with good freeze-thaw stability.

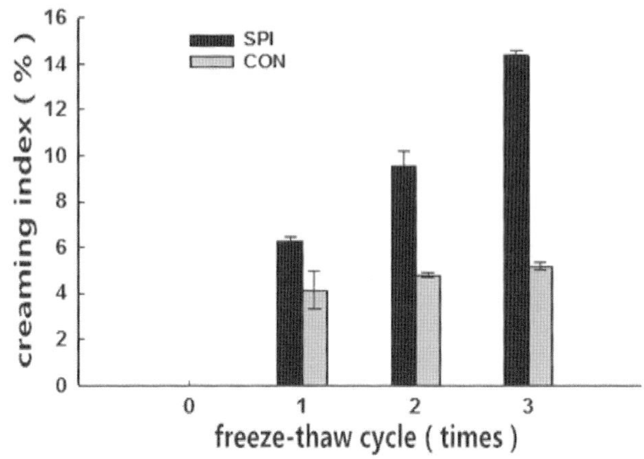

Fig. 5. Impact of freeze-thaw cycling on creaming index of O/W emulsions prepared with SPI-MD conjugate

The creaming index of SPI and SPI-MD emulsion after three freeze-thaw processes is showed in Fig. 5. SPI emulsion happen a clear stratification after freeze-thaw, which can be attributed to the fact that the emulsion in the process of freezing and thawing, flocculation or coalescence occurs between the oil droplets. For SPI-MD emulsion, creaming phenomenon is not obvious, remains a stable homogeneous state.

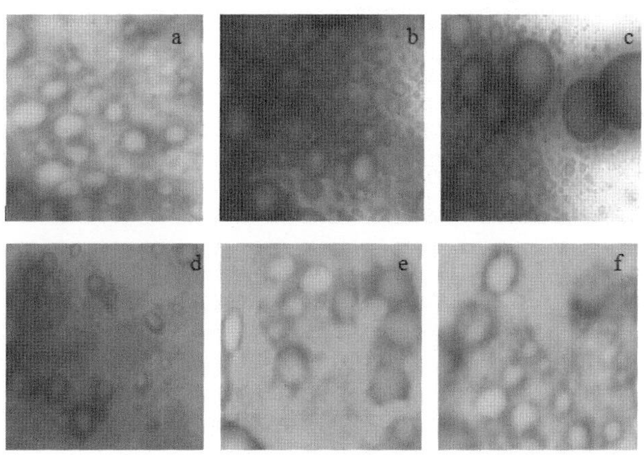

a, b, c for SPI by 1, 2, 3 times of freeze-thaw cycle respectively
d, e, f for SPI-MD by 1, 2, 3 times of freeze-thaw cycle respectively

Fig. 6. TEM analysis of SPI and SPI-MD conjugate emulsions after three freeze-thaw cycles

Figure 6 reflects the micrographs of SPI (a, b, c) emulsion and SPI-MD (d, e, f) emulsion after three times freeze-thaw process. SPI emulsion particle size after freeze-thaw was significantly larger, after freezing and thawing three times (c) have significant coalescence phenomenon, uneven distribution in the observation area, indicating the destruction of emulsion, occurrence of flocculation, coalescence. In contrast, SPI-MD emulsion did not show this phenomenon, droplet particles are evenly distributed in the emulsion.

Conclusions

In this experiment, the freeze-thaw stability of SPI-MD emulsion within the range of pH3.0-10.0 was significantly better than SPI emulsion, less particle size change, and emulsion with no obvious stratification and flocculation phenomenon. In addition, the SPI-MD emulsion shows the characteristics of the salt tolerance, the emulsion in 0-500mmol/L NaCl concentrations have a better freeze-thaw stability. TEM results show that SPI-MD complexes can maintain the stability of the emulsion after three times of freeze-thaw cycles without obvious aggregation and creaming.

Acknowledgments

This study was supported by Natural Science Foundation of China (Grant no. 31071493).

References

1. P. Walstra. Physical Chemistry of Foods Marcel Decker [M], New York, 2003.
2. M.A. Bos, T. van Vliet. Interfacial rheological properties of adsorbed protein layers and surfactants: a review Adv[J]. Colloid Interface Sci., 2001(91): 437–471.
3. P. Wilde, A. Mackie, F. Husband, P. Gunning, V. Morris Proteins and emulsifiers at liquid interfaces Adv[J]. Colloid Interface Sci., 2004, 108(9): 63–71
4. Chiu T H, Chen M L, Chang H C. Comparisons of emulsifying properties of Maillard reaction products conjugated by green, red seaweeds and various commercial proteins[J]. Food Hydrocolloids, 2009, 23 :2270–2277.
5. Supratim Ghosh, John N. Coupland. Factors affecting the freeze–thaw stability of emulsions[J]. Food Hydrocolloids, 2008, 22: 105–111.
6. McClements, D.J. Protein-stabilized emulsions[J]. Current Opinion in Colloid and Interface Science, 2004, 9: 305-313.
7. Yeshajahu Pomeranz. Functional properties of food components[M], 1985.

Schizandrin B relieves repression of NF1 tumor suppressor by the AML1-ETO fusion protein

Wen-Yue Zhuang and Jian-Guang Chen*
Medical Department, Beihua University,
Jilin, China
E-mail: wenyuezhuang@163.com
**Corresponding author*

Schizandrin B isolated from the fruit of *Schisandra chinensis*, has been reported as an anti-cancer substance. In this study, we investigated the effects of Schizandrin B on the tumor suppressor gene NF1 in AML1-ETO-expressing cell lines. We found that AML1-ETO-expressing cell lines displayed a low level of NF1 gene. Chromatin immunoprecipitation assays demonstrated that NF1 was the direct transcriptional target of AML1-ETO. The universal binding of AML1-ETO to genomic DNA resulted in recruitment of methyl-CpG binding protein 2 (MeCP2), indicating that AML1-ETO induced heterochromatic silencing of NF1. Our findings indicate that Schizandrin B can relieve NF1 suppression in AML1-ETO-expressing cells through a mechanism that involves inversion of epigenetic alterations.

Keywords: Schizandrin B; NF1; AML1-ETO.

1. Introduction

The NF1 gene on chromosome 17q11.2 is a classic tumor suppressor gene. Its product, neurofibromin is an important negative regulator of the Ras cellular proliferation pathway [1, 2]. Individuals with NF1 are at increased risk of developing various tumors, including MPNST, pheochromocytoma, leukemia, glioma, and rhabdomyosarcoma [3] due to increased Ras signaling [4]. The Ras protein family is involved in signal transduction, and their activation leads to growth, differentiation and cellular survival. NF1 has been recently identified as important in the development of acute myelogenous leukemia (AML) [5], the same form of leukemia that is associated with exposure to benzene, an established human leukemogen.

The t(8;21) is the second most common chromosomal abnormality in AML, accounting for 10-15% of cases with discernible translocations, and characteristically induces a leukemia with the French-American-British (FAB) M2 phenotype. A simple model of AML1-ETO function in leukemogenesis reflects its dominant negative effects on AML1 target genes, to a large extent via the aberrant recruitment of epigenetic modifiers such as HDACs and DNA methyltransferases (DNMTs). A direct transcriptional regulation by AML1-ETO through the AML1 DNA-binding activity has been demonstrated for a few genes, notably the tumor suppressor gene NF1.

Several phytochemicals, including dietary plant products, have been shown to have anti-tumor properties. Schizandra chinensis belongs to the Magnoliaceae family and is called omiza in Korean, gomishi in Japanese, and wuweizi in Chinese because it has five tastes: salty, sweet, sour, astringent, and bitter. In Chinese and Korean traditional medicine, Schizandra chinensis has been used to treat diabetes, palpitation, insomnia, nocturnal enuresis, dysentery, cough, asthma, phlegm, and jaundice [6]. Because of its various biologically activities, many chemicals have been isolated from Schizandra chinensis, including schizandrin, gamma-schizandrin, gomisins A-F, schindilactones A-C, schintrilactones A and B, wuweizidilactones A-F [6-9]. It was reported that the lignans isolated from Schisandra chinensis showed an anti-proliferative effect in A549 humanlung cancer cells, inhibited humanprostate cancer proliferation, and induced apoptosis in human leukemia U937 cells [10-12].

In this study, using AML1-ETO-expressing cell line U937-A/E as an *in vitro* model, we performed chromatin immunoprecipitation (ChIP) assays to investigate how the binding of AML1-ETO affected the chromatin structure of its target gene NF1 and thus caused deregulated gene expression. Furthermore, we investigated the molecular mechanisms of the anti-tumor effect of Schizandrin B in U937-A/E cells.

2. Materials and methods

2.1. *Cell culture and treatment*

Human myeloid U937 cells were maintained in RPMI-1640 medium supplemented with 10% fetal bovine serum (FBS) at 37°C in a humidified incubator with an atmosphere of 5% CO_2. U937-Mock and U937-A/E cells were maintained in RPMI-1640 medium supplemented with 10% FBS and 0.5 mg of G418/ml. Selected cell lines were cultured with 5.0μM of Schizandrin B (Shanghai Hui Cheng Biological Technology Co., Ltd) daily for 2 consecutive days.

2.2. *RNA extraction and analysis*

Total RNA was extracted from cells using the TRIzol RNA isolation system (Invitrogen). Quantitative reverse transcription-PCR (qRT-PCR) was carried out by using the 7500 Sequence detector ABI, and the PCR products were detected using SYBR green I chemistry (Applied Biosystems). The following primers were used: NF1, primer F: 5′-GGACACTGCTCAATATCGCATTAC-3′; primer R: 5′-AGGTACAAGTTAAGGCAC ACAGAAGA-3′.

2.3. *ChIP assay*

ChIP was performed as described previously following standard procedures (Supplementary Information) [13]. Promoter sequences were detected with PCR primers for NF1 promoter, forward: 5′-GGCCTGAAGTTTGGGTGTCTTA-3′ and reverse: 5′-TGTAGGGAAGAAGATCAGGGAGATAG-3′.

2.4. *Statistical analysis*

All values in the figures are expressed as means±SD. To determine statistical significance, the values were compared using two-group *t*-tests with differences considered significant for *P* values less than 0.05.

3. Results

3.1. *The AML1-ETO fusion protein represses NF1 gene expression*

AML1-ETO (also known as RUNX1-MTG8) represses genes that contain RUNX1 binding sites [14]. A quantitative reverse transcription PCR (qRT-PCR) assay was used to assess the mRNA expression of NF1 and GAPDH. This assay was tested on U937 AML1-ETO-expressing cells and U937 non-expressing cells. We observed that AML1-ETO-expressing cells contained markedly reduced levels of NF1 as compared with control-transfected cells or wild-type cells ($P<0.001$; Fig. 1).

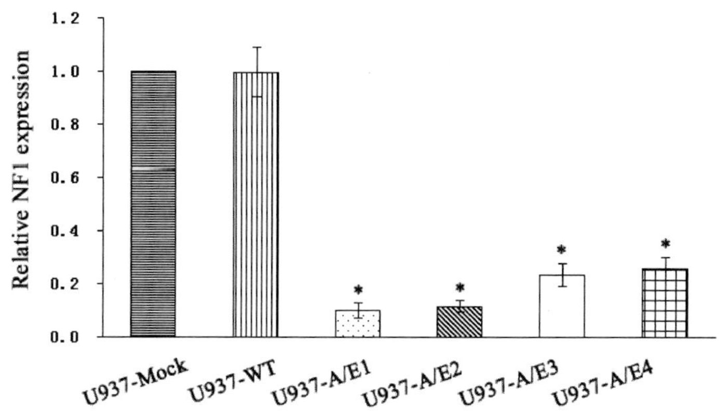

Fig. 1. AML1-ETO positive cells downregulate NF1 mRNA. Relative qRT-PCR quantization of NF1 level in AML1-ETO-expressing cells and non-expressing cells. The results represent the average of three independent evaluations ± SD.

3.2. *AML1-ETO binds the NF1 promoter in vivo*

AML1-ETO maintains the ability of AML1 to bind the consensus sequence TGT/cGGT on target gene promoters and acts as a dominant-negative repressor of AML1 targeting genes. ChIP using primers that encompassed the NF1 promoter region was performed with anti-ETO antibody to verify these AML1-ETO targets enriched within the AML1-ETO bound genomic sequences in U937-AML1-ETO cells and as a negative control in U937-empty vector. All comparisons in the cell lines were made between AML1-ETO expressing and non-expressing cells. DNA sequences specifically precipitated by anti-ETO antibody in AML1-ETO expressing cells (but not in AML1-ETO negative cells) most likely represent the AML1-ETO specific targets. We have detected the NF1 promoter sequences in anti-ETO immune complexes, but not in control immune complexes (Fig. 2), indicating that NF1 is a direct and specific target of the t (8; 21) fusion protein.

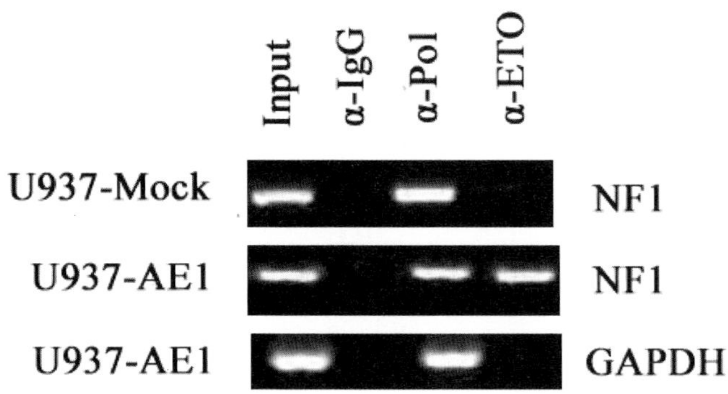

Fig. 2. Verification of AML1-ETO target (NF1) in AML1-ETO-positive cells. AML1–ETO bound to the NF1 promoter. ChIP assays were performed using specific antibodies for ETO, RNA polymerase II (Pol II), as well as non-immune IgG. Immunoprecipitated chromatin was analyzed by PCR with primers specific for the regions of the NF1 promoter. Input showed the amplification from sonicated chromatin. Amplification of GAPDH DNA was a control for nonspecific precipitated sequences. Data are expressed as fold-differences relative to control conditions (in which IgG is used instead of specific antibodies in the ChIP). The results represent the average of three independent evaluations ± SD.

3.3. *Treatment of Schizandrin B partially reverses NF1 suppression*

We next treated the AML1-ETO-positive and -negative U937 cells with Schizandrin B. Schizandrin B increased the expression of NF1 by ~2- to 3-fold (P<0.001) (Fig. 3A). In addition, Schizandrin B impaired the ability of anti-MeCP2 antibody to immunoprecipitate naked DNA surrounding the region of AML1 binding sites on NF1, gene promoter (P<0.001) (Fig. 3B).

A

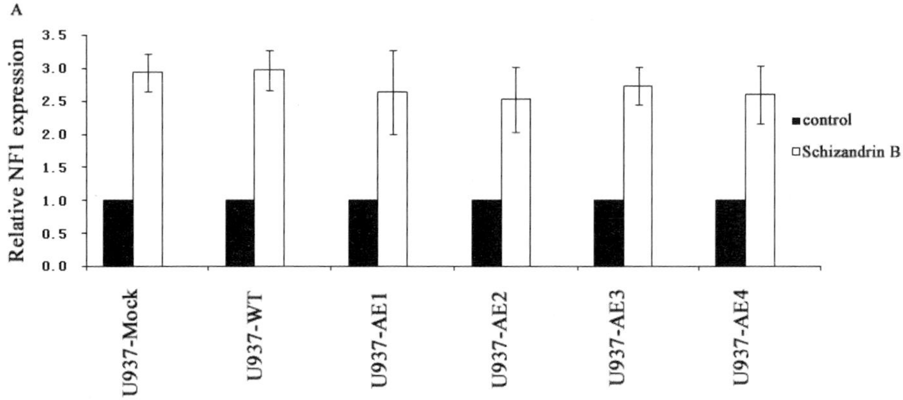

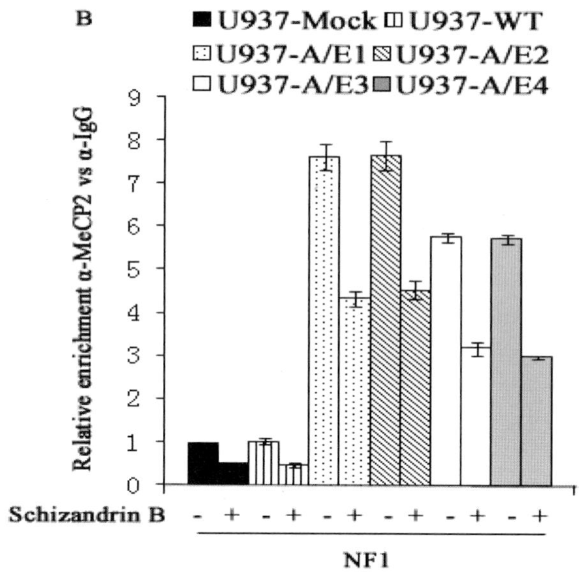

Fig. 3. Schizandrin B treatment induces NF1 expression. (A) Relative level of mRNA expression of NF1 in Schizandrin B -exposed and non-exposed U937 cells. (B) ChIP assay performed with antibodies for MeCP2 following Schizandrin B treatment in AML1-ETO-positive cells. IgG isotype control was used. The results represent the average of three independent evaluations ± SD. *P<0.001 compared respectively with untreated U937 cells.

4. Discussion

Neurofibromatosis type 1 (NF1), also referred to as von Recklinghausen disease, is an autosomal dominant inherited disease that affects approximately one in 3,000 individuals [15]. It is well established that the incidence of tumors in patients with NF1 is high compared with the normal population, which is the main reason for the reduced life span of NF1 patients [16]. NF1 is a potent suppressor of RAS oncogenic signaling, and its inactivation has been shown to induce a myeloproliferative phenotype [17].

Cancer is a genetic and epigenetic disease. The contribution of epigenetic mechanisms for a correct cell function is highlighted by the effects of their deregulation in cooperation with genetic alterations leading to the establishment and progression of tumors. Heterochromatic gene silencing represents an alternative oncogenic mechanism to gene mutation or deletion for the transcriptional repression of tumor suppressor genes [18].

In our study, we showed that the tumor suppressor gene NF1 was specifically downregulated in U937 AML1-ETO-expressing cells. Here, we found that the recruitment of MeCP2 to NF1 chromatin in AML1-ETO-expressing cell lines resulting in the silence of the NF1 gene. We showed that AML1-ETO could markedly downregulate the expression of NF1 by inducing repressive chromatin structure at its promoter. This finding is noteworthy as DNA hypermethylation is often associated with decreased gene expression.

Natural products are well recognized as sources of drugs in several human ailments including cancers [19]. Schizandrin B, for example, has several pharmacological activities, including anticancer and cancer chemopreventive potential in animal models [20]. Our findings indicate that Schizandrin B can relieve NF1 suppression in AML1-ETO-expressing cells through a mechanism that involves inversion of epigenetic alterations. Therefore, this study broadens our understanding of the molecular mechanisms by which Schizandrin B reverse the inhibition of myeloid-specific genes. In addition, our results highlight that Schizandrin B is a promising nutritional compound for developing potential multi-targeted therapies to re-establish a normal differentiation program. If so, these mechanisms may be potential targets for therapeutic strategies based on the reversal of epigenetic silencing in t(8;21)-positive leukemias.

Acknowledgments

This study was supported by grants from the Jilin Health and Family Planning Commission of China (No. 2014Q039).

References

1. Cawthon, R. M., R. Weiss, G. F. Xu, D. Viskochil, M. Culver, J. Stevens, et al. A major segment of the neurofibromatosis type 1 gene: cDNA sequence, genomic structure, and point mutations. Cell 1990, 62:193–201.
2. Xu, G. F., P. O'Connell, D. Viskochil, R. Cawthon, M. Robertson, M. Culver, et al. The neurofibromatosis type 1 gene encodes a protein related to GAP. Cell 1990, 62:599–608.
3. Brems, H., E. Beert, T. de Ravel, and E. Legius. Mechanisms in the pathogenesis of malignant tumours in neurofibromatosis type 1. Lancet Oncol. 2009, 10:508–515.
4. Cichowski K, Jacks T: NF1 tumor suppressor gene function: narrowing the GAP. Cell 2001, 104(4):593–604.
5. Parkin B, Ouillette P, Wang Y, Liu Y, Wright W, Roulston D, Purkayastha A, Dressel A, Karp J, Bockenstedt P, et al: NF1 inactivation in adult acute myelogenous leukemia. Clin Cancer Res 2010, 16(16):4135–4147.
6. Panossian, A.; Wikman, G. Pharmacology of Schisandra chinensis Bail: an overview of Russian research and uses in medicine. J. Ethnopharmacol. 2008, 118, 183.
7. Huang, S. X.; Li, R. T.; Liu, J. P.; Lu, Y.; Chang, Y.; Lei, C.; Xiao, W. L.; Yang, L. B.; Zheng, Q. T.; Sun, H. D. Isolation and characterization of biogenetically related highly oxygenated nortriterpenoids from Schisandra chinensis. Org. Lett. 2007, 9, 2079.
8. Huang, S. X.; Yang, J.; Huang, H.; Li, L. M.; Xiao, W. L.; Li, R. T.; Sun, H. D. Structural characterization of schintrilactone, a new class of nortriterpenoids from Schisandra chinensis. Org. Lett. 2007, 9, 4175.
9. Huang, S. X.; Yang, L. B.; Xiao, W. L.; Lei, C.; Liu, J. P.; Lu, Y.; Weng, Z. Y.; Li, L. M.; Li, R. T.; Yu, J. L.; Zheng, Q. T.; Sun, H. D. Wuweizidilactones A-F: novel

highly oxygenated nortriterpenoids with unusual skeletons isolated from Schisandra chinensis. Chem. Eur. J. 2007, 13, 4816.

10. Min, H. Y.; Park, E. J.; Hong, J. Y.; Kang, Y. J.; Kim, S. J.; Chung, H. J.; Woo, E. R.; Hung, T. M.; Youn, U. J.; Kim, Y. S. Antiproliferative effects of dibenzocyclooctadiene lignans isolated from Schisandra chinensis in human cancer cells. Bioorg. Med. Chem. Lett. 2008, 18, 523.
11. Vandyke, K.; White, M. Y.; Nguyen-Khuong, T.; Ow, K.; Luk, S. C. W.; Kingsley, E. A.; Rowe, A.; Pang, S. F.; Walsh, B. J.; Russell, P. J. Plant-derived MINA-05 inhibits human prostate cancer proliferation in vitro and lymph node spread in vivo. Neoplasia 2007, 9, 322.
12. Park, C.; Choi, Y. H.; Hyun, S. K.; Kwon, H. J.; Hwang, H. J.; Kim, G. Y.; Choi, B. T.; Kim, B. W.; Choi, I. W.; Moon, S. K.; Kim, W. J.; Choi, Y. H. Induction of G1 arrest and apoptosis by schisandrin C isolated from Schizandra chinensis Baill in human leukemia U937 cells. Int. J. Mol. Med. 2009, 24, 495.
13. Zhuang, W.Y., et al., Epigenetic silencing of Bcl-2, CEBPA and $p14^{ARF}$ by the AML1-ETO oncoprotein contributing to growth arrest and differentiation block in the U937 cell line. Oncol Rep, 2013. 30(1): p. 185-92.
14. Lutterbach, B., and S. W. Hiebert. Role of the transcription factor AML-1 in acute leukemia and hematopoietic differentiation. Gene 2000, 245:223–235.
15. Brems H, Beert E, de Ravel T and Legius E: Mechanisms in the pathogenesis of malignant tumours in neurofibromatosis type 1. Lancet Oncol 10: 508-515, 2009.
16. Seminog OO and Goldacre MJ: Risk of benign tumours of nervous system, and of malignant neoplasms, in people with neurofibromatosis: population-based record-linkage study. Br J Cancer 2013, 108: 193-198.
17. Le, D.T., Kong, N., Zhu, Y., Lauchle, J.O., Aiyigari, A., Braun, B.S., Wang, E., Kogan, S.C., Le Beau, M.M., Parada, L., and Shannon, K.M. Blood 2004, 103, 4243–4250
18. Sharma S, Kelly TK and Jones PA: Epigenetics in cancer. Carcinogenesis 2010, 31: 27-36,.
19. Ye Y, Wang H, Chu JH, Chou GX, Chen SB, Mo H, Fong WF, Yu ZL. Atractylenolide II induces G1 cell-cycle arrest and apoptosis in B16 melanoma cells. J. Ethnopharmacol. 2011, 136, 279–282.
20. Kim SH, Kim YS, Kang SS, Bae K, Hung TM, Lee SM. Antiapoptotic and hepatoprotective effects of gomisin A on fulminant hepatic failure induced by D-galactosamine and lipopolysaccharide in mice. J. Pharmacol. Sci. 2008, 106, 225–233.

Application of DNA fingerprint based on SSR in rice adulteration detection and origin traceability

Hong Liu[1], Jie Cui[1*], Ying Ma[1], Cuihong Dai[1], Dongjie Zhang[2] and Lili Qian[2]
[1]School of Food Science and Engineering, Harbin Institute of Technology, Harbin, China, 150090;
[2]Heilongjiang Bayi Agricultural University, Daqing, China, 163319
E-mail: hgdliuhong@163.com; cuijie@hit.edu.cn(Corresponding author)[*]

This paper briefly introduced mainstream DNA molecular marker techniques and their advantages, disadvantages and application, mainly focuses on the SSR technique in DNA fingerprinting construction of rice adulteration detection. The problems, the new development and application of SSR DNA fingerprint technology were reviewed in this paper. A fast and accurate new detection technology which base on SSR molecular marker technique to construct rice DNA fingerprint database were hoped to be established to replace the traditional rice adulteration detection method.

Keywords: Detection of Rice Adulteration; DNA Molecular Marker; SSR; DNA Fingerprinting Database

1. Preface

Rice is one of the major crops in China, sharing 40% of the total grain output per year. Rice is also the staple food of most people in the world and nearly 2/3 of the population in China eat rice[1,2]. Color, taste, aroma and price are variable in accordance with the origin and varieties of rice. Unscrupulous traders often adulterated in the high quality rice in order to get more benefit from it. Adulteration of commercial available rice in domestic market is serious so far, such as the adulteration of Wuchang rice in Xi'an grain and oil trading market, 2010. Influence spread throughout the country, resulting in the over storage of many small brands of Wuchang rice and a huge loss to consumers and producers, seriously disrupting the order of the grain market[3]. Therefore, in order to protect the origin brand of rice and the interests of massive consumers and farmers, establishing a rapid and effective new rice adulteration detection technology is urgent.

A crediable reference and a feasible means of detection must be established to detect the adulteration of rice. Traditional test methods for rice varieties and purity are mainly based on the chemical properties of rice and the human senses. High error rate is often caused by the subjectivity of artificial sense; the artificial chemical analysis method can accurately detect the difference of the components of rice in different varieties, but it is difficult to be promoted because it takes too much time and efforts[4]. DNA molecular marker analysis technology developed from the 1980s is a kind of ideal genetic marker technology[5], which developed after morphological markers, cell markers and biochemical markers. Grain can be identified from the gene level by DNA molecular marker, not influenced by external factors, individual biological developmental stages and organ tissue[6]. It has the characteristics of high accuracy, specificity and simplicity, and obvious advantages compared with traditional method. In 2005, the International Plant Variety Rights Protection Organization (UPOV) set the Sequence Repeats Simple (SSR) and Nucleotide Polymorphisms Single (SNP) and other molecular markers as the average methods of the construction of DNA fingerprint database in the BMT test guide[7]. Based on these advantages, the application of molecular marker technology to build a rice variety specific fingerprint database has extended to breeding and production guidance[8]. The mainstream of DNA molecular marker technology are summarized in this paper, method based on SSR molecular marker technology is selected to construct rice fingerprint database, this technology is expected to replace the traditional identification method to solve the problem of current adulteration of commercially available rice.

2. DNA molecular marker technology

DNA molecular marker technology is a technique to detect the polymorphism of DNA molecules(mechanisms such as the absence, shift, inverse position, rearrangement of bases, or the existence of different length and arrangement of repeated sequences), which can directly reflect the difference of DNA and genetic polymorphism in DNA level between individual and population through the electrophoretic patterns. DNA molecular marker technology has been widely used in crop genetics breeding, genome mapping, gene mapping, identification genetic relationship of plants, gene cloning, etc[6]. It is worth noting that, with the foundation of PCR, RFLP technology and repeat sequences, dozens of molecular marker technologies have emerged so far, such as AFLP, SSR, RAPD, SNP and so on, but no one suits for all cases still. Each molecular marker technology is different in the aspects of gene detecting position, polymorphism level, site specificity, repeatability, and the difficulty of technical operation, and the cost of experiment and so on.

2.1 RFLP molecular marker technology

RFLP (restriction fragment length polymorphism) is the first generation of DNA marker for genetic studies. Advantages: quite a lot of allele genes exist in the same RFLP locu and there is no interaction between different locus[8]; RFLP can detect variation of the whole genome and identify the genotype of the individual in the isolated population. Disadvantages: clone the probe which can reveal DNA polymorphism of different genomes is difficult, low detection sensitivity; the complex experimental operation and high cost is not suitable for large-scale application of breeding and variety identification; the usage of radioactive isotopes largely limited the application of RFLP technology. Although RFLP is the first generation of molecular marker technology but still is widely used in the genetic map of 10 kinds of plants, such as maize, wheat, rice, and so on. It is especially favorable for the establishment of RFLP gene linkage map of rice, but there are still a lot of blank regions[6].

Songwen Wang used 28 RFLP indica-japonica specific probes to analysis a plurality of rice varieties, the bioinformatics analysis showed that, these markers were different in genomic level and divided these rice varieties into Indica and japonica rice[9]. Yanbo Jiang used 45 indica-japonica specific probes to analysis indica-japonica differentiation of 58 varieties of hybrids rice through the methods of indica- japonica specific RFLP markers. Morphological index classification was also used to analysis the indica-japonica differentiation and study the relationships between the parents and heterosis. The results show that the two classification methods are basically the same. The hybrids and their parents were divided into two groups- Indica and japonicarice[10]. Zhang P J[11] used 12 rice parents to design 32 F1 hybrids with NC II design methods, in order to explore the possibility to predict heterosis using RFLP markers with the control of Shanyou 63. The results of cluster analysis showed that, Indica and japonica is easily separated, and easy to separate the ordinary japonica from nuda rice, java. Bao Ying[12] et al. come up with the analysis of the restriction fragment polymorphism (RFLP) of the ribosomal ITS, which can quickly and accurately distinguish the species of the genus CD.

2.2 RAPD molecular marker technology

RAPD (Random Amplified Polymorphism DNA), is the second generation of molecular markers technology which based on non-repetitive sequence. Advantages: simple and quick with high detection efficiency, just a small amount of DNA sample is required and a set of primers can be used in the analysis of different genomes of organisms. Disadvantages: there exists the issues of co-migration, electrophoresis can not separate the same molecular weight but different sequence DNA fragment; the uneven distribution in the genome and the limited number of RAPD markers greatly limit its application in gene mapping; heterozygous and homozygous can not be discriminated through RAPD and the genetic information provided was incomplete; many of factors affect RAPD technique, therefore, the stability and reproducibility of the experiment is not ideal. RAPD technology was widely used in rice gene mapping and genetic variability analysis in the early time but has been declining currently because it is not suitable for breeding and species identification[7].

Yunhai Li[13] et al. screened 14 RAPD primers from 100 RAPD primers to analyze 30 rice varieties, which can effectively identify all the sterile lines and the restorer lines. Jianing Mao[14] used RAPD technology to screen 13 primers which can amplify 43 stable polymorphic fragments between three-lines hybrid rice and their parents for trial. Sterile lines, restorer lines, maintainer lines, F1 and their parents were distinguished effectively and their genetic relationship were showed clearly. Shihua Duan[15] et al. used 13 RAPD primers to analyze restorer lines of 35 major hybrid rice, and 93 polymorphic fragments with good repeatability were detected, which can effectively distinguish all the restorer lines for trial. Cluster analysis showed that the genetic difference among Chinese hybrid rice is small, the genetic background is relatively unitary, and the utilization of heterosis of rice in China is largely limited by above factors. Taihe Xiang[16] screened 8 primers from 500 RAPD primers, which can effectively distinguish Xianyou 63,Xianyou 64,Xianyouwan 3 and their parents. Zhongming Chen [17] used 100 RAPD primers to analyze the polymorphism of Liangyoupei 9 and their parents, filtered out stable amplified polymorphic primer OPN-9 to identify their purity and the results showed that the primer could be used to identify the artificial mixture with sterile lines. Yuancheng Peng[18] extracted DNA for RAPD amplification from the parents of 5 combination of two-line hybrids rice and F1 seeds. It was found that the BW97277 can be used to find the difference in the test materials, which indicates that the RAPD technique can be used for the identification of the purity of the seeds.

2.3 AFLP molecular marker technology

AFLP(Amplified Fragment Length Polymorphism), is the second generation of molecular marker technology which based on the restriction enzyme digestion and PCR. Advantages: AFLP is more reliable and economical and have the ability to reveal the level of polymorphism of the species more effective than RFLP, RAPD, SSR, and have the advantages of randomness, convenience of RAPD as well as the specificity and reliability of RFLP[19]. Disadvantages: the cost is expensive, the requirements of purity and quality of genomic DNA are high; AFLP also have the mutual defectiveness such as dominant inheritance of multi copy markers, results of electrophoretic bands between different experiments are difficult to be compared for the subjectivity, thus AFLP can not be applied to variety identification neither[20]. In addition, a small number of markers are different from tightness of gene map when AFLP is used for genetic mapping[6]. AFLP experiment needs high precision instruments and equipments, and had applied for patents, so it is difficult to popularize its use in varieties purity identification. AFLP not only has the advantages of RAPD and RFLP, but also avoids the shortcomings of the two. It is considered as the most effective molecular marker technology [21,22]. It is widely applied in all application aspects of general molecular markers, such as genetic diversity and germplasm identification, molecular marker breeding, gene mapping, gene orientation, construction of linkage map and so on[13]. In theory, no matter how complex the genome DNA is, the AFLP method can detect polymorphism between any DNA.

Wang[23] et al. analysed 10 varieties of peanut with AFLP(the 10 varieties are from the International Institute for Tropical Crops), 78 polymorphic bands were gained, the result showed that AFLP was the effective method to construct the peanut fingerprint. Chunqing Zhang[24] et al. screened the primer combination M21P87 and M73P17 to construct the rice DNA fingerprint to distinguish the different varieties. Shibing Gao[25] et al. analysed AFLP fingerprint of 19 maize inbred lines using the primer combination EcoR I- AGG Mse I- CAA , maize inbred lines were able to be identified from the level of genome, AFLP can be used for the identification of maize inbred lines.

2.4 SNP molecular marker technology

SNP (Single Nucleotide Polymorphism) is the third generation molecular marker techniques for the variation of single nucleotides in the genome. Advantages: the number of SNP is large, the density is higher than SSR, the polymorphism is high, and it can quickly detect the single nucleotide difference and only a small insertion, deletion, etc. between different alleles, the genetic stability is high. Disadvantages: primer design is difficult, the information provided by SNP is limited and the cost is high. SNP markers play an important role in crop breeding[26], construction of rice genetic map, gene cloning and functional genomics research, species evolution[27], molecular genetics, pharmaco genetics, forensic science and disease diagnosis and treatment and so on[28]. Although still no reports about usage of SNP molecular marker technology in rice purity identification[29], but due to the high genetic stability, plenty of locus and wide distribution (there appears a SNP every 232 bp in rice), clear distinction between female parent and male parent and hybrid genotypes, with dimorphic and allelic gene characteristics of SNP markers, these characteristics offset the deficiency of two generations of molecular marker method above, and has the advantages of high throughput and a high degree of automation, SNP detection and analysis technology will make a difference in rice varieties of seed purity identification for gradually maturity and continue improvement.

Jingjiao Lu[30] took advantage of SNP to analyse genetic differentiation of 104 parents of southern two-line indica hybrid rice. The results showed large genetic distance variation between the 104 indica two-line hybrid rice parents pairwise, and the average genetic distance between sterile lines and male parents> genetic distance between sterile lines> genetic distance between male parents. The analysis of genetic differences among rice varieties by SNP was consistent with the results of pedigree analysis. 51 inbred lines of maize were classified into seven heterotic group with genotype analysis by using 1041 SNP locus by Jinfeng Wu[31], inter group genetic distance classification result basically was consistent with pedigree. Yaling Li et al[32]. developed a A/T SNP locus in the Mi-1 gene of tomato, and the SNP assortment of the 438 F2 groups was carried out then. The results showed that the genotype of F2 population conform to the ratio of 1:2:1 with 72 homozygous T/T genotype: 175 homozygous A/T genotype: 70 homozygous A/A genotype besides unexplained absence of 121 F2 individual.

2.5 SSR molecular marker technology

SSR (Simple Sequence Repeat) is the second generation of molecular marker technology[33,34], which is based on specific primers PCR markers. Advantages: the number of SSR is rich, revealing high polymorphism, dominant inheritance, it can identify the heterozygous and homozygous; stable amplification of electrophoresis strip, technology maturity, detection time is short, not affected by environmental impact, easy to operate, only need less quantity of DNA template, etc. Disadvantages: development of SSR primer is difficult, DNA sequence information at both ends of the repeat sequence must be known. Otherwise, sequence the whole genome of rice must be carried out firstly. Fortunately, this problem has been solved.

N Nandakumar[35] et al. used 10 SSR markers to analyse 11 rice varieties and their parents, 9 SSR markers can be used in the construction of the fingerprint, and 4 SSR markers (RM216, RM258, RM263 and RM206) can be used to identify all the hybrids and purity protection and variety identification. 24 SSR primers were used to construct the DNA fingerprinting of 19 high quality conventional japonica rice varieties in Taihu by Bing Luo[36]. The results showed that the genetic diversity of conventional japonica rice varieties in Taihu region was narrow, and it need to be further strengthened. Chengliang Guo[37] et al. identified the purity of 30 Liangyou 287 hybrid rice by field planting investigation method, and the specificity of the whole 10 species in the field was verified by RM21. The results showed that the identification results of the two methods were basically consistent, and the higher number of the samples examined by SSR, the higher the consistency is with the results of field survey. Anning Zhang [38] used 600 SSR markers which is distributed on 12 chromosomes of rice to analyse genetic devisity of parents of 13 hybrid combination, and 3 markers RM3, RM475, RM304which were able to detect polymorphism between all the parents were chosen to identify the purity of seeds of 13 hybrid rice. Ronglin Chen[39] et al. screened 12 pairs of primers which showed polymorphism among parents of super hybrid rice Guiliangyou 2 from 24 pairs of SSR primers. RM17 was chosen to identify the purity of 4 batches of Guiliangyou 2 seeds for its most obvious complementary electrophoresis strip, and the results were compared with the results of field cultivation. The results showed that the purity variation amplitude among the 4 batches of Guiliangyou 2 were 97.0% ~ 97.4%, 0.1~0.4 percentage point higher than field cultivation respectively. 5 hybrid strains were detected by SSR marker identification and field planting identification method. Mutant plant cultivated in the field with similar agronomic characters but whose growth period is 3 days longer, SSR detection showed normal complementary bands of parents. This was the main reason why the sensitivity of SSR is higher than field cultivation method in purity identification.188 portions of Yangliangyou 6 seeds were identified by using SSR detection and field cultivation respectively by Qinghua Yang et al[40]. The results showed that the detection of purity of hybrid rice Yangliangyou 6 by SSR molecular marker method was feasible, and the results were accurate and reliable.

2.6 Features of mainstream DNA molecular marker technology

The characteristics of mainstream DNA molecular markers, such as distribution, heredity, polymorphism, and so on, were compared and summarized in this paper to provide convenience for the researchers (Table 1).

Tab. 1 The characteristics of mainstream DNA molecular markers

Character of Molecular Mark	RFLP	RAPD	SSR	AFLP	SNP
Distribution	ubiquitous	ubiquitous	ubiquitous	ubiquitous	ubiquitous
Genetic characteristics	codominanc	mostly codominance	codominanc	mostly codominance	codominanc
Polymorphism	medium	higher	high	Very higher	high
Technical difficulty	medium	high	Very higher	high	high
Repeatability	high	medium	high	high	high
DNA sample	2-30	1-100	50-100	100	Little (ng)
Experimental period	long	short	short	short	short

Genomic region	Low copy sequence	genome	genome	genome	non encoding area
Radioactivity	usually	no	no	usually	No
Cost	high	low	low	high	Low
Application	physical map	gene tags	gnetic diversity	gene tags	
Test base	molecular hybridization	random PCR	specific PCR	specific PCR	specific PCR
Probe	DNA fragment	random primer	specific primer	specific primer	specific primer
Genetic diversity detection	little	ubiquitous	little	medium	little
Difficulty of band statistics	easy	medium	easy	difficult	
Site detection frequency	1~2	5~20	1	20~100	

2.7 Advantages of SSR technology to construct the DNA fingerprint of rice

The domestic and foreign studies have generally experienced two stages on how to construct fingerprint database by using molecular markers: the first period is specific markers(searching for molecular markers that can specifically label crop varieties), the second period is the combination of different molecular markers (combining different molecular markers to distinguish the varieties of crops)[41]. The method used for the construction of DNA fingerprint database domestic at present is mainly based on the international standard-the SSR marking method[42]. SSR labeling bear all the genetic advantages of RFLP method and avoid its disadvantages of using radioactive isotopes, its repetition rate and credibility are higher than RAPD, simpler experimental operation and lower cost and the quality of template DNA required is not high compared with AFLP method, more suitable than SNP method for the detection of large scale samples and multi locus and so on. SSR molecular marker technology is widely used in genetic diversity analysis, genetic map construction, target gene orientation, molecular marker assisted breeding, seeds purity and authenticity identification and many other research[22], it has become the most popularized genetic marker technology present[43]. There is rarely application in the construction of rice DNA fingerprinting with RAPD, AFLP, RFLP, STS, SNP and so on currently. Due to the rapid development of DNA extraction technology and PCR, grain, rice and plant leaves can be directly used for DNA extraction, greatly speeding up the detection process, high performance and stability of PCR technique ensure the detection accuracy of the results, so the application of SSR molecular markers in DNA fingerprinting construction is rapidly expanding[44].

2.8 Research and application of DNA fingerprint database in rice based on SSR Technology

The SSR DNA fingerprinting is mainly applied in the screening and location of the crop excellent genes, identification of old and new cultivars, intellectual property protection of new varieties, identification of variety authenticity and purity, detection of adulteration in commercial available rice and so on. There is an inevitable trend in the identification of new varieties of plants but has not yet formed a standard SSR fingerprint detection system in rice[45]. Japan, South Korea, the United States had already complete the construction of SSR DNA fingerprinting of rice varieties, and SSR DNA fingerprinting research of crops in China had just started shortly before. Rice Institute of Chinese Academy of Agricultural Sciences constructed SSR DNA fingerprinting of rice in north and south of China which attend the national rice variety regional test respectively, other provinces and cities such as Sichuan Province has also participated in the construction of SSR DNA fingerprinting, the result has a great impact on the production, operation, usage and management of seed. China has completed acquisition of DNA fingerprint data of 900varieties of rice[46], that providing technology and basic data support for the new rice variety test and the construction of database of known species, and also providing basis for the identification of hybrid rice varieties.

In the research of rice variety SSR DNA fingerprinting, Jieyun Zhuang et al. ascertain 24 core SSR markers to identify China's main rice varieties with 58 candidate SSR markers and 63 representative rice varieties. The national agricultural industry standard (NY/T 1433-2007) has been developed, which is the most representative work in DNA fingerprint research. A new "rice varieties identification with SSR mark method" NY/T 1433-2014 is pending approval on the website of Ministry of agriculture, the new standard will replace the old one. 48 pairs of SSR primers are obtained in the new standard, the discriminability will greatly improve compared with the old standard, but whether it is applicable to identify the different rice varities all over the country remains to be tested in practice. It has great significance to construct the fingerprint of rice of the whole country and the local varieties by using SSR molecular marker technology, and it is worthy to be promoted.

3. Detection of rice adulteration based on SSR Technology

3.1 Process to build rice DNA fingerprint database with SSR technology

The SSR-PCR amplification results with core SSR primers were used to construct a binary DNA fingerprint database consisting of “0” and “1” according to the strip information. The digital results "0 and 1" are arranged in order with the range of number magnitude and short chromosome arm then long chromosome arm as well as rice varieties arrangement. Each kind of rice will get a unique sequence of number comprised by “0,1”, which is the specific rice SSR fingerprint for each kind of rice or called molecular identity.

XuzhongLu[47] et al. constructed the molecular identity of 127 rice varieties. The ID number is composed of three parts: the first part is the commodity code, which represents basic commodity information of the rice, including rice varieties, region and time of breeding (or validation). The second part is the molecular fingerprint code, showed the molecular fingerprint information of the rice varieties obtained by 12 SSR markers (one marker per chromosome). The third part is the specific gene identification code which represented specific gene information. Huang Liang[48] et al. took advantage of 4 pairs of ISSR primers and their combinations to obtain the ISSR characteristic band of 18 rice varieties and then construct their ISSR finger prints to identify them effectively through ISSR molecular marker technology. Yingchun Han[49] et al. used 38 pairs of SSR primers which distributed on 12 rice chromosomes uniformly to analyse genetic diversity of 10 rice varieties in Henan area, screening 7 pairs of core SSR primers with high polymorphism to construct SSR fingerprint to distinguish 10 upland rice varieties.

3.2 Research on the construction of DNA fingerprint which is based on SSR technology to detect adulteration of rice

According to national standard NY/T 1433-2007 and NY/T 1433-2014, using the same core SSR primers to construct the fingerprint of commercial available rice then compare with the fingerprint of the corresponding pure varieties of rice. If it is found that the number of differential primers in the fingerprints is more than or equal to 2, it will be judged as “different species”, the commercial available rice can be regarded as adulteration; if the number of differential primers in the fingerprints is less than or equal to 1, it will be judged as "similar species”. It’s not sure that whether there is adulteration in the commercial available rice, other methods can be applied to explore further. If there is no different primers, it can be judged as “the same or very similar species”, the commercial available rice can be regarded as no adulteration.

4. Problems in SSR DNA fingerprinting technology and the application of new technology

4.1 Problems of DNA SSR fingerprinting technology

DNA fingerprinting technology based on SSR technology has been developed in recent years, constructing DNA fingerprinting technology by SSR technology has been widely used in variety identification and protection of intellectual property, detection of purity and authenticity of varieties, genetic relationship and classification of varieties, genetic mapping, heterosis prediction and molecular marker selection breeding in rice and other crops[50]. But most of the research still have considerable distance from the large-scale breeding application, there are still many problems in the technical level. The current DNA fingerprinting technology still exists four key issues: (1) The current DNA extraction methods suit only for laboratory research basically, the problem of extracting DNA from samples quickly and effectively still unsolved; (2) DNA fingerprint detection system is relatively cumbersome and detection efficiency is low. It is difficult for effective integration and accurate comparison among large-scale and different batches of DNA fingerprinting data, that limit the large-scale application of this technology in breeding; (3) The majority varieties of rice in China have low genetic diversity, limited number of samples and sources of rice in most studies makes it difficult to filter out the representative core SSR primers as standard [44]; (4) Since the study of DNA fingerprinting crop has just started in China, there is no unified standard for the construction of rice DNA fingerprinting among various national and local research institutions. As a result, the same varieties of rice has several incompatible and unconvertible DNA fingerprint database. It's a waste of resources. A single database has low capacity, low coverage, insufficient resolution, instability and other shortcomings. That will hinder the construction and development of the rice DNA fingerprint database, and also hinder its deep research and application in rice molecular breeding [42].

4.2 Application of new technologies in SSR DNA fingerprinting

Despite these problems, with the development of artificial intelligence and the emergence of new and sophisticated molecular biology instruments, many foreign research institutions and breeding units already have some testing equipment with high degree of automation. From sample adding to data analysis, automated operation in the entire process improve the efficiency and accuracy of DNA fingerprinting test greatly. With the development of new technologies, SSR fluorescence labeled and capillary electrophoresis technology is increasingly used in DNA fingerprinting techniques, and will replace the PAGE silver staining eventually. The current SSR fluorescence labeled and capillary electrophoresis are mainly used in canola, corn, wheat, tobacco and other crops, research on rice is less. Previous studies in corn and rapeseed and other crops showed that, compared with conventional electrophoresis method, results of SSR fluorescence labeled and capillary electrophoresis test are more accurate, sensitive, efficient and more suitable for detection and analysis of high-volume varieties[51] (as shown in Table 2). It is believed that with the development of modern biotechnology breakthroughs in key technologies, DNA fingerprinting technology will be widely used in all aspects of crop germplasm resources.

Tab. 2. Advantages and disadvantages of detection methods of SSR-PCR amplification product

Polymorphism detection method	Advantages and Disadvantages
Agarose gel electrophoresis	Advantages: simple operation, low cost, less electrophoresis pattern background interference; horizontal comparison can achieved between different batches of results. Disadvantages: low resolution, maximum detectable amplification fragment is about 8 bp; unable to show the estimated value of the amplified fragment; difficult to achieve automation and direct comparisons between the results of the same batch.
PAGE	Advantages: medium resolution, 3 bp or more difference in the length is distinguishable theoretically; according to the amplified bands of representative varieties, horizontal comparison can be achieved between different batches of results. Disadvantages: complicate operation, technical difficulties, multiple experimental factors, silver staining is easy to be interfered by the presence of background; unable to show the estimation value of amplified fragment length; a direct comparison among different batches of results and automation is not easy to achieve.
Capillary electrophoresis	Advantages: High resolution, 1 bp difference in length can be distinguished in theory; an estimate of amplified fragment length can be given combined with standard varieties; automation and results can be directly compared between different batches. Disadvantages: requires high quality of the template DNA and PCR amplification specificity; the cost of equipment and reagents is high.

Acknowledgments

Research supported by the Application technology research and development major projects in Heilongjiang Province(GA14B104).

References

1. S. H. Duan, P. S. Hu, Chinese rice science and technology development strategy, *Chinese Rice Science. Paris* **22**, 3 (2008).
2. Y. H. Wang, *Wuchang rice breakout*, Vol 3 (Company and Industry, 2011), pp. 031.
3. J. Sun, X. M. Jin, H. P. Mao et al, Hyperspectral imaging technology in rice adulteration detection application, *Agricultural Engineering Journal. Paris* **30**, 21 (2014).
4. Y. X. Zhang, Q. Guan, X. L. Xu et al, Research progress of DNA molecular markers and several new molecular markers technology, *Heilongjiang Agricultural Sciences*. 1 (2008).
5. Y. L. An, S. C. Guo, S. P. Li et al, Application progress of DNA molecular markers in crop germplasm resources, *North China Agricultural Journal. Paris* **22**, S3 (2007).
6. W. Li, T. Wang, Development of DNA molecular markers and its main technology, *Technology Information*. 30 (2008).
7. J. Y. Zhuang, Characteristics and application of DNA molecular markers in rice, *China Rice. Paris* **14**, 1 (2008).
8. S. Li, G. F. Zhao, Molecular markers of AFLP and its application, *Northwest China Plant. Paris* **23**, 5 (2003).
9. S. W. Wang, X. Liu, Y. Wang et al, Polymorphism of indica japonica genome revealed by RFLP, *Chinese Agricultural Science. Paris* **39**, 5 (2005).
10. Y. B. Jiang, C. Q. Sun, R. H. Li et al, Study on classification of two-line hybrid rice and their parents by RFLP markers, *China Agricultural Science. Paris* **32**, 6 (1999).
11. P. J. Zhang, H. W. Cai, P. R. Yuan et al, Relationship between RFLP genetic distance and heterosis in rice, *Hybrid Rice. Paris* **16**, 5 (2001).
12. Y. Bao, S. Ge, DNA analysis of the ITS sequence of nuclear RFLP identified the species of the genus CD, *Botany Gazette. Paris* **45**, 7 (2003).
13. Y. H. Li, Q. Qian, RAPD identification and genetic relationship of main hybrid rice parents in China, *Journal of crops. Paris* **26**, 2 (2000).
14. J. N. Mao, S. H. Duan, S. Q. Li et al, Genetic analysis and identification of rice and its parents by RAPD molecular markers in three groups of three lines, *Heredity. Paris* **24**, 3 (2002).
15. S. H. Duan, J. N. Mao, Y. G. Zhu, DNA polymorphism of the major restorer lines of hybrid rice in China by RAPD molecular markers, *Journal of Wuhan University of Technology (SCIENCE EDITION). Paris* **47**, 4 (2001).
16. T. H. Xiang, X. F. Wang, L. Li et al, To distinguish and identification of indica hybrid rice and its parents in China by using RAPD markers, *Journal of Crops.*

Paris **26**, 3 (2000).
17. Z. M. Chen, X. E. Wang, Y. Zhao et al, Rapid identification of Liangyoupei 9 by RAPD, *Hybrid Rice. Paris* **18**, 4 (2003).
18. Y. C. Peng, X. F. Wang, Comparative study of SDS-PAGE, isozyme isoelectric focusing and RAPD in identification of two-line hybrid rice F_1 seed purity, *Biotechnology. Paris* **16**, 5 (2006).
19. P. Vos, AFLP: a new technique for DNA fingerprinting, *Nucleic Acids Research. Paris* **23**, 21 (1995).
20. M. L. Weng, B. Wang, Principle and application of AFLP, *Hybrid Rice.* 5 (1996).
21. Y. Wu, Z. B. Li, Y. Wang et al, SNP marking technique and its application in crop breeding, *Journal of Liaoning Agricultural College. Paris* **12**, 3 (2010).
22. Y. Tian, Y. Xie, X. Y. Lu, Single nucleotide polymorphism and its application in rice, *Subtropical Plant Science. Paris* **41**, 1 (2012).
23. X. J. Wang, G. Santosh, N. Nigan, AFLP fingerprinting in groundnut, *Eighteenth International Congress of Genetics Abstract.* (Beijing, China, 1998) pp.154.
24. C. Q. Zhang, J. Z. Jia, Studies on the primers screening for AFLP Fingerprintsof rice cultivars, *Agricultural Sciences in China. Paris* **9**, 1 (2002).
25. W. C. Li, S. B. Gao, T. Z. Rong et al, Identification of AFLP fingerprinting in maize inbred line, *Journal of Sichuan Agricultural Uniersity. Paris* **19**, 2 (2001).
26. L. Q. Tang, C. L. Xiao, W. P. Wang, Research and application progress of SNP molecular markers, *Journal of Chinese Agricultural Science. Paris* **28**, 12 (2012).
27. J. Qiao, L. Y. Wang, Y. Shi, Research progress of SSR on purity identification of rice varieties, *Tianjin Agricultural Sciences. Paris* **17**, 4 (2011).
28. J. Y. Zhuang, S. T. Peng, Q. C. Yan et al, SSR markers and purity identification of main hybrid rice combinations and their parents in China, *Chinese Rice Science. Paris* **17**, 1 (2003).
29. Y. Shi, M. Li, P. He et al, Single Nucleotide Polymorphism and its utilization in plant research, *Hans Journal of Agricultural Sciences. Paris* **2014**, 2014.
30. J. J. Lu, Z. Y. Yuan, B. Zhou et al, Genetic difference analysis of parents of southern two-line indica hybrid rice based on SNP markers, *Hybrid Rice. Paris* **29**, 5 (2014).
31. J. F. Wu, W. Song, R. Wang et al, The group classification of 51 maize inbred lines using SNP markers, *Maize Sciences. Paris* **22**, 5 (2014).
32. Y. L. Li, J. F. Li, L. G. Kang et al, SNP genotyping of tomato Mi-1 gene, *Journal of Northeast Agricultural University. Paris* **41**, 10 (2010).
33. X. F. Wu, R. L. Lv, M. C. Liu, Molecular marker technology and application in rice genetic studies, *Journal of Chinese Agricultural Science. Paris* **25**, 4 (2009).
34. Y. Y. Xin, Y. Zhang, Y. P. Xiong et al, Identification of super hybrid rice combinations and their purity by SSR molecular markers, *Chinese Rice Science. Paris* **19**, 2 (2005).
35. N. N. andakumar, A. K. Singh, R. K. Sharma et al, Molecular fingerprinting of hybridsand assessment of genetic purity of hybrid seeds in rice using microsatellite markers, *Euphytica. Paris* **136**, 2004.

36. B. Luo, H. Y. Sun, Z. G. Yang et al, DNA fingerprint construction and genetic similarity analysis of Taihu conventional japonica rice based on SSR markers, *Journal of Southern Agriculture. Paris* **46**, 1 (2015).
37. L. G. Cheng, B. Y. Guo, M. G. Yu et al, Using SSR markers to identify two-line hybrid rice seed purity, *The Chinese Seed Industry*. 11 (2011).
38. A. N. Zhang, J. H. Wang, X. Q. Yu et al, Application of SSR markers in the purity identification of hybrid rice, *Shanghai Journal of Agricultural Sciences. Paris* **29**, 6 (2013).
39. R. L. Chen, Y. C. Huang, M. Tang et al, Study in the detection of purity of super hybrid rice Guiliangyou 2 by SSR markers, *Hybrid Rice. Paris* **4**, (2013).
40. Q. H. Yang, X. Q. Li, M. H. Lu, Comparison of SSR molecular detection and field planting purity identification of Yangliangyou 6, *Chinese Seed Industry*. 7 (2014).
41. Z. F. Xiao, Construction of hybrid rice DNA fingerprint database by SSR markers, Master's thesis, Central South University (2011).
42. W. G. Kong, J. F. Wang, H. F. Ding, Application of DNA fingerprint database in the main crop germplasm resources in China, *Shandong Agricultural Sciences. Paris* **46**, 4 (2014).
43. Y. L. Wang, Y. S. Jiang, Application of SSR markers in crop germplasm, *Shandong Agricultural Sciences. Paris* **44**, 10 (2012).
44. H. Wang, M. B. Li, J. J. Bai et al, Research progress on the construction of DNA fingerprinting of rice varieties using SSR molecular markers, *Chinese Rice. Paris* **17**, 1 (2011).
45. S. M. Zuo, N. N. Zhou, Z. X. Chen et al, Application and prospect of DNA fingerprinting in rice varieties identification, *Jiangsu Agricultural Sciences. Paris* **42**, 12 (2014).
46. Y. Zhan, D. D. Cao, Y. Wang et al, Comparative study on seed purity identification of hybrid rice, *Zhejiang Agricultural Sciences. Paris* **1**, 5 (2015).
47. Z. L. Xu, J. L. Ni, L. Li et al, Rice ID constructed by SSR molecular fingerprinting and product information, *Journal of Crops, ACTA AGRONOMICA SINICA. Paris* **40**, 5 (2014).
48. L. Huang, Y. X. Li, DNA fingerprinting construction and inheritance of common rice varieties, *Grain and Oil. Paris* **27**, 10 (2014).
49. C. H. Ying, J. Z. Li, X. D. Yan, Construction of SSR fingerprint and analysis of genetic diversity of Henan main upland rice varieties, *Molecular Plant Breeding. Paris* **12**, 2 (2014).
50. Y. Z. Zhang, C. H. Wang, Y. S. Xia et al, Progress in the application of SSR molecular markers in Hybrid Rice Research, *Liaoning Agricultural Sciences*. 6 (2009).
51. J. H. Xia, B. Y. Cheng, J. Y. Gong et al, Application of SSR fluorescence markers and capillary electrophoresis in identification of rice DNA fingerprinting, *Chinese Rice Science. Paris* **25**, 6 (2011).

Ferric pyrophosphate: A versatile and alternative iron fortification compound

Liuqin Ge[1], Meisheng Xia[1,*], Zhitong Yao[2,3]* and Qingping Sun[3]
[1]*Ocean College, Zhejiang University, Zhoushan, 316021, China*
[2]*College of Materials Science and Environmental Engineering, Hangzhou Dianzi University, Hangzhou, 310018, China*
[3]*Yangzhou SinoMicro Biotech Co., Ltd., Yangzhou, 225000, China*
gliuqin@163.com, msxia@zju.edu.cn, sxyzt@126.com, sqp@163.com

Iron deficiency is the most common micronutrient deficiency in the world, with many adverse health effects on human beings and the animals. The success of an iron fortification program depends largely on the iron compounds. In this work, the various iron fortification compounds were investigated and their advantages and disadvantages were discussed. There are three stages in the development of iron fortification compounds: inorganic iron (e.g. ferrous sulfate), organic iron (e.g. iron fumarate) and chelating iron (e.g. iron-glycine chelate). Among them, the ferric pyrophosphate does not modify organoleptic characteristics of foods, and can have a similar bioavailability with ferrous sulfate through micronization or solubilization, making it a potential and alternative iron fortification compound.

Keywords: Ferric Pyrophosphate; Bioavailability; Iron Fortification; Iron Compound; Food.

1. Introduction

Iron deficiency is considered to be the most common and widespread micronutrient deficiency in the world, affecting 1.5 to 2 billion people, among whom 500 million have iron deficiency anemia [1]. Iron is an essential trace element of humans, naturally present in cytochromes, enzymes, haemoglobin and myoglobin [2]. It plays a crucial role in various biochemical processes, including oxygen transport and storage, electron transfer, and catalysis [3]. Iron deficiency has adverse health effects on growth rate, cognitive performance, pregnancy outcome, immune status, and work capacity [4]. This deficiency is partly induced by inadequate iron intake and absorption, such as plant-based diets. In addition, it also results from increased iron requirements during fast growth, and excessive blood losses. Consequently, iron deficiency is quite prevalent in infants, adolescents, and women of reproductive age [5].

There are three main strategies for combating iron deficiency in populations. The first one is dietary modification or diversification, which has the disadvantages that change of dietary preferences is difficult and foods with highly bioavailable iron are expensive. The second one is iron supplementation, usually in high doses of iron provision without food. It is highly efficacious to increase Fe status, but limited by its low compliance due to gastrointestinal discomfort [6]. The last one, iron fortification of foods, is often regarded as the most practical, sustainable, and cost-effective approach to reduce prevalence of iron deficiency at the national level. Effective iron fortification of foods still remains a challenge. Iron is the most problematic essential mineral to add to foodstuffs. On one hand, the water-soluble iron compounds with high bioavailability often cause undesirable color and flavor changes in the food vehicles, because they release free iron ions that are highly reactive with food components [7]. On the other hand, insoluble compounds may be too poorly absorbed to have enough nutritional benefit, although they are much more stable and cause few sensory changes.

Therefore, the success of an iron fortification program depends largely on the Fe compound. Based on their solubility in water or dilute acid, iron fortification compounds can be divided into three categories, including freely water-soluble compounds, compounds poorly soluble in water but soluble in dilute acid, and compounds insoluble in water and also poorly soluble in dilute acid. The first compounds can dissolve instantaneously in the gastric juice during digestion and thus have the highest bioavailability among conventional iron compounds. The second group have a similar or slightly lower bioavailability depending on how fast they dissolve in gastric juice. The final group have a low and variable bioavailability, as their solubility in gastric juice depends on the composition of the meal and the physical characteristics of the fortificant particles, such as size, shape, and surface area.

2. Conventional iron compounds

Ferrous sulfate is a common freely water-soluble iron compound, which has been designated as the reference standard iron source with a relative bioavailability value (RBV) of 100. In practice, it is used to fortify infant formula, bread, pasta, and wheat flour stored for short periods [8]. Although ferrous sulfate has a high bioavailability, it is so highly reactive with food composition that sensorial problems may arise in some vehicles. It is not suitable to fortify salt, infant cereal, cocoa products, fish sauce and soy sauce, as well as cereal flours stored for long periods [8].

Ferrous fumarate, the representative of the compounds poorly soluble in water but soluble in dilute acid, causes less organoleptic changes in food vehicles than ferrous sulfate does. It is usually added to infant cereals, chocolate drink powders, complementary foods for infants and young children, with a bioavailability equivalent to ferrous sulfate [8, 9]. Even though its poor solubility in water makes it less reactive, it can still provoke rancidity because of hygroscopicity [10]. Moreover, since its absorption requires to be dissolved in the gastric juice, it may not be so well absorbed in children, or in populations whose gastric acid secretion is impaired owing to bacterial infections or nutrient deficiencies [8, 11].

Elemental iron powders are insoluble in water and poorly soluble in dilute acid, which are particularly used to fortify cereal products. These fortificants consist of five different types based on their manufacturing processes, which are electrolysis, hydrogen (H)-reduction, carbon monoxide (CO)-reduction, atomization, and carbonylation.

Electrolytic iron is the only one demonstrated to be a useful fortificant, but its absorption is just about half as ferrous sulfate at best [8].

3. Novel iron fortificant

With the great progress in synthesis and preparation method, more and more novel iron fortificants are under development and investigation. Those fortificants include conventional iron compounds modified by microencapsulation, micronization, or solubilization, new iron compounds synthesized through chelation reactions (e.g. sodium iron ethylenediaminetetraacetec acid, and ferrous bisglycinate), and heme iron extracted from animal resources. Unlike those conventional iron compounds, they not only own high availability but also have desirable application performance in food vehicles.

3.1. *Microencapsulated iron compound*

Microencapsulation technique has been used in iron fortification usually for ferrous sulfate and ferrous fumarate. It provides a barrier between iron and food matrix, thus reducing undesirable sensory changes and reactions in food vehicles [8, 12-14]. In addition, it has little influence on the bioavailability of iron compounds [8, 13], or even contributes to increasing iron bioavailability [12] because it counteracts inhibitors of iron absorption (e.g. phytic acid, polyphenols, and calcium) [15] and protects iron from oxidation [12]. Microencapsulated ferrous sulfate or ferrous fumarate can be used to fortify milk [2, 12, 16] and iodized salt [17-20], respectively. They can be produced by various techniques, including fatty acid ester method, liposome method, emulsification method [16], spray-drying [15, 21], and so on [12, 22]. Problems involved in the development of microencapsulated iron compound include heat instability of the capsule and relatively high cost of microencapsulation [8, 13].

3.2. *Ferrous bisglycinate*

Ferrous bisglycinate is an iron amino acid chelate, which consists of one Fe^{2+} atom bound to two molecules of glycine [23-25]. This structure provides a protection to ferrous iron from binding to inhibitors in food, making ferrous bisglycinate two to four times more bioavailable than ferrous sulfate [24-26]. Furthermore, chelate also reduces the reactivity of iron and the possibility to cause sensory changes in food vehicles [24, 25]. It has been demonstrated to be an effective iron fortificant from various efficacy studies [23-27] and the US Food and Drug Administration has acknowledged it as a safe iron fortificant [24, 25]. However, its high cost and tendency to promote fat oxidation in stored cereals [8, 13], make it unsuitable for many food vehicles.

3.3. *Heme iron*

Heme iron concentrate is prepared from animal hemoglobin through enzyme hydrolysis and purification [28]. Compared to non-heme iron, heme iron has a greater bioavailability, mainly because that heme iron is absorbed as an intact porphyrin ring [29], which protects it from inhibitors of iron absorption. It has been reported from animal and human studies that bioavailability of heme iron was about 23% higher than that of ferrous sulfate [30, 31]. Besides, an efficacy trial in adolescent girls confirmed that biscuit fortified with heme iron concentrate was more effective in restoring serum hemoglobin than that with iron sulfate [32]. In addition, the ingestion of heme iron is not associated with side-effects, such as gastric irritation [32]. However, its widespread use is limited due to many factors, including the elevated cost, intense red-brown color, low iron content (1.38%), and technical difficulties in collection, drying and storage [29].

4. Ferric pyrophosphate

Ferric pyrophosphate is an important kind of compounds insoluble in water and also poorly soluble in dilute acid. It does not modify organoleptic characteristics of foods because of its white colour and low reactivity, making it suitable for many difficult-to-fortify food vehicles, such as salt, cereals, infant formulas, and rice. It has been demonstrated to be an effective iron fortificant from efficacy studies [8, 13, 32]. However, the poor solubility of ferric pyrophosphate in gastric juice limits its bioavailability, which furthermore reduces its nutritional value. The relative bioavailability value (RBV) of ferric pyrophosphate was reported to be only about 50% [8, 34] from human studies.

4.1. *Micronized ferric pyrophosphate*

The iron absorption from poorly soluble iron compounds largely depends on the particle size. As to ferric pyrophosphate, particle size reduction not only contributes to improving its bioavailability in solid foodstuffs, but also makes it an alternative fortificant for liquid foods [35]. The mean particle size (MPS) of ferric pyrophosphate can be decreased to 2–3 μm through conventional grinding, while further reduction in MPS to < 1 μm requires aqueous-phase synthesis with emulsifiers to prevent particle agglomeration [36]. Micronized dispersible ferric pyrophosphate (MDFP) can be prepared from ferric chloride and sodium pyrophosphate using dispersion techniques [34], with monoglycerides and diglycerides used as the emulsifiers [37].

The effect of reducing particle size of ferric pyrophosphate on its bioavailability has been tested in many studies. Wegmüller et al. [36] studied the bioavailability of ferric pyrophosphate with different particle size in rats. The RBV was 59 % for the regular ferric pyrophosphate (MPS $\approx$ 21 μm), 69% for micronized ferric pyrophosphate (MPS $\approx$ 2.5 μm), and 95 % for MDFP (MPS $\approx$ 0.5 μm). It is also found that the bioavailability of MDFP varied with food matrix. Fidler et al. [34] measured Fe absorption from MDFP (MPS $\approx$ 0.3 μm) fortified in wheat-based infant cereal and yoghurt drink in healthy women. The RBV of MDFP was discovered to be 83% from wheat-based infant cereal and 94% from yoghurt drink.

Salt is a good potential vehicle for iron fortification, because it is one of the few regularly consumed food items particularly in poor regions. Water-soluble iron compounds react with moisture and impurities in salt, leading to color changes and iodine losses [38]. As to ferric pyrophosphate, it has negligible influence on the sensory characteristic and iodine stability of iodized salt [38]. A efficacy trial for 6–15-y-old Moroccan children reported that salt fortified with potassium iodate and micronized ferric pyrophosphate (MPS = 2.5 μm) increased mean hemoglobin by 16 g/L, improved iron status and body iron stores, and decreased the prevalence of iron deficiency anemia from 30 % at baseline to 5 % [38]. A trial in 5–15-y-old Indian children also indicated that iodized salt with micronized ferric pyrophosphate (MPS = 2.5 μm) was efficacious in reducing the prevalence of iron deficiency and anemia [39].

Another promising food vehicle for iron fortification is rice, as it is the leading staple food in developing world such as South Asia. Addition of water-soluble iron compounds to rice grains often causes oxidation and unacceptable sensory changes, while rice grains fortified with micronized ferric pyrophosphate have excellent organoleptic characteristics [4]. Besides, studies in Indian children have demonstrated that micronized ferric pyrophosphate fortified in rice was effective in increasing iron stores and reducing iron deficiency [40, 41].

4.2. *Soluble ferric pyrophosphate*

Soluble ferric pyrophosphate (SFP), with its iron core chelated to citrate and pyrophosphate ligands, is highly soluble in aqueous solutions over a wide range of pH (pH 2-8) [42]. Due to the relatively low price, food-grade SFP is allowed to be used as an iron fortificant in Europe [43]. Similarly to ferric pyrophosphate, SFP does not change the sensory characteristic of fortified foods as well [44].

An early test in rats indicated that RBV was 58% for ferric pyrophosphate, 83% for ferric pyrophosphate solubilized with ammonium citrate, and 103% for ferric pyrophosphate solubilized with sodium citrate [45]. Another rat test reported that RBV of ferric pyrophosphate stabilized and solubilized with glycine was 106% when fortified in water and 125% when fortified in yoghurt [46]. An in vitro experiment on Caco-2 cells indicated that the iron bioavailability of SFP was the same high or higher and the effect of iron absorption inhibitors on its iron smaller, as compared to that of $FeSO_4$ [43]. A newly efficacy study in rats showed that SFP-fortified milk significantly increased haemoglobin, plasma ferritin and plasma transferrin content as well as iron content in liver [42].

5. Conclusions

Iron fortification of foods is the most cost-effective approach to control iron deficiency. Conventional iron compounds don't have satisfactory performance either in bioavailability or stability when fortified in foodstuffs. Novel iron fortificants include chelating iron compounds, heme iron, and conventional iron compounds modified by microencapsulation, micronization, or solubilization. Ferric pyrophosphate has remarkable advantages that it does not interact with the food matrix or change the sensory characteristics of foods, making it a versatile fortificant. Through micronization or solubilization, ferric pyrophosphate can have improved bioavailability similar to or even greater than ferrous sulfate. It can be a promising alternative iron source for food fortification. To realize its widespread use, more absorption and efficacy studies particularly on humans are still needed.

References

1. A. Ahmed, F. M. Anjum, M. A. Randhawa, U. Farooq, S. Akhtar and M. T. Sultan, Effect of multiple fortification on the bioavailability of minerals in wheat meal bread, *J. Food Sci. Tech. Mys.* **49**, 737 (2012).
2. S. Abbasi and S. Azari, Efficiency of novel iron microencapsulation techniques: fortification of milk, *Int. J. Food Sci. Tech.* **46**, 1927 (2011).
3. G. Papanikolaou and K. Pantopoulos, Iron metabolism and toxicity, *Toxicol. Appl. Pharm.* **202**, 199 (2005).
4. D. Moretti, T. C. Lee, M. B. Zimmermann, J. Nuessli and R. F. Hurrell, Development and evaluation of iron-fortified extruded rice grains, J. Food Sci. **70**, S330 (2005).
5. S. R. Pasricha, H. Drakesmith, J. Black, D. Hipgrave and B. A. Biggs, Control of iron deficiency anemia in low- and middle-income countries, Blood **121**, 2607 (2013).
6. R. Blanco-Rojo, A. M. Perez-Granados, L. Toxqui, C. Gonzalez-Vizcayno, M. A. Delgado and M. P. Vaquero, Efficacy of a microencapsulated iron pyrophosphate-fortified fruit juice: a randomised, double-blind, placebo-controlled study in Spanish iron-deficient women, *Brit. J. Nutr.* **105**, 1652 (2011).
7. Y. M. van Leeuwen, K. P. Velikov and W. K. Kegel, Colloidal stability and chemical reactivity of complex colloids containing Fe3+, Food Chem. **155**, 161 (2014).
8. R. F. Hurrell, Fortification: Overcoming technical and practical barriers, J. Nutr. **132**, 806s (2002).
9. M. Harrington, C. Hotz, C. Zeder, G. O. Polvo, S. Villalpando, M. B. Zimmermann, T. Walczyk, J. A. Rivera and R. F. Hurrell, A comparison of the bioavailability of ferrous fumarate and ferrous sulfate in non-anemic Mexican women and children consuming a sweetened maize and milk drink, Eur. J. Clin. Nutr. **65**, 20 (2011).
10. M. J. Salgueiro and J. Boccio, Ferric Pyrophosphate as an Alternative Iron Source for Food Fortification, in Handbook of Food Fortification and Health: From Concepts to Public Health Applications, eds. V.R. Preedy, Nutrition and Health, Vol.1, pp. 91-97.
11. R. Hurrell, Use of ferrous fumarate to fortify foods for infants and young children, Nutr. Rev. **68**, 522 (2010).
12. C. Gupta, P. Chawla, S. Arora, S. K. Tomar and A. K. Singh, Iron microencapsulation with blend of gum arabic, maltodextrin and modified starch using modified solvent evaporation method - Milk fortification, Food Hydrocolloid **43**, 622 (2015).
13. R. Hurrell, How to ensure adequate iron absorption from iron-fortified food, Nutr. Rev. **60**, S7 (2002).

14. H. Majeed, H. J. Qazi, W. Safdar and Z. Fang, Microencapsulation Can Be a Novel Tool in Wheat Flour with Micronutrients Fortification: Current Trends and Future Applications - a Review, Czech J. Food Sci. **31**, 527 (2013).
15. M. Moslemi, H. Hosseini, M. Erfan, A. M. Mortazavian, R. Mazaheri, N. Fard, T. R. Neyestani and R. Komeyli, Characterisation of spray-dried microparticles containing iron coated by pectin/resistant starch, Int. J. Food Sci. Tech. **49**, 1736 (2014).
16. C. Gupta, P. Chawla and S. Arora, Development and evaluation of iron microencapsules for milk fortification, Cyta-J. Food **13**, 116 (2015).
17. Y. O. Li, L. L. Diosady and A. S. Wesley, Iodine stability in iodized salt dual fortified with microencapsulated ferrous fumarate made by an extrusion-based encapsulation process, J. Food Eng. **99**, 232 (2010).
18. Y. O. Li, D. Yadava, K. L. Lo, L. L. Diosady and A. S. Wesley, Feasibility and optimization study of using cold-forming extrusion process for agglomerating and microencapsulating ferrous fumarate for salt double fortification with iodine and iron, J. Microencapsul **28**, 639 (2011).
19. D. Romita, Y. L. Cheng and L. L. Diosady, Microencapsulation of Ferrous Fumarate for the Production of Salt Double Fortified with Iron and Iodine, Int. J. Food Eng. **7**, 1556 (2011).
20. D. Yadava, Y. O. Li, L. L. Diosady and A. S. Wesley, Optimisation of polymer coating process for microencapsulating ferrous fumarate for salt double fortification with iodine and iron, J. Microencapsul **29**, 729 (2012).
21. L. L. D. Bittencourt, C. Pedrosa, V. P. de Sousa, A. P. T. Pierucci and M. Citelli, Pea Protein Provides a Promising Matrix for Microencapsulating Iron, Plant Food Hum. Nutr. **68**, 333 (2013).
22. B. M. Ding, X. M. Zhang, K. Hayat, S. Q. Xia, C. S. Jia, M. Y. Xie and C. M. Liu, Preparation, characterization and the stability of ferrous glycinate nanoliposomes, J. Food Eng. **102**, 202 (2011).
23. P. Ferrari, A. Nicolini, M. L. Manca, G. Rossi, L. Anselmi, M. Conte, A. Carpi and F. Bonino, Treatment of mild non-chemotherapy-induced iron deficiency anemia in cancer patients: Comparison between oral ferrous bisglycinate chelate and ferrous sulfate, Biomed. Pharmacother. **66**, 414 (2012).
24. J. F. Haro-Vicente, D. Perez-Conesa, F. R. Braquehais and G. Ros, Iron absorption and haemoglobin status of rats fed a ferrous bisglycinate-fortified growing-up milk, J. Sci. Food Agr. **89**, 2107 (2009).
25. M. E. van Stuijvenberg, C. M. Smuts, P. Wolmarans, C. J. Lombard and M. A. Dhansay, The efficacy of ferrous bisglycinate and electrolytic iron as fortificants in bread in iron-deficient school children, Brit. J. Nutr. **95**, 532 (2006).
26. N. Milman, L. Jonsson, P. Dyre, P. L. Pedersen and L. G. Larsen, Ferrous bisglycinate 25 mg iron is as effective as ferrous sulfate 50 mg iron in the prophylaxis of iron deficiency and anemia during pregnancy in a randomized trial, J. Perinat. Med. **42**, 197 (2014).

27. L. H. S. Miglioranza, T. Matsuo, G. M. Caballero-Cordoba, J. B. Dichi, E. S. Cyrino, I. B. N. Oliveira, M. S. Martins, N. M. Polezer and I. Dichi, Effect of long-term fortification of whey drink with ferrous bisglycinate on anemia prevalence in children and adolescents from deprived areas in Londrina, Parana, Brazil, Nutrition **19**, 419 (2003).
28. M. Aleman, C. D. Nuchi, R. Bou, A. Tres, J. Polo, F. Guardiola and R. Codony, Effectiveness of antioxidants in preventing oxidation of palm oil enriched with heme iron: A model for iron fortification in baked products, Eur. J. Lipid Sci. Tech. **112**, 761 (2010).
29. S. Miret, S. Tascioglu, M. van der Burg, L. Freinken and W. Klaffke, In Vitro Bioavailability of Iron from the Heme Analogue Sodium Iron Chlorophyllin, J. Agr. Food Chem. **58**, 1327 (2010).
30. G. Gonzalez-Rosendo, J. Polo, J. J. Rodriguez-Jerez, R. Puga-Diaz, E. G. Reyes-Navarrete and A. G. Quintero-Gutierrez, Bioavailability of a Heme-Iron Concentrate Product Added to Chocolate Biscuit Filling in Adolescent Girls Living in a Rural Area of Mexico, J. Food Sci. **75**, H73 (2010).
31. A. G. Quintero-Gutierrez, G. Gonzalez-Rosendo, J. Sanchez-Munoz, J. Polo-Pozo and J. J. Rodriguez-Jerez, Bioavailability of heme iron in biscuit filling using piglets as an animal model for humans, Int. J. Biol. Sci. **4**, 58 (2008).
32. A. G. Quintero-Gutierrez, G. I. Mariaca-Gaspar, J. Villanueva-Sanchez, J. Polo, C. Rodriguez and G. Gonzalez-Rosendo, Acceptability and use of heme-iron concentrate product added to chocolate biscuit filling as an alternative source of a highly available form of iron, Cyta-J. Food **10**, 112 (2012).
33. L. Davidsson, S. A. Sarker, K. A. Jamil, S. Sultana and R. Hurrell, Regular consumption of a complementary food fortified with ascorbic acid and ferrous fumarate or ferric pyrophosphate is as useful as ferrous sulfate in maintaining hemoglobin concentrations > 105 g/L in young Bangladeshi children, Am. J. Clin. Nutr. **89**, 1815 (2009).
34. M. C. Fidler, T. Walczyk, L. Davidsson, C. Zeder, N. Sakaguchi, L. R. Juneja and R. F. Hurrell, A micronised, dispersible ferric pyrophosphate with high relative bioavailability in man, Brit. J. Nutr. **91**, 107 (2004).
35. M. A. Roe, R. Collings, J. Hoogewerff and S. J. Fairweather-Tait, Relative bioavailability of micronized, dispersible ferric pyrophosphate added to an apple juice drink, Eur. J. Nutr. **48**, 115 (2009).
36. R. Wegmuller, M. B. Zimmermann, D. Moretti, M. Arnold, W. Langhans and R. F. Hurrell, Particle size reduction and encapsulation affect the bioavailability of ferric pyrophosphate in rats, J. Nutr. **134**, 3301 (2004).
37. D. Moretti, M. B. Zimmermann, R. Wegmuller, T. Walczyk, C. Zeder and R. F. Hurrell, Iron status and food matrix strongly affect the relative bioavailability of ferric pyrophosphate in humans, Am. J. Clin. Nutr. **83**, 632 (2006).
38. M. B. Zimmermann, R. Wegmueller, C. Zeder, N. Chaouki, F. Rohner, M. Saissi, T. Torresani and R. F. Hurrell, Dual fortification of salt with iodine and micronized

ferric pyrophosphate: a randomized, double-blind, controlled trial, Am. J. Clin. Nutr. **80**, 952 (2004).

39. M. Andersson, P. Thankachan, S. Muthayya, R. B. Goud, A. V. Kurpad, R. F. Hurrell and M. B. Zimmermann, Dual fortification of salt with iodine and iron: a randomized, double-blind, controlled trial of micronized ferric pyrophosphate and encapsulated ferrous fumarate in southern India, Am. J. Clin. Nutr. **88**, 1378 (2008).
40. M. S. Radhika, K. M. Nair, R. H. Kumar, M. V. Rao, P. Ravinder, C. G. Reddy and G. N. V. Brahmam, Micronized ferric pyrophosphate supplied through extruded rice kernels improves body iron stores in children: a double-blind, randomized, placebo-controlled midday meal feeding trial in Indian schoolchildren, Am. J. Clin. Nutr. **94**, 1202 (2011).
41. D. Moretti, M. B. Zimmermann, S. Muthayya, P. Thankachan, T. C. Lee, A. V. Kurpad and R. F. Hurrell, Extruded rice fortified with micronized ground ferric pyrophosphate reduces iron deficiency in Indian schoolchildren: a double-blind randomized controlled trial, Am. J. Clin. Nutr. **84**, 822 (2006).
42. B. Sachdeva, R. Kaushik, S. Arora and S. Kapila, Bioavailability of iron in multiple fortified milk, J. Food Sci. Technol. **52**, 6017 (2015).
43. L. Zhu, R. P. Glahn, D. Nelson and D. D. Miller, Comparing Soluble Ferric Pyrophosphate to Common Iron Salts and Chelates as Sources of Bioavailable Iron in a Caco-2 Cell Culture Model, J. Agr. Food Chem. **57**, 5014 (2009).
44. B. Sachdeva, R. Kaushik, S. Arora and K. P. Indumathi, Impact of fortification with iron salts and vitamin A on the physicochemical properties of laboratory-pasteurised toned milk and bioaccessibility of the added nutrients, Int. J. Dairy Technol. **68**, 253 (2015).
45. R. F. Hurrell, D. E. Furniss, J. Burri, P. Whittaker, S. R. Lynch and J. D. Cook, Iron Fortification of Infant Cereals - a Proposal for the Use of Ferrous Fumarate or Ferrous Succinate, Am. J. Clin. Nutr. **49**, 1274 (1989).
46. M. J. Salgueiro, S. Arnoldi, M. A. Kaliski, H. Torti, E. Messeri, R. Weill, M. Zubillaga and J. Boccio, Stabilized-Solubilized Ferric Pyrophosphate as a New Iron Source for Food Fortification. Bioavailability Studies by Means of the Prophylactic-Preventive Method in Rats, Biol. Trace Elem. Res. **127**, 143 (2009).

The effect of polysaccharides from platycodonis radix on rat airway smooth muscle cells proliferation

Yu Sheng[1], Guangchen Liu[2,*], Yiyao Gao[3], Leichao Zhang[4], Zuying Lv[5]

[1]*College of Pharmacy, Beihua University,*
Jilin 132013, Jilin, Peoples R China
[2]*Department of Orthopaedics and Traumatology, The First Hospital of Jilin University,*
Changchun 130021, Jilin, Peoples R China
[3]*Department of Scientific research center, China-Japan Union Hospital of Jilin University,*
Changchun 130033, Jilin, Peoples R China
[4]*Department of pathology, China-Japan Union Hospital of Jilin University,*
Changchun 130033, Jilin, Peoples R China
[5]*School of Pharmaceutical Sciences, Jilin University,*
Changchun 130021, Jilin, Peoples R China
*shengyushirley@163.com, *lgcdo2011@163.com, xiaoyao19860202@163.com,*
baobaopig1986@163.com, 731859728@qq.com
**Corresponding author*

To observe the effect of Polysaccharides from Platycodonis Radix (PRPS) on the rat airway smooth muscle cells (ASMCs) proliferation in vitro. **Methods** ASMCs were obtained from adult SD rats by trypsin-collagenase digestion method.In vivo experiment, the cells proliferating model of rat ASMCs were established by ET-1 and the intervention of PRPS. ASMCs were treated with PRPS at different concentrations and the growth of the cells was analyzed by MTT assay. The expression of PCNA in cells was detected by immunocytochemical method. **Results** At the end of the experiment, the growth curve of high dosage group of PRPS was significantly lower than that in the model group (0.1521±0.0053). The expression of PCNA in this group was also lower than the model group and slightly higher in the dexamethasone treated group. **Conclusion** PRPS can inhibit the proliferation of ASMCs induced by endothelin-1 in vitro, the mechanism has correlation with down- regulation of the expression and PCNA.

Keywords: PRPS, ASMCs, Cell Proliferation, MTT, PCNA.

1. Introduction

Platycodonis Radix is the dry root of *Campanulaceae Platycodon grandiflorum (Jacq.) A.DC*[1]. It is a famous medicinal plant in China. It has been used in traditional oriental medicine as an expectorant and antitussive and used to treat coughs, colds, upper respiratory infections, sore throats, tonsillitis, and chest congestion. According to reports, the polysaccharides of Platycodonis Radix (PRPS) has immunomodulatory and anti-tumor effects. In recent years, PRPS has become the hotspot of the research of Platycodonis Radix.Studies have found that PRPS could significantly decrease airway hyperresponsiveness (AHR) in asthma[2]. Airway smooth muscle cells (ASMCs) was the main effector cell of airway constriction which plays a crucial role in the course of airway remodeling[3]. The airway remodeling could result in persistent bronchial hyperresponsiveness (BHR)[4]. In this study, we make ASMCs to be the target of PRPS.

We research the effects of PRPS on the proliferation of ASMCs induced by endothelin-1 in vitro.

2. Material and Method

2.1 Material

2.1.1 Experimental medicine

Platycodonis Radix purchased from Changchun Guo Ao Pharmaceutical Co., Ltd. They were identified as the dried root of Campanulaceae plants of Platycodon grandiflorum (Jacq.) A.DC.

2.1.2 Experimental animal

Male SD rats were purchased from Jilin university school of public health laboratory animal center.

2.1.3 Reagents

DMEM medium (high sugar), DMSO:Gibco Company, US; collagenase IV, MTT, PCNA mAb(anti-rat):sigma company, US; α-smooth muscular actin α-SMA):Santa company; FBS: HYCLONE company; DAB coloration kit:State Maixin Biological Technology Co.,Ltd.; SABC reagent: Wuhan Dr. de bio engineering Co., Ltd.; Anhydrous ethanol, acetone, ether, etc:Beijing chemical reagent.

2.2 Method

2.2.1 The isolation and culture of rat ASMCs in vitro

Rat was put to death by dislocation after etherization. Removed the trachea and lung tissue after opened chest under sterile conditions. Put them into D-Hanks. Isolated main trachea and primary bronchial without ectotrachea. They were cut into 1 mm ×1 mm × 1 mm tissue block. Then put tissue block in a petri dish which contained the digestive juices (0.25% trypsin : 0.4% collagenase IV= 1:1) at 37 °C for 30 minutes. Poured the broth through a strainer with mesh. Put cell suspension in culture bottle after centrifugal.

2.2.2 The purification of rat ASMCs

The ASMCs were purified by method of differential adhesion. When passaging cells, put the cell suspension solution in the culture bottle, in the 5% CO2 incubator for 10 ~ 15 minutes, were observed under inverted microscope that part of the cells has attached, the cells had not attached were put into the second culture bottle, let them stand for 10 minutes, then put the cells had not attached into the third culture bottle. The Solution in the third bottle were the pure ASMCs. The solution in the second bottle were mixed cells, they can be used to cultivate, the smooth muscle cells and fibroblasts can be separated when the cells were purefied by passaging cells againand again.

2.2.3 The authentication of rat ASMCs

The form of cells were observed under inverted microscope. Third-generation cells were identified by smooth muscle cell specific alpha actin immunochemical staining.

2.2.4 MTT method

The proliferation of ASMCs was detected by MTT colorimetric assay at 12h, 48h, 24h, 72h. The cells proliferating model of rat ASMCs were established by ET-1 (10-9mol/L). Divided the cells into Control group, model group, high dose group (PRPS 0.6mg/ml), Middle dose group (PRPS 0.4mg/ml), low dose group (PRPS 0.2mg/ml), DXM group (0.4mg/ml).

2.2.5 Expression of PCNA protein in ASMCs

The single cell suspension (1.25×104/ml) was seeded into 24-hole culture plates which had been pre-placed with coverslips. They were incubated in CO2 incubator for 24h until cells had completely stick wall stretching. Then they were incubated in non-serum medium DMEM for 24h. Divided the cells into control group, model group, high dose group (PRPS 0.6mg/ml), middle dose group (PRPS 0.4mg/ml), low dose group (PRPS 0.2mg/ml), DXM group (0.4mg/ml). ASMCs were cultured on coverslips, and were stained of PCNA by the S-P method after cultivation for 48 hours. PCNA were located in cell nuclei. Cellular nuclear was stained yellow-brown in PCNA positive group. Under a microscope randomly selected five high power field of vision, each view counted 100 cells, computed the rate of PCNA positive cells (positive nuclei number/total number of the nucleus).

3. Results

3.1 The identification of rat ASMCs

Before cell fusion, ASMCs were spindle or polygona, had 1~2 cell nuclei, and had one or several foot processes. After cell fusion, ASMCs arranged in fascicular clusters, growed in the "peak and valley" mode under microscope, with the big cell nucleus located in the central of the cells (Fig. 1). Immunohistochemistry Assay announced strong positive expression of a-actin, cell purity reached more than 95%, and there had no abnormal changes in cell morphological (Fig. 2).

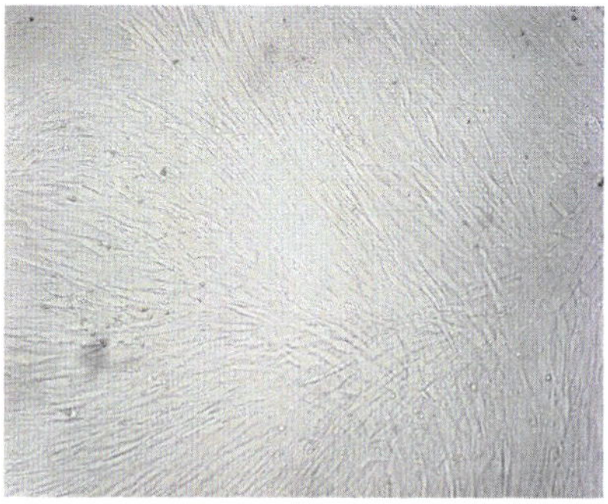

Fig. 1 The cells grew into typical "hill and valley" pattern (×100)

Fig. 2 The positive immunohistochemistry staining with anti-α-sarcomeric antibody(×100)

3.2 The proliferation of ASMCs

The determination results showed that high dose group compared with model group, significant difference ($n = 6$, $P < 0.05$), indicated that the high dose of PRPS can obviously inhibit in ASMCs proliferation. Middle dose and low dose group of PRPS compared with the DXM group, proliferation reaction is higher, but lower than the model group, still can inhibit histamine ASMCs proliferation of excitation. The results are shown in Table 1, and the Proliferation trend is shown in Fig. 3.

Tab. 1 OD values of cells in different PRPS groups($\overline{x}$ ±s, n=6)

Group	n	OD values			
		12h	24h	48h	72h
Control	6	0.0529±0.0023	0.0602±0.0039	0.0758±0.0026	0.1259±0.0034
Model	6	0.0946±0.0032##	0.1031±0.0035##	0.1276±0.0055##	0.2063±0.0052##
High dose	6	0.0695±0.0049*	0.0783±0.0035*	0.0976±0.0043*	0.1521±0.0053*
Middle dose	6	0.0783±0.0045	0.0891±0.0043	0.1064±0.0049	0.1636±0.0047
Low dose	6	0.0852±0.0039	0.0942±0.0044	0.1153±0.0042	0.178±0.0043
DXM	6	0.0622±0.0048**	0.0684±0.0050**	0.0872±0.0044**	0.1367±0.0049**

##: P<0.01 vs Control group *: P<0.05 vs Model group; **: P<0.01 vs Model group.

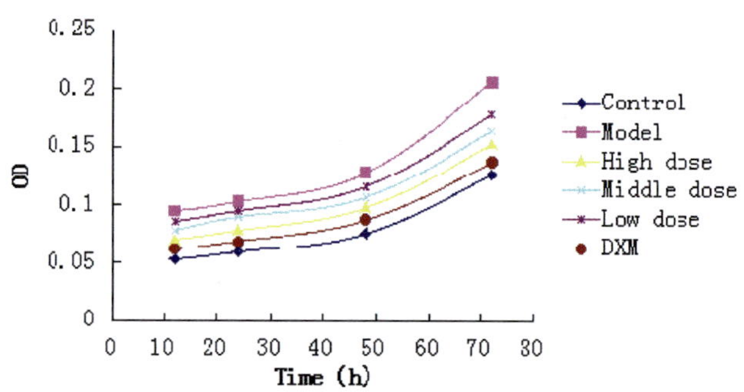

Fig. 3 The growth curves of ASMCs with different PRPS. Values were presented as mean value(n = 6).

3.3 The positive rate of PCNA

The positive rate of PCNA in high dose group was 57.84±2.65% (n=6, P<0.05 vs Model group), but lower than the DXM group, still had obvious inhibitory effect on ASMCs proliferation of excitation. The results are shown in Table 2.

Tab. 2 The effect of PRPS on proliferation of rat ASMCs

Group	Positive Rate of PCNA(%)
Control	27.33±2.58
Model	73.62±3.21
High dose	57.84±2.65*
DXM	43.18±2.43

*: P<0.05 vs Model group

4. Discussion

Airway remodeling is the important pathological features of asthma. It can stimulate the continuous and development of airway inflammation. It also can lead to continuous airway hyperresponsiveness and irreversible airflow obstruction. In recent years, research on the mechanism and treatment increased gradually. The way to prevent airway remodeling has become a hotspot of asthma in the domestic and foreign research. This experiment made the airway reconstruction as the breakthrough point, studied the influence of PRPS on ASMCs proliferation, so as to determine the PRPS effect of the treatment of asthma.

The experiment observed the effect of PRPS on the proliferation model of ASMCs established by ET-1. MTT assay showed that the high dose of PRPS can obviously inhibit in ASMCs proliferation. At the same time, high dose of PRPS can downregulate the expression of PCNA in ASMCs. These results suggest that PRPS can inhibit the proliferation of ASMCs induced by ET-1, thus it can delay the process of airway remodelling and reduce airway responsiveness in asthma.

As the main component of Chinese traditional medicine, PRPS has advantages for the treatment of asthma, because of having anti-inflammatory and immune regulation effect. Otherwise it can inhibit the proliferation of ASMCs. These effects can coordinate with each other, the common fight against airway reconstruction, thus effectively for the treatment of asthma. For a long time, people think that Platycodonis Radix has antitussive and expectorant effects mainly depends on the characteristics of another major component called platycodon grandiflorum saponins [5]. This experiment provides theory basis for PRPS which can treatment of asthma.

Acknowledgments

This is study was supported by a grant from Jilin Science and Technology Project (Excellent Youth Talents Training Plan), Jilin city, China (No.20156415) and a grant from Beihua University, China.

References

1. Pharmacopoeia Commission. The 2010 edition of the Pharmacopoeia of the people's Republic of China [M]. Beijing: China Medical Science and Technology Press, 2010, 259.
2. Jae Ho Choi, Yong Pil Hwang , Hyun Sun Lee, et al. Inhibitory effect of Platycodi Radix on ovalbumin-induced airway inflammation in a murine model of asthma [J]. Food and Chemical Toxicology, 2009, 47:1272–1279.
3. Redington A E, Howarth P H. Thorax [J], 1997, 52(3): 310-312.
4. Ting Yang. The mechanism of airway remodeling and its influence on the physiological function of bronchial asthma[J]. Section of Respiratory System Foreign Medical, 2002, 22(5) : 258-260.
5. Lee KJ, You H J, Park S J, et al. Hepatoprotective effects of Platycod on, grandiflorum on acetaminophan-induced liver damage in mice[J]. Cancer Lea, 2001, 174(1): 73-81.

Study on extracting tannin from waste recycle after the persimmon used to brew persimmon wine

Sha Li, Weihua Liu, Renbang Zhao*, Yaqing Zhang, Mengying Sun, Yang Wang, Yuanyuan Huang
Department of Food Science and Technology, Agricultural University of Hebei, Baoding, 071000, P.R. China
**E-mail: zhaorenbang@sina.com*

Marc from the process of brewing persimmon wine was selected as research material to extract tannin. The changes of content of tannin during ripping period and different part of persimmon were also determined. Based on the single-factor experiments, three independent variables including ethanol concentration, extraction time and extraction temperature were selected as affecting factors to Response Surface Tannin. The optimized of extracting conditions of tannin are as follow: the 1:20 of liquid-to-solid ratio, the ethanol concentration of 30%, extraction temperature of 90°C, extraction time of 2.9h. The predicted tannin extraction yield reached 20.3mg/g under the optimized conditions, while the actual extraction yield is 22.7mg/g with the relative error of 2.4%.

Keywords: Persimmon; Fruits Wine; Persimmon Residue; Tannin; Pectin.

1. Introduction

Tannin, also known as tannic acid, is a bitter plant polyphenolic compound that binds to and precipitates proteins and various other organic compounds including amino acids and alkaloids. The concentration of tannin accounts for 0.13% - 1.54% of fresh persimmon fruit which has the highest of tannin concentration in all fruits.

Persimmon (Diospyros kaki Linn) is originated from East Asia (SUGIURAA 1984) and has 2000 years of cultural history. Persimmons are cultivated both in southern China and northern China, and could account for 90% of the world and its yield makes up 76% of the world, which makes China become the largest country of persimmon cultivation and production at present (Ai Chengxiang *et al.* 2014). Persimmon could be classified into 4 categories, which are sweet persimmon, non-astringent persimmon, not astringent Persimmon and astringent persimmon, based on the research of Kajiura (Kajiura M 1946) that the relationships among the possibility of self-deastringency persimmon tree, the amount of seeds and the color of flesh. 90% of Chinese persimmons are astringent persimmon (Luo Zhengrong and Wang Renzi 2008). Moreover, each area has its local flavour varieties, such as Yue persimmon, Mopan persimmon and Xiaofang persimmon (Yang Yong and Li Gaochao 2009). There are six kinds of persimmons which can be called as the Chinese famous persimmons, which are Dapan persimmon from North China, Lianhua persimmon and Jingmian persimmon from Ji and Lu areas, Jixinhuang persimmon and Jian persimmon of Shanxi Province and Fang persimmon of Zhejiang Province (Ma Qi *et al.* 2005).

Persimmon could be used to produce wine and the problems of persimmon wine production which are having deep color, returnable astringent taste and liquor turbidity, were solved in our laboratory. However, 40%-46% of the total persimmon, most of which were peel and marc, was left over in the process of brewing wine. During the process of brewing persimmon wine in our laboratory, tannin was filtered out and remained in the pomace. Hence, the residue was rich in tannin. To solve the follow-up problems of persimmon product, the residue was used as material to extract tannin. This not only avoids the waste of persimmon fruit resources but also solves the environmental pollution caused by the persimmon debris decomposition.

Freudenberg (K. Freudenberg 1920), according to characteristics of the chemical structure of tannin, divided tannin into hydrolyzed tannin and condensed tannin. Condensed tannin is the main type of persimmon tannin, and the main material in the astringency.

Tannin is easy to be oxidized to brown quinones, which could be added into food as natural pigments. In addition, tannin combined with oral saliva protein precipitate, can change the oral feeling, and make the products formed a kind of unique taste. Tannin with the activity of scavenging free radicals and antioxidant capacity can be used as food preservative. According to the antibacterial properties of tannin, tannin can mothproof the teeth and fresh the air. Tannin also can inhibit leukemia, transform nitroso compound, reduce blood press, remove free radicals, dispel the effect of ethanol, and inhibit snake venom, so those are used to prevent disease of heart head blood-vessel, tumor and cancer (Song J M *et al.* 2005, Gorinstein S *et al.* 2001) in pharmaceutical industry. Tannin can also be used in cosmetics. Because of a large number of phenolic hydroxyl groups in tannin molecule, which easily absorbs the water in the air, tannin can be used as moisturizer. Moreover, having strong absorption of ultraviolet ray, tannin is used as natural additive which makes the cosmetics have sunscreen and whitening effect. Tannin can also be used in other areas, such as flocculant which can absorb heavy metal ions in waste water to purify water. Finally, as it is well known, tannin is used for converting animal hides to leather, owing to their ability to interact with and precipitate proteins (He Q. *et al.* 2007, Mateus N. *et al.* 2004).

Solvent extraction can be employed directly with water, organic solvents or both for the extraction of tannin. Some experimental data on tannin extraction from natural matrix, at a fixed temperature, are reported.

Gu Haifeng (Gu Haifeng *et al.* 2007) used anhydrous ethanol containing 1% hydrochloric acid as extracting agent to extract for 40 min under 90°C with the liquid-to-solid ratio of 1:12, and got 16.9 mg/g condensed tannin from fresh persimmon pulp.

Novel extraction methods including ultrasound-assisted extraction (Yu Xianchun *et al.* 2011) and microwave-assisted extraction (Jiang Ni *et al.* 2007) are fast and efficient for extracting chemicals from solid plant matrixes. Furthermore, the possibility of utilizing non-toxic, environmentally safe, cheap solvent, such as carbon dioxide, has been investigated by Readel (Readel K. *et al.* 2001) and Luengthanaphol et al. (Luengthnaphol S. *et al.* 2004).

Although the extraction of tannin from the natural matrix has been reported, extraction process is nowadays performed by empirical methods, and industrially no optimization of extraction process has been carried out.

Due to plenty of tannin was remained in the marc, during the process of brewing persimmon wine in our groups, the extraction of tannin from marc of process of persimmon was proposed. The change of tannin in the entire mature process and different spot of persimmon was also determined. In order to maximize the recovery of tannin from persimmon marc, the influence of concentration of solvent, temperature and extraction time, and solvent to solid ratio were studied. The results obtained can be considered a helpful tool for the optimization of the existing extraction processes of tannin from solid matrix and for the design of production plants.

2. Experiment

2.1 Materials

The persimmons used for experiment were picked in Tangxian county, Baoding, Hebei province.

2.2 The marc from the process of brewing persimmon wine

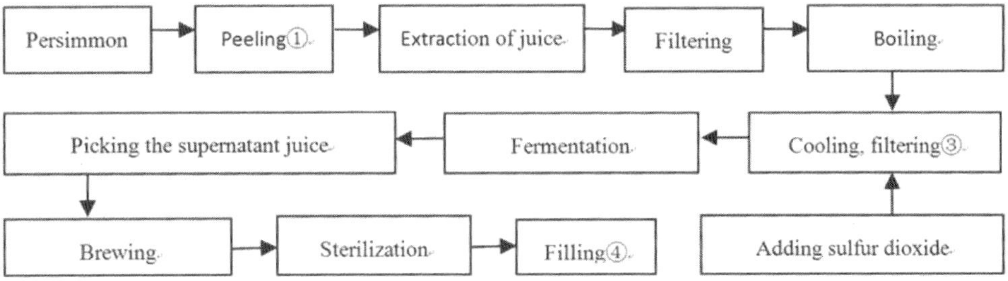

The marc comes from ①, ②, ③, ④ steps in the process of brewing persimmon wine. There are 353.5mg/g tannin in marc of ①, ② and ③ steps, and 120.7mg/g tannin in marc of ④ step.

2.3 Apparatus

Electronic balance CP114 (Ohaus instrument Shanghai Co., Ltd.); Super thermostatic water bath HH-601 (JintanHengfeng Instrument Manufacturing Co. Ltd.); Ultraviolet visible spectrophotometer UV-5200PC (Shanghai Element Analysis Instrument Co., Ltd.); Food mixer JLL28-B (Shunde City, Foshan Branch Shun plastic electrical appliance industry Co., Ltd.); DZF-6020 vacuum drying oven (Department of Beijing science and Technology Co Ltd -); Rotary evaporator RE-52A (Shanghai YaRong biochemical instrument factory); Circulating water type vacuum pump SHD- III (Baoding high tech Zone Sunshine science and Technology Instrument Factory).

2.4 Apparatus

Ammonium oxalate, absolute ethyl ethanol, the soda ash, the sodium tungstate, sodium molybdate, lithium sulfate, Phosphoric acid and Gallic acid are all analytically pure.

2.5 Sample treatments

5 g of persimmon marc was filtered to extracting solution, which was performed in triplicates.

2.6 Selection of extracting reagent

In order to determine the suitable extracting reagent, anhydrous ethanol, water and 50% ethanol aqueous solution were employed to compare the extraction effect on tannin of the persimmon residue.

2.7 Single factor experiment of extracting tannin

Four dependent variables including ethanol concentration, solid to liquid ratio, extraction time and extraction temperature were selected as affecting factors were showed in Table 1.

Tab. 1. Single factor experiment design of tannin ethanol concentration(%)

ethanol concentration(%)	Liquid-to-solid ratio(g/100ml)	time(h)	temperature(°C)
20	1:05	1.0	50
30	1:10	2.0	60
40	1:15	3.0	70
50	1:20	4.0	80
60	1:25	5.0	90

2.8 Optimization of extraction of tannin using response surface methodology

According to central composite design the three factors, three levels experiments (showed in Table 2) were carried out with the tannin extraction yield as evaluation index.

Tab. 2. The response surface analysis of tannin

level	ethanol concentrationX_1(%)	tempratureX_2(°C)	timeX_3(h)
-1	20	50	2.0
0	40	70	4.0
1	60	90	6.0

2.9 Determination of tannin content

Referring to the agricultural industry standard of NY/T 1600-2008, the content of tannin in persimmon and its products was determined by colorimetry.

3. Results and Discussion

3.1 Changes of tannin content in different parts during the same period

As shown in Figure 1, when the extractant was distilled water and ethanol, tannin yield was less than aqueous ethanol, therefore, extraction reagent was aqueous ethanol.

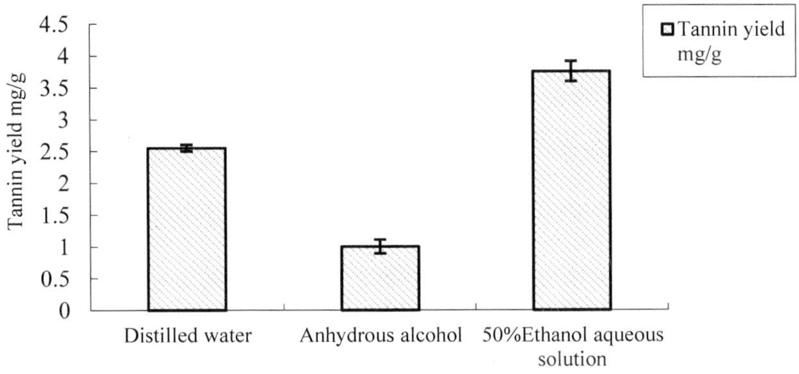

Fig. 1. The effect of different extracting agent on tannin

3.3 Single factor experiment of extracting tannin

3.3.1 Effect of ethanol concentration

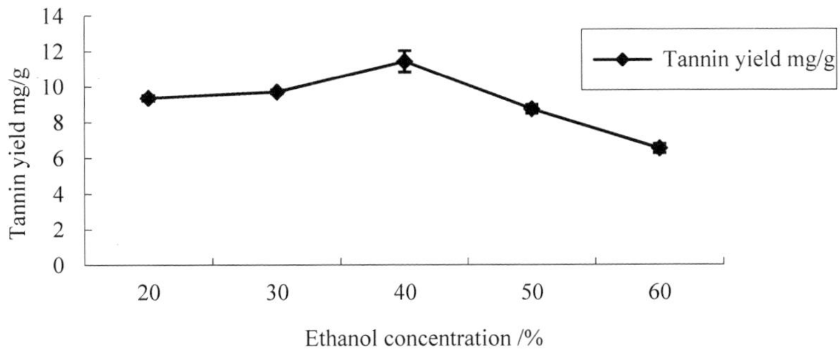

Fig. 2. The effect of ethanol concentration on tannin

According to Figure 2, tannin yield increased rapidly with increasing concentration of ethanol aqueous solution, reaching a peak at 40%, where the yield was 11.4mg/g. And then, the yields slowly decreased. The reason was that a change in polarity of ethanol aqueous could be compatible with water-soluble and ethanol-soluble tannin. When ethanol concentration improved continually, the difference of polarity between solvent and tannin decreased (Zhang Yanping *et al.* 2009). Hence, 40% ethanol concentration was employed in the following experiments.

3.3.2 Effect of liquid-to-solid ratio

According to Figure 3, increment of liquid-to-solid ratio increased the tannin yield which reached a plateau at a value around 12% at liquid-to-solid ratio of 1:20. Low liquid-to-solid ratio has low solvent and impacted mass transfer rate. Soluble tannin couldn't dissolve leading to low tannin yield (Su Ning *et al.* 2012). Therefore, tannin dissolved completely when liquid-to-solid ratio reaching 1:20.

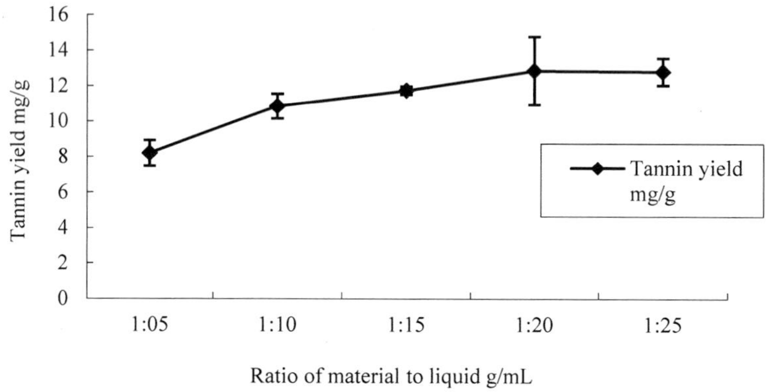

Fig. 3. The effect of liquid-to-solid ratio on tannin

3.3.3 Effect of time on tannin

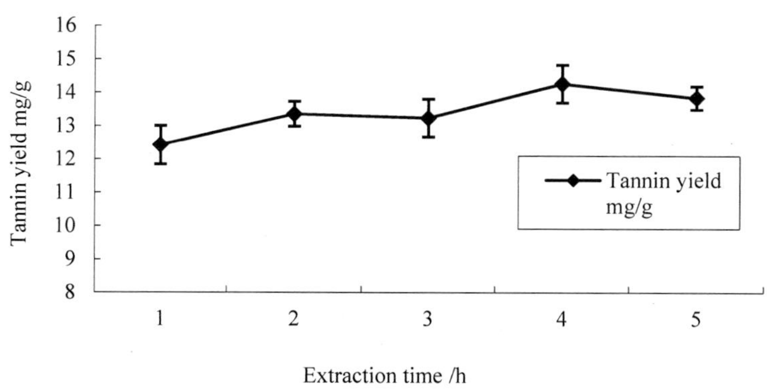

Fig. 4. The effect of extraction duration on tannin

With 1:20 of liquid-to-solid ratio, 40% ethanol and 70°C of water bath, the extracting time was 1h, 2h, 3h, 4h and 5h respectively. As shown in Figure 4, tannin yield increased with the extracting time increasing, which reached the peak at 4.0h and fell down afterwards. The reason might be that tannin was oxidized with long bath time (Chen Guogang *et al.* 2007). So 4.0h of extracting time was employed in the following experiments.

3.3.4 Effect of temperature

As shown in Figure 5, tannin content increased with the rise of temperature. The tannin yield was 17.9% higher in 60°C than that in 50°C. Afterwards, the tannin yield increased with the change of temperature. The reason might be that bond of tannin and protein was broken with high temperature, leading to soluble tannin yield increasing. So 90°C of extracting temperature was employed in the following experiments.

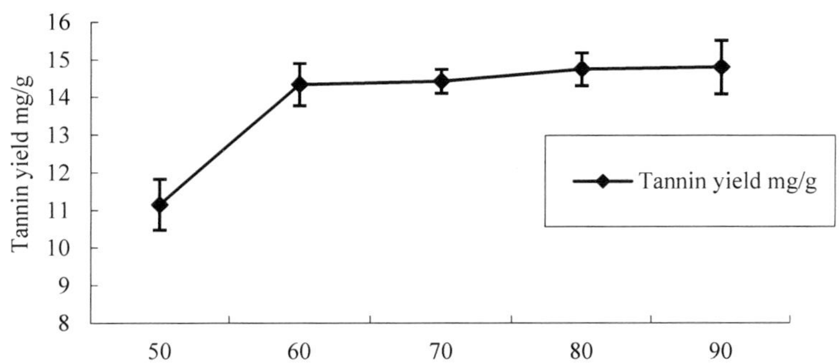

Fig. 5. The effect of extraction temperature on tannin

3.3.5 Optimization of extraction of tannin using response surface methodology

The ethanol concentration, temperature and time and extraction of tannin field fitting equation was obtained.

Tannin yield = $16.11-1.86X_1+5.30X_2+1.14X_3-1.53X_1X_2-0.044X_1X_3-0.22X_2X_3-3.02X_1^2-1.63X_2^2+0.38X_3^2$. The result that extraction temperature and ethanol concentration had a significant effect on the extraction of tannin was obtained by analysis of variance model.

The coefficient of determination was 0.8963 and the model was statistically significant ($P<0.05$), indicating a goodness of fitting the model can be used as a measure of precision in predictions for extraction condition.

The water bath temperature was the maximum effect on extracting tannin, and the extracting time was the minimum one. This result is consistent with Wang Chengdong (Wang Chengdong *et al.* 2004). The optimization results analyzed from response was that ethanol concentration 30%, 90°C of extraction temperature for 2.9 h. Under these conditions, the extraction of tannin was 20.3mg/g. According to the suitable extraction conditions given the model, verification experiment has been done. In this way, the yield of tannin was (22.7±0.3)mg/g. Compared with the theory of model, the difference was (2.4±0.3)mg/g.

4. Conclusions

There are plenty of tannin remained in the marc from the process of brewing persimmon wine. To make reasonable and full use of persimmon resources, the marc was used to extract tannin. The optimized extraction conditions for tannin is: material liquid ratio of 1:20, ethanol concentration of 30%, extraction temperature of 90 °C, time of 2.9 h. The result was proved by actual extraction yield, which was (22.7±0.3)mg/g.

Acknowledgement

This work was supported by Science and Technology Support Project of Hebei, China (Research and demonstration of ecological restoration of heavy metal wastewater in typical coastal zone and estuary wetland NO.13273301D; Research on the key technology of pectin extraction and processing of persimmon juice NO.14236810D-3).

References

1. SUGIURAA. Origin and varieties evolution of persimmon. Recent Advances in Breeding, 25, 30-37. (Japan. 1984).
2. A. Chengxiang, Q. Zhihua and X. Li, China persimmon industry development report 2013. China Fruit Vegetable, 34(2), 10-13. (China. 2014).
3. Kajiura M. Persimmon cultivars and their improvement 2. Breeding Horticulture, 1, 175-182. (Japan. 1946).
4. L. Zhengrong and W. Renzi. Persimmon in China: Domestication and traditional utilization of genetic resources. Advances in Horticultural Science, 22(4), 239-243. (China. 2008).
5. Y. Yong and L. Gaochao. Persimmon cultivars and commodity Characters. 2009 of the Fourth National Symposium on scientific research and production of persimmon, 55-66. (China. 2009).
6. M. Qi, Z. Qiang, Q. Tao. Status of research and development resources of persimmon in China. Shaanxi Journal of Agricultural Sciences, (6), 56-57. (China. 2005).
7. K. Freudenberg. Die Chemie der Naturliche Gerbstoffe. Springer, Berlin. (1920).
8. Song J M *et al.* Antiviral effect of catechins in green tea on influenza virus. Antiviral Research, 68(2), 66-74. (2005).
9. Journal of Agricultural and Food Chemistry, total phenolics, and minerals in persimmons and apples. Journal of Agricultural and Food Chemistry, 49(2), 952-957. (2001).
10. He Q. *et al.* Biodegradability of tannin-containing wastewater from leather industry. Biodegradation, 18, 465-472. (2007).
11. Mateus N. *et al.* Influence of the tannin structure on the disruption effect of carbohydrates on protein tannin aggregates, Anal. Chim. Acta 513, 135-140. (2004).
12. G. Haifeng, L. Chunmei and X. Yujuan. Study on Preparation and antioxidant activity of tannin from persimmon. Transactions of the Chinese Society of

Agricultural Engineering, 23(5), 241-245. (2007).

13. Y. Xianchun, S. Dilin and L. Xiangsu. Study on ultrasonic assisted extraction of tannin from persimmon leaves. Chemical Research and Application, 23(3), 345-349. (China. 2011).
14. J. Ning, L. Xiaopeng, Y. Xianhui, Y. Song and Z. Xiaojiang. Research on gallnut tannic acid in microwave-assisted extraction. China Practical Medicine, 2(25), 81-83. (China. 2007).
15. Mateus N. *et al.* Influence of the tannin structure on the disruption effect of carbohydrates on protein tannin aggregates, Anal. Chim. Acta 513, 135-140. (2004).
16. Luengthnaphol S. *et al.* Extraction of antioxidants from sweet Thai tamarind seed coat–preliminary experiments, J. Food Eng. 63, 247-252. (2004).
17. Z. Yanping F. Xiaohang and Y. yuanzhi. Optimization of extraction process of polyphenols from Chestnut. Food Science and Technology, 34(9), 187-191. (2009).
18. S. Ning, W. Chao, W. Changtao, L. Xiaopeng, L. Juan, Z. Hongyan and L. Nan.Extraction and efficacy studies of hydrolysis tannin from emblica. Food Science and Technology, 37(10), 191-195. (2012).
19. C. Guogang, D. Zhixiang, J. Ying and Z. Xinhua. Extraction and Content Determination of tannin in the grape stems. Journal of Shihezi University, 25(6), 749-752. (2007).
20. W. Chengdong, Z. Zhenwen and S. Shiren. Study on extraction method of grape polyphenols. Acta Botanica Boreali-Occidentalia Sinica, 24(11):2131-2135. (China. 2004).

An investigation of pathogens associated with porcine respiratory disease complex and interactions analysis in Sichuan

Ping Li[a], Lun Liu[b], Peng-Juan Liu[c], Bo Guo[d], Fan Yang[e], He-Jun Fang[f] and Ling Zhu[g]

College of Veterinary Medicine. Sicau, Sichuan Agricultural University, Chengdu, 611130, China

E-mail: [a]lipinga104@163.com; [b]liulun1@outlook.com; [c]1210695671@qq.com; [d]657907821@qq.com; [e]511075708@qq.com; [f]373290661@qq.com; [g]abtczl72@126.com

In the current study, 212 samples of Porcine Respiratory Disease Complex (PRDC) with typical respiratory desease symptoms, were collected from 41 large-scale pig farms in 12 cities of Sichuan to detect pathogens to discover the rules of development of PRDC. 11 species were aimed at, including Porcine reproductive and respiratory syndrome virus (PRRSV), Streptococcus pneumoniae (S.pn), porcine circovirus type 2 (PCV2), Pasteurella multocida (Pm), Mycoplasma hyopneumoniae (MH), Haemophilus parasuis (HPS), Actinobacillus pleuropneumoniae (APP), Porcine pseudorabies virus (PRV), Bordetella bronchiseptica (Bb), Classical swine fever virus (CSFV) and Swine influenza virus (SIV), and the infection rates were respectively 58.02%, 57.55%, 56.60%, 54.72%, 49.06%, 47.17%, 20.75%, 16.98%, 9.43%, 3.78% and 1.89%. Moreover, Listeria (LM) and Pseudomonas aeruginosa (PM) were isolated from respiratory passages and had higher infection rates, as 16.98% and 13.21%. For co-infection by two, three and four kinds of pathogens, S.pn - PRRSV, S.pn - PRRSV - PCV2 and S.pn -PRRSV- PCV2 - HPS had the highest infection rates in their types, respectively as 49.06%, 33.96% and 16.98%. All these above showed that PRRSV, S.pn, PCV2, Pm, MH and HPS were major pathogens of PRDC in Sichuan, having high infection rates over 40.00%, and pathogens above were easy to co-infected to worsen the situation, but it was limited, which indicated that there was a certain mechanism between pathogens to help them inhibit each other to gain growth advantage. And the investigation can help research on the mechanism of inhibition and interaction between pathogens, which may provide insight.

Keywords: Porcine Respiratory Disease Complex; co-infections; Sichuan; respiratory system microecology

1. Introduction

Porcine Respiratory Disease Complex (PRDC) is a complex disese caused by many panthgons[1], including virus, mycoplasma, bacteria, parasites and management. The main clinical characteristics of it include a wide range of pathogens, far-ranging coverage of onset age and higher prevalence, causing the prevention and control of PRDC to be a penal servitude. In the intensive, scale hoggery, the clinical and pathological features of PRDC demenstrated a variety of patterns, mainly influenced by large-scale pig field rearing environment, nutritional status, immune state and epidemic pathogens[2,3]. The infectious pathogens main include Porcine reproductive and respiratory syndrome (PRRSV), Classical swine fever virus (CSFV), Swine influenza virus (SIV), Porcinecircovirus 2 (PCV2), Porcine pseudorabies virus (PRV), Haemophilus parasuis (HPS), Actinobacillus pleuropneumoniae (APP), Pasteurella multocida (Pm), Bordetella bronchiseptica (Bb), Streptococcus pneumoniae (S.pn) and Mycoplasma pneumoniae (MH).

That the relationship and interaction between pathogens, normal flora and the host is the basis of the study of animal micro ecology. Sullivan A. believes that normal microbial flora can play a role in preventing potential pathogenic microorganisms and preventing the growth of opportunistic bacteria that already exist[4]. Currently, microecological preparation mainly and widely used in intestina, including new drugs, health food, feed additives, plant growth accelerator etc[5]. But the study on the microbial ecology of the respiratory tract disease are much less[6,7,8].

This study took an investigation of pathogens associated with PRDC to analyze the regularity and characteristics of PRDC in Sichuan, compared with other areas, to formulate effective prevention and control meseaures. Moreover, it is helpful to understand infection condition of interaction between pathogens to further contribute to the research of animal micro ecology of respiratory system in pathology situation of infection.

2. Material and methods

This investigation was taken to understand the infection of pathogens associated with PRDC in Sichuan large-scale pig farms from April 2014 to April 2015. And collected samples with classical symptom of PRDC from 12 representative cities, where pig husbandries were flourishing, to analyze the species of major pathgen and co-infections. 212 samples, having symptoms of respiratory tract diseases, including the lung tissue, antiades, nasal swabs and weasand swabs, were collected from 41 large-scale farms in Sichuan, locating in 12 cities, including Suining (14.15%), Ya'an (12.74%), Leshan (8.96%), Chengdu (8.96%), Guangyuan (8.49%), Xinjin (8.02%), Meishan (7.55%), Neijiang (7.08%), Nanchong (7.08%), Guang'an (6.13%), Mianyang (5.66%), Deyang (5.19%). The number of samples in different cities were listed in Table 1. Primers[a], using in this study, were cited from some methods that has been established and made functional verification[9,10,11,12,13,14]. These primers were specific and located in their conservation genome region. And the universal primer 16SrRNA (27F/1492R) was used for bacterial identification.The sequences of the primers were listed in Table 2.

Bacterial pathogens and MH were isolated by culturing on proper medium[b], and identifed by biochemical identification, serotype identification, pathogenic identification and nucleic acid identification. And virus pathogen, causing PRDC, should be divided into DNA virus and RNA virus to detected. The former, having DNA genetic material, could be detected by PCR, including PCV2 and PRV. And the latter, having RNA genetic material, needed RT-PCR, including PRRSV, CSFV and SIV. The PCR production were sequenced and using the tool BLAST in the NCBI were processed to assure the PCR results.

3. Results

After the horizontal comparision of infection situation of these pathogens, here we showed the infection rate of each pathogen. Through isolation and identification, some pathogenic bacterias were found, and they were S. Pn, Pm, HPS, APP, LM, PA and Bb, the infection rate of which were 57.55%, 54.72%, 47.17%, 21.23%, 16.98%, 13.21% and 8.96%. Virus, aimed at, were detected successfuly, including PRRSV, PCV2, PRV, CSFV and SIV, the infection rate of which were 58.02%, 56.60%, 17.45%, 3.78% and 1.89%. The PRRSV had the highest infection rate, and following with S.pn. And the infection rate of PCV2 and Pm were also in a high level, all above 50%. The resule datas was showed in Table 3.

The test compiled the data to get the co-infection condition of the samples mentioned above. In the co-infection pattern of 2 pathogens, S. pn-PRRSV co-infection rate was the highest, approaching to 49.06% and the second were S. pn-PCV2 co-infection and PRRSV-PCV2 co-infection, both 47.17%. The other kinds of co-infection, such as Pm-PRRSV, HPS-S.pn, Pm-PCV2 and HPS-PRRSV, were also at a rather high level. And it was certain that there were other pairs with lower co-infection rates, and all of them were showed in figure 1. Comparing the data in Table 3 and Table 4, it showed that pathogens with high infection rate, like PRRSV and S. pn, were prone to be in co-infection condition and had higher co-infection rate.

In the statistics of co-infection of 3 pathogens, with its co-infection rate of 33.96%, the S. pn-PRRSV-PCV2 was the highest. And there were other conditions that were all above 20.00%, of which the co-infection patterns were S. pn-PCV2-HPS, PRRSV-PCV2-HPS, PRRSV-PCV2-MH and S. pn-PRRSV-MH.The result was showed in Table 5.

Through analyzing structure of quadruplex co-infection, S. pn-PRRSV-PCV2-HPS had the highest co-infection rate, 16.98%. And for S. pn-PRRSV-PCV2-MH and Pm-PCV2-HPS-S. pn, the co-infection rates were 15.09% and 11.32%. For the others, all of their co-infection rates were within 10.00%. The result was showed in Table 6.

4. Discussion

PRDC, in recent years, has become a major desease, threatening the development of swine industry. Multi-factors infection is the difficult point of prevention, control and treatment. The study, through statistics and analysis of various pathogens mixed infection, found that the major pathogens of PRDC in Sichuan large-scale pig farms were PRRSV, S.pn, PCV2, Pm, MH and HPS, all of which had infection rates about 40.00%. Especially, the infection rate of PRRSV was up to 58.02%. But for Bb, CSFV and SIV, the infection rates were within 10.00%, which showed that these deases were controled relatively well in Sichuan. Notably, in this investigation, more often than not, the LM and PA were able to be isolated from the respiratory tract, and both of them had higher infection rates, which widen the awareness of respiratory pathogens.

From the result, PRRSV having the highest infection rate, a conclusion could be drawed that, it was the key point of PRDC and in prevalence in many parts of Sichuan. And some investigations were taken in other areas of China. Liu reported that the mian pathogens of PRDC in Guangxi, from May 2007 to May 2009, were PRRSV and PCV2, the infectin rates of which were 51.36% and 36.54%, and CSFV also had a higher infection rate[15]. Nearly at the same time, Shu did the similar from February 2007 to June 2008, and the result showed that PCV2, PRV and PRRSV were the top three, and their infectin rates respectively were 60.99%, 45.15% and 27.31%[16]. All of the above showed that PRDC in different areas had different epidemiological characteristics and rules, and PRRSVand PCV2 were the major.

What caused PRRSV and PCV2 to be the main pathogens of PRDC in Sichuan large-scale pig farms, the core effects could be the immunosuppression action and the evolution of these virus. PCV2 causes immunosuppression by decreasing and making immune cells apoptosis, and changing expression level of some cytokines. It also can promote the development of immunosuppression by a protein, encoded by ORF3, a important virulent gene of it[17]. ORF2, a virulent gene with high mutation rate, can encod Cap, a main structural protein with dominant antigen determinants of PCV2. Wu got that ORF2 had a higher mutation rate with significant difference by comparing the mutation rate of ORF2 with ORF1 and ORF2 in Sichuan[18]. And for PRRSV, the Nsp2, which has high mutation rate in the PRRSV genome and has an important immunological function, may not be a molecule marker to define virus strains in China[19,20]. Futher more, nucleotide deletion and other kinds of mutation was also common in the rest areas in its genemo, which means that mutation and the prevalence of the PRRSV will become increasingly difficult to control. In PRDC, PRRSV was almost always mixed infections, relating to its immunosuppressive effect. With a strong tissue tropism for the mononuclear phagocyte system, PRRSV mainly invade macrophages in blood and lung[21], and antibody dependent enhancement (ADE) can significantly enhance their ability to infect the macrophage, resulting in apoptosis of macrophage, reduce the nonspecific immunity, provide the conditions for secondary infection[22,23,24]. And there is a point should be noticed that the recessive infection with PRRSV can not help other pathogens attack the body, only if there was symptom expression. And it help to correct lyunderstand the pathogenic mechanism of PRRSV.

From all the multiple infection samples, there was no unlimited mixed infection among pathogens, which indicated that a certain mechanism between pathogens should be existing, to help them inhibit each other to gain growth advantage. Some beneficial microorganisms can release antibacterial substances, compete nutrient, adhere epitopes and stimulate the immune system to improve the level of body's immune. And microecological preparation take clever use of the interaction relationship between flora to achieve the purpose of conditioning microflora. Therefore, sartificially adding beneficial microorganism to prevent and treat diseases was widly used.In the analysis of respiratory diseases, bacterial pathogens had important status. In general, most of these bacteria are normal resident flora in the respiratory tract and often play a pathogenic role in econdary infections. Respiratory system, digestive system and urogenital system has their own particular habitats and microecological environment. Using corresponding microecological preparation can help to maintain the ecological balance of the original habitat, and prevent diseases repeated attacks. It has very broad prospect of development and application.

5. Tables

Tab. 1. Samples distribution.

Areas	Samples	Rates
Suining	19/212	14.15%
Ya'an	15/212	12.74%
Leshan	17/212	8.96%
Chengdu	27/212	8.96%
Guangyuan	12/212	8.49%
Xinjin	15/212	8.02%
Meishan	19/212	7.55%
Neijiang	16/212	7.08%
Nanchong	13/212	7.08%
Guang'an	30/212	6.13%
Mianyang	18/212	5.66%
Deyang	11/212	5.19%

Tab. 2. Primers of PRRSV, CSFV, PCV2, SIV, PRV, MH and 16SrRNA.

Primer category	Primer sequences (5'~3')	sequence length (bp)
PRRSV	PRRSV-1: AATTCCCGCACCTCGCGGAACTGT PRRSV-2: AGTTCCTGCACCGCGTAGAACTGT	180
CSFV	CSFV-1: GGAATGGCTAAAGACAAGA CSFV-2: TCTGTGAGACCKRTGCTG	270
PCV2	PCV2-1: CGGATATTGTAGTCCTGGTCG PCV2-2: ACTGTCAAGGCTACCACAGTCA	481
SIV	SIV-1: CAGCGTGGTGGTAATCTTC SIV-2: GTTCTTGTGCCTTTCTTGC	242
PRV	PRV-1: ATGGGCATCGGCGACTACCT PRV-2: CCACCGCCACAAAGAACACG	178
MH	MH-1: TTACAGCGGGAAGACC MH-2: CGGCGAGAAACTGGATA	472
16SrRNA	27F: AGAGTTTGATCCTGGCTCAG 1492R: TACGGCTACCTTGTTACGACTT	1500

Tab. 3. Single pathogen infection rate.

Pathogens	Infection rate [n/N(%)]
Porcine reproductive and respiratory syndrome virus (PRRSV)	123/212 (58.02)
Streptococcus pneumoniae (S. pn)	122/212 (57.55)
Prcine circovirus type 2 (PCV2)	120/212 (56.60)
Pasteurella multocida (Pm)	116/212 (54.72)
Mycoplasma hyopneumoniae (MH)	104/212 (49.06)
Haemophilus parasuis (HPS)	100/212 (47.17)
Actinobacillus pleuropneumoniae (APP)	45/212 (21.23)
Porcine pseudorabies virus (PRV)	37/212 (17.45)
Listeria monocytogenes (LM)	36/212 (16.98)
Pseudomonas aeruginosa (PA)	28/212 (13.21)
Bordetella bronchiseptica (Bb)	19/212 (8.96)
Classical swine fever virus (CSFV)	8/212 (3.78)
Swine influenza virus (SIV)	4/212 (1.89)

Tab. 4. Two pathogens co-infection rates.

Two pathogens co-infection	Infection rate [n/N(%)]	Two pathogens co-infection	Infection rate [n/N(%)]
S.Pn+PRRSV	104/212 (49.06)	PA+PCV2	15/212 (7.08)
S.Pn+PCV2	101/212 (47.64)	LM+PRRSV	14/212 (6.60)
PRRSV+PCV2	100/212 (47.17)	Bb+PRRSV	14/212 (6.60)
Pm+PRRSV	84/212 (39.62)	Bb+MH	13/212 (6.13)
HPS+S.Pn	68/212 (32.07)	PRV+MH	13/212 (6.13)
Pm+PCV2	65/212 (30.66)	HPS+PA	13/212 (6.13)
HPS+PRRSV	64/212 (30.19)	HPS+LM	12/212 (5.66)
Pm+S.Pn	63/212 (29.72)	HPS+APP	12/212 (5.66)
S.Pn+MH	56/212 (26.42)	Pm+Bb	12/212 (5.66)
HPS+PCV2	52/212 (24.21)	S.Pn+Bb	12/212 (5.66)
HPS+MH	49/212 (23.11)	PRRSV+CSFV	11/212 (5.19)
Pm+MH	48/212 (22.64)	LM+MH	11/212 (5.19)
PRRSV+MH	45/212 (21.23)	APP+PRV	10/212 (4.72)
HPS+Pm	40/212 (18.87)	PA+LM	8/212 (3.77)
S.Pn+PRV	37/212 (17.45)	Bb+PCV2	6/212 (2.83)
LM+PCV2	29/212 (13.68)	Pm+CSFV	6/212 (2.83)
APP+PRRSV	28/212 (13.21)	S.Pn+CSFV	6/212 (2.83)
PCV2+MH	27/212 (12.74)	Bb+CSFV	5/212 (2.36)
PCV2+PRV	27/212 (12.74)	MH+CSFV	5/212 (2.36)
APP+S.Pn	27/212 (12.74)	S.Pn+SIV	4/212 (1.89)
S.Pn+LM	24/212 (11.32)	APP+SIV	3/212 (1.42)
APP+MH	22/212 (10.38)	MH+SIV	3/212 (1.42)
Pm+PRV	21/212 (9.91)	PA+MH	3/212 (1.42)
PRRSV+PRV	20/212 (9.43)	CSFV+MH	2/212 (0.94)
APP+Pm	20/212 (9.43)	SIV+MH	2/212 (0.94)
S.Pn+PA	20/212 (9.43)	PA+PRV	2/212 (0.94)
PA+PRRSV	18/212 (8.49)	LM+PRV	1/212 (0.47)
HPS+PRV	18/212 (8.49)	HPS+Bb	1/212 (0.47)
Pm+PA	16/212 (7.55)	Bb+LM	1/212 (0.47)
Pm+LM	16/212 (7.55)		

Tab. 5. Three pathogens co-infection rates.

Three pathogens co-infection	Infection rate [n/N(%)]	Three pathogens co-infection	Infection rate [n/N(%)]
S.Pn+PRRSV+PCV2	69/212 (32.55)	Pm+S.Pn+Bb	8/212 (3.77)
S.Pn+PCV2+HPS	56/212 (26.42)	Pm+S.Pn+PA	8/212 (3.77)
PRRSV+PCV2+HPS	51/212 (24.06)	Pm+S.Pn+LM	8/212 (3.77)
PRRSV+PCV2+MH	49/212 (23.11)	S.Pn+PRRSV+Bb	8/212 (3.77)
S.Pn+PRRSV+MH	44/212 (20.75)	S.Pn+PRRSV+PA	7/212 (3.30)
S.Pn+PRRSV+HPS	42/212 (19.81)	S.Pn+PRRSV+LM	7/212 (3.30)
S.Pn+PCV2+MH	40/212 (18.87)	PRRSV+PCV2+LM	7/212 (3.30)
PRRSV+PCV2+Pm	40/212 (18.87)	Pm+PRRSV+Bb	7/212 (3.30)
Pm+PRRSV+MH	38/212 (17.92)	Pm+PRRSV+PRV	6/212 (2.83)
Pm+PCV2+HPS	36/212 (16.98)	HPS+S.Pn+APP	6/212 (2.83)
S.Pn+PRRSV+Pm	32/212 (15.09)	HPS+S.Pn+PA	6/212 (2.83)
S.Pn+PCV2+Pm	32/212 (15.09)	HPS+S.Pn+LM	6/212 (2.83)
Pm+PRRSV+HPS	32/212 (15.09)	Pm+PCV2+PA	6/212 (2.83)
HPS+S.Pn+Pm	29/212 (13.68)	S.Pn+MH+APP	5/212 (2.36)
HPS+S.Pn+MH	28/212 (13.21)	S.Pn+MH+Bb	5/212 (2.36)
HPS+PRRSV+MH	28/212 (13.21)	S.Pn+MH+PRV	5/212 (2.36)
HPS+PCV2+MH	27/212 (12.74)	HPS+MH+LM	5/212 (2.36)
S.Pn+PCV2+LM	25/212 (11.79)	Pm+MH+Bb	5/212 (2.36)
S.Pn+PCV2+PRV	24/212 (11.32)	PRRSV+MH+APP	5/212 (2.36)
Pm+PCV2+MH	24/212 (11.32)	PRRSV+MH+PRV	5/212 (2.36)
HPS+MH+Pm	23/212 (10.85)	S.Pn+PRRSV+CSFV	4/212 (1.89)
S.Pn+PRRSV+PRV	20/212 (9.43)	Pm+PRRSV+LM	3/212 (1.42)
PRRSV+PCV2+PRV	19/212 (8.96)	Pm+PRRSV+CSFV	3/212 (1.42)
Pm+S.Pn+MH	18/212 (8.49)	HPS+PRRSV+PA	3/212 (1.42)
Pm+S.Pn+PRV	17/212 (8.02)	HPS+PRRSV+PRV	3/212 (1.42)
S.Pn+PRRSV+APP	16/212 (7.55)	S.Pn+MH+SIV	2/212 (0.94)
Pm+PRRSV+APP	16/212 (7.55)	HPS+PCV2+Bb	2/212 (0.94)
HPS+S.Pn+PRV	15/212 (7.08)	HPS+PCV2+PRV	2/212 (0.94)
Pm+PCV2+PRV	14/212 (6.60)	HPS+MH+APP	2/212 (0.94)
Pm+S.Pn+APP	12/212 (5.66)	HPS+MH+Bb	2/212 (0.94)
S.Pn+PCV2+PA	12/212 (5.66)	HPS+MH+PA	2/212 (0.94)
Pm+PRRSV+PA	11/212 (5.19)	Pm+MH+ PA	1/212 (0.47)
Pm+PCV2+LM	11/212 (5.19)	Pm+MH+ CSFV	1/212 (0.47)
HPS+PCV2+PA	11/212 (5.19)	PRRSV+MH+Bb	1/212 (0.47)
HPS+PCV2+LM	11/212 (5.19)	PRRSV+MH+CSFV	1/212 (0.47)

Tab. 6. Four pathogens co-infection rates.

Four pathogens co-infection	Infection rate [n/N(%)]	Four pathogens co-infection	Infection rate [n/N(%)]
S.Pn+PRRSV+PCV2+HPS	23/212 (10.85)	PRRSV+PCV2+PRV+MH	4/212 (1.89)
S.Pn+PRRSV+PCV2+MH	22/212 (10.38)	Pm+S.Pn+MH+Bb	4/212 (1.89)
Pm+PCV2+HPS+S.Pn	21/212 (9.91)	S.Pn+PCV2+HPS+PA	4/212 (1.89)
HPS+S.Pn+MH+PCV2	19/212 (8.96)	S.Pn+PCV2+LM+PA	3/212 (1.42)
S.Pn+PCV2+PRV+PRRSV	18/212 (8.49)	S.Pn+PCV2+Pm+PA	3/212 (1.42)
S.Pn+PRRSV+PCV2+PRV	17/212 (8.02)	Pm+PRRSV+MH+Bb	3/212 (1.42)
Pm+PRRSV+MH+HPS	15/212 (7.08)	Pm+PRRSV+MH+PA	3/212 (1.42)

Pm+PRRSV+MH+PCV2	15/212 (7.08)	Pm+PRRSV+MH+CSFV	3/212 (1.42)
S.Pn+PCV2+Pm+PRV	14/212 (6.60)	Pm+PCV2+HPS+PA	2/212 (0.94)
HPS+S.Pn+MH+PRRSV	13/212 (6.13)	Pm+PCV2+HPS+LM	2/212 (0.94)
HPS+PRRSV+MH+PCV2	13/212 (6.13)	Pm+PCV2+HPS+PRV	2/212 (0.94)
HPS+MH+Pm+PRRSV	13/212 (6.13)	S.Pn+PRRSV+Pm+Bb	2/212 (0.94)
PRRSV+PCV2+Pm+MH	13/212 (6.13)	S.Pn+PRRSV+Pm+PA	2/212 (0.94)
Pm+PCV2+HPS+PRRSV	13/212 (6.13)	Pm+PRRSV+HPS+PA	2/212 (0.94)
Pm+PCV2+HPS+MH	12/212 (5.66)	HPS+S.Pn+Pm+APP	2/212 (0.94)
S.Pn+PRRSV+Pm+PCV2	11/212 (5.19)	HPS+S.Pn+Pm+PA	2/212 (0.94)
S.Pn+PCV2+Pm+LM	11/212 (5.19)	HPS+S.Pn+Pm+LM	2/212 (0.94)
HPS+S.Pn+Pm+MH	10/212 (4.72)	HPS+S.Pn+MH+APP	2/212 (0.94)
HPS+MH+Pm+S.Pn	9/212 (4.25)	HPS+S.Pn+MH+PRV	1/212 (0.47)
HPS+MH+Pm+PCV2	8/212 (3.77)	HPS+PRRSV+MH+PA	1/212 (0.47)
Pm+PRRSV+MH+S.Pn	8/212 (3.77)	HPS+PCV2+MH+Bb	1/212 (0.47)
S.Pn+PRRSV+Pm+HPS	7/212 (3.30)	HPS+PCV2+MH+LM	1/212 (0.47)
S.Pn+PRRSV+Pm+APP	7/212 (3.30)	S.Pn+PCV2+LM+PRV	1/212 (0.47)
S.Pn+PRRSV+Pm+PRV	7/212 (3.30)	S.Pn+PCV2+PRV+PA	1/212 (0.47)
S.Pn+PCV2+Pm+Bb	7/212 (3.30)	HPS+MH+Pm+PA	1/212 (0.47)
S.Pn+PCV2+Pm+MH	6/212 (2.83)	S.Pn+PRRSV+PRV+HPS	1/212 (0.47)
HPS+S.Pn+Pm+PRV	6/212 (2.83)	PRRSV+PCV2+PRV+HPS	1/212 (0.47)
S.Pn+PCV2+LM+HPS	6/212 (2.83)	Pm+S.Pn+PRV+APP	1/212 (0.47)
S.Pn+PCV2+LM+PRRSV	5/212 (2.36)	Pm+S.Pn+PRV+PA	1/212 (0.47)
S.Pn+PCV2+PRV+HPS	5/212 (2.36)	Pm+S.Pn+PRV+LM	1/212 (0.47)
S.Pn+PCV2+PRV+MH	4/212 (1.89)	S.Pn+PRRSV+MH+APP	1/212 (0.47)
S.Pn+PRRSV+PRV+MH	4/212 (1.89)	S.Pn+PRRSV+MH+Bb	1/212 (0.47)
PRRSV+PCV2+PRV+Pm	4/212 (1.89)		

Acknowledgments

The author(s) disclosed receipt of the following financial support for the research, authorship, and/or publication of this article. This work was supported by New Century Excellent Talents in University of Ministry of Education of China (Project No: NCET-11-1059), Sichuan Youth Waterfowl Disease Control and Prevention Science and Technology Innovative Research Team (Project No: 2013TD0015) and Ministry of science and technology of the international science and technology cooperation projects (Project No: 2014DFA31260).

References

1. Hulsen J. Mycoplasma and PRDC. Seminar Working Group Medicine of the Pig in cooperation with Boehringer Ingelheim, J. *Tijdschrift Voor Diergeneeskunde,* **127**, 731(2002).
2. Zhang C Q. Pathogenesis and prevention measures of porcine respiratory disease complex syndrom, J. *The breeding and feed.* **9**, 41(2008).

3. Hansen M S, Pors S E, Jensen H E, Bille-Hansen V. An investigation of the pathology and pathogens associated with porcine respiratory disease complex in Denmark, J. *Journal of comparative pathology.* **143**, 120(2010).
4. Zhao. Establishment and Application of Four Pathogenic Baeterias on PRDC by Complex PCR, Anhui Agricultural University (2012).
5. Sullivan Å, Edlund C., Nord C E. Effect of antimicrobial agents on the ecological balance of human microflora. *The Lancet infectious diseases,* J. **1**, 101(2001).
6. Kim M. J., Lee D K., Park J E., Park I H. Antiviral activity of Bifidobacterium ado lescentis SPM1605 against Coxsackievirus B3, J. *Gastroenterol Res Pract.* **28**, 681(2014).
7. Forestier C., Guelon D., Cluytens V, et al. Oral probiotic and prevention of Pseudomonas aeruginosa infectio-ns: a randomized, double-blind, placebo-controlled pilot study in intensive care unit patients, J. *Crit Care.* **12**, p. R69(2008).
8. Esposito S., Rigante D., Principi N. Do children's upper respiratory tract infections benefit from probiotics? J. *BMC Infect Dis.* **14**: p.194(2014).
9. Gao, Z., Kang Y., Yu J., Ren L. Human pharyngeal microbiome may play a protective role in respiratory tract infections, J. *Genomics Proteomics Bioinformatics.* **12**, p. 144(2014).
10. Yang X. Molecular Epidemiological Investigation of Porcine Reproductive and Respiratory Syndrome Virus in Sichuan, Sichuan Agricultural University. (2013).
11. Huang H. Investigation on epidemiological of classical swine fever in Leshan district of Sichuan province, Sichuan Agricultural University (2013).
12. Wu J J. Molecular epidemiology porcine circovirus type2 In Sichuan, Sichuan Agricultural University (2013).
13. Sun Z Y., Guo W Z., Huan G Q. Isolation and Identification of Swine Influenza A Subtype H3N2 Strain and Sequencing of the Virus Genom, J. Chinese Journal of Animal and Veterinary Sciences. 39, 195(2008).
14. Yang X., Zhu L., Shi X H. A study of PCR assay for detection of pseudorabies based on PRV gE gene, J. Heilongjiang Animal Science and Veterinary Medicine. 12, 10 (2010).
15. Liu M J., Li D M., Shao G Q. Establishment of PCR Method for the Identification of Mycoplasma hyopneumoniae, J. Journal of Jinling Institute of Technology. 24, 99(2008).
16. Liu C Q. Aetiological investigation, control and prevetion measure of procine rispiratory disease complex in large scale pig farms in Guangxi, Nanjing Agricultural University (2011).
17. Shu X H., Yin G F., Yang Z L. An Epidemiological Survey on PCV-2, CSFV, PRRSV and PRV that Play as Main Pathogens of PRDC in Some Herds of Swinein Yunnan Areas, J. Journal of Yunnan Agricultural University. 26, 54(2011).
18. Hong J S, Kim N H, Choi C Y, Lee J S. Changes in cellular microRNA expression induced by porcine circovirus type 2-encoded proteine, J. Veterinary research. 46, 39(2015).

19. Thevenot C, Fanget M, Fayol M. Retrieval or nonretrieval strategies in mental arithmetic? An operand recognition paradigm, J. Memory & Cognition. 35, 1344(2007).
20. Zhou L, Yang X, Tian Y. Genetic diversity analysis of genotype 2 porcine reproductive and respiratory syndrome viruses emerging in recent years in China, J. BioMed research international (2014).
21. Kimman T G., Cornelissen L A., Moormanm R J., Rebel J M.Challenges for porcine reproductive and respiratory syndrome virus (PRRSV) vaccinology, J. Vaccine. 8, 3704(2009).
22. Yoon K J., Wu L L., Zimmerman J J., Platt K B. Antibody dependent enhancement (ADE) of porcine reproductive and respiratory syndrome virus (PRRSV) infection in pigs, J. Viral Immunol. 9, p.51(1996).
23. Cheng Q. Isolation, Identification and Pathogenicity Analyses of Porcine Reproductive and Respiratory Syndrome Virus SHxx13/2013 Strain, Chinese Academy of Agricultural Sciences (2014).
24. Salguero, F.J., Frossard J P., Rebel J M., Stadejek T. Host-pathogen interactions during porcine reproductive and respiratory syndrome virus1 infection of piglets. Virus Res. 202, p. 135(2015).

Adsorption of Pb^{2+} and Cd^{2+} onto chestnut shell combined with ^{60}Co-γ irradiation

Renbang Zhao, Yaqing Zhang, Weihua Liu, Sha Li, Yang Wang, Mengying Sun, Yuanyuan Huang

Department of Food Science and Technology, Agricultural University of Hebei, Baoding, 071000, P.R. China

E-mail: zhaorenbang@sina.com

The ^{60}Co-γ irradiation combined with chestnut shell to absorb the Pb^{2+} and Cd^{2+} ions in stimulated water and real waste water was studied. Compared to the Heating absorption, Irradiation absorption has the same removal rate of Pb^{2+} in simulated waste water, and has the low removal rate of Cd^{2+} in simulated waste water. After Irradiation absorption, addition heating reduced both removal rate of Pb^{2+} and Cd^{2+} ions. In most of the real waste water, the Irradiation absorption has higher removal rate than the Heating absorption. Hence, the ^{60}Co-γ irradiation combined can replace the heating as energy source to combine with chestnut shell to absorb the Pb^{2+} and Cd^{2+} ions from the real waste water.

Keywords: Absorption, Chestnut Shell, Heavy Metal, ^{60}Co-γ Irradiation.

1. Introduction

Chinese chestnut, Castanea mollissima, is a species of chestnut native to china and the tree is widely cultivated in the province of Hebei, Hubei, Hunan, etc. The nuts are edible and sweet, and provide a significant food source for human and wildlife. The tradition products of chestnut were canned, frozen and sugar fried chestnut. Some of products were exported to East Asian, and the main by-product of chestnut was chestnut shell, which was burned in the north of China as fuel for heating source, resulting in a waste in the resource. Recently, the recycling chestnut shell has been notice, and the main ingredients were determined (G. Vázquez *et al.* 2008, N. Jung-Ran *et al.* 2010, G. Vázquez *et al.* 2010).

Many works have been done to use the chestnut as absorbent to remove the heavy metal ions from the water. Wang *et al.* (W. Guohui *et al.* 2009) used the chestnut shell to remove Cr(VI), and the removal rate of Cr(VI) was 99.7% when pH=2. Qi. *et al.* (Q. Jianhua *et al.* 2009) use the chestnut shell to absorb the Cu(II) and found that the adsorption processes followed the pseudo-second order kinetic model and the equilibrium data were satisfactorily fitted to the Langmuir models. Deng *et al.* (D. Zhenli *et al.* 2012) prepared the poron carbons from chestnut shell and remove Cr^{6+} from the aqueous solution.

Gonzalo *et al.* (G.Vázquez *et al.* 2009) used acid formaldehyde pre-treated chestnut shell as adsorbent to optimize Pb^{2+}, Cu^{2+} and Zn^{2+} removal from aqueous solutions. The maximum pre-treated chestnut shell adsorption capacity was obtained for Pb^{2+} ions, $8.5 mg\cdot g^{-1}$, and the order of cation affinity was $Pb^{2+}>Cu^{2+}>Zn^{2+}$. Moreover, Gonzalo *et al.*

(G. Vázquez *et al.* 2012) prepared the chestnut shell by an alkaline pre-treatment in order to increase its adsorption capacity on cadmium ions adsorption. Didem *et al.* (D. Özçimen *et al.* 2009) prepared activated carbons from chestnut shell and grape seed, and removed Cu (II) ions from aqueous solutions.

From the current reports, we can found that the chestnut shell has a significant ability to absorb the heavy metal ions from the water, and the absorption process is an endothermic process, where the heating energy was needed. According to our research, the favorite condition of adsorption for Pb^{2+} was at 54°C of temperature, 6.5 h of contact time. And for Cd^{2+} was at 54°C, 7 h of contact time. It is obvious that the absorption process is energy and time consumption.

^{60}Co-γ irradiation is the most commonly used source of gamma radiation for radiation technology, using for environment (Z. Renbang *et al.* 2009, Y. Jing *et al.* ,Z. Renbang *et al.* 2009), pharmaceuticals (B.P. Fairand *et al.* 2002), food products (world health organization, R.A. Molins 2001), and material modification (J.G. Drobny 2003). In the case of application of ^{60}Co-γ irradiation, such advantages include: product can be used immediately after treated; the product temperature rise minimally during the process; the irradiation has high penetrability; it is easy to control the process (only dose to be controlled); and the treatment process is very precise and reproducible treatment process.

In this work, ^{60}Co-γ irradiation has been adopted to combine with the absorption of Pb^{2+} and Cd^{2+} onto chestnut shell in order to obtain a higher removal rate of Pb^{2+} and Cd^{2+} in acid solution, and achieve the purpose of saving energy and time. The real waste water, polluted by heavy metal, from the Hebei province was used to find out the removal rate of Pb^{2+} and Cd^{2+} by the combination of ^{60}Co-γ irradiation and absorption of chestnut shell.

2. Experiment

2.1. *Materials*

2.1.1. *Chestnut*

The chestnut, purchased from locate market, were peeled first, and the shell was washed with distilled water, finally crushed and sieved through with 40 mesh size.

2.1.2. *Water sample and adsorption assay*

Water samples were collected from Hengshuihu lake and Baiyangdian River. The water 10cm below the surface of water was taken, and filtrated to remove precipitate. After that the pH and the concentration of Pb^{2+} and Cd^{2+} were determined.

2.2. *Preparation of simulated heavy metal waste water*

Stock solutions of lead nitrate and cadmium nitrate were prepared at 500.0mg/L in nitric acid solution with pH 5.0 and stored in dark. Aliquots of stock solution will be diluted in nitric acid solution to prepare working solution of heavy metal ions (Pb^{2+} 40mg / L, Cd^{2+} 30mg / L). The working solution will be prepared individually.

2.3. *The adsorption experiment*

2.3.1. *Heating absorption*

1.0g chestnut shell were added into the working solution and real waste water, respectively, and then placed in conical flash for water bath at 50°C (Control experiment was placed at room temperature) for 5h. The conical flash was shocked once every hour. The solution was filtered, and the liquid-phase was collected.

2.3.2. *Irradiating absorption*

1.0g chestnut shell were added into the working solution and real waste water, respectively, and then placed in conical flash for irradiation (the absorb dose was 2kGy, 4kGy and 6kGy, respectively). 1mL of aliquots solutions were picked up to determine the concentration of heavy metal ions after irradiation, and the remaining solutions was used in Section 2.3.3.

2.3.3. *Irradiation & heating absorption*

The solution in Section 2.3.3 were placed in water bath at 50°C for 5h. The conical flash was shocked once every hour. The solution was filtered, and the liquid-phase was collected.

2.4. *Removal rate*

The removal rates of heavy metal ions were obtained according to the following equation:

The adsorption rate=q^e/q^0 (100%)(1)

Where q^e was the amount of lead (II) or cadmium (II) absorbed onto chestnut shell. q^0 was the initial amounts of lead (II) or cadmium (II) before adsorption.

3. Results and discussion

3.1. *Irradiation combined with chestnut shell to the adsorption of Pb^{2+} and Cd^{2+} in acidic solution*

The removal rate of Pb^{2+} in working solution of Section 2.3.2 and 2.3.3 were given in the Fig. 1. And the removal rate of Pb^{2+} in Heating experiment and Control experiment is 81.23% and 97.26%, respectively.

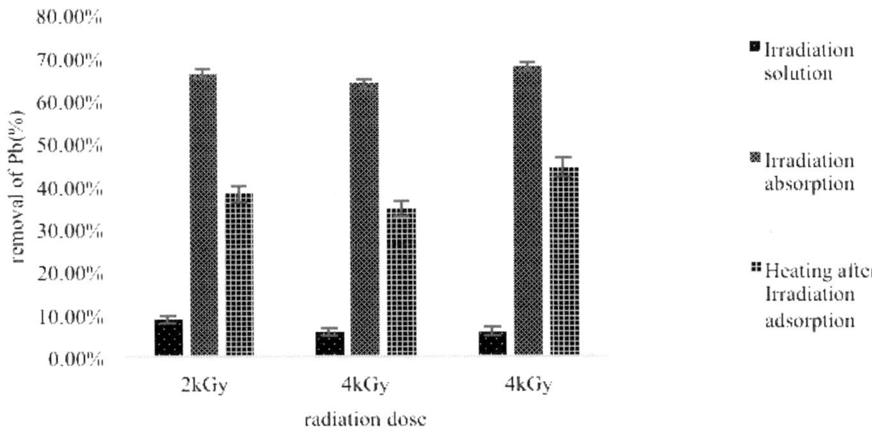

Fig. 1. Removal rate of Pb^{2+} in stimulated water

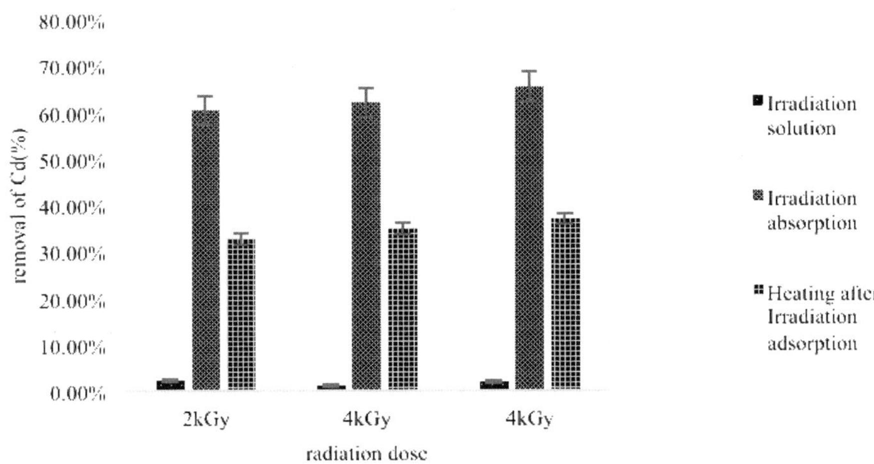

Fig. 2. Removal rate of Cd^{2+} in stimulated water

According to the Fig. 1, the removal rate of Pb^{2+} at 2kGy of absorb dose is 97%, which is larger than that of Control experiment, and similar to the Heating absorption, but with the raise the absorb dose the removal rate decreased. However, after irradiation absorption, the addition heating would cause the removal rate low. Hence, the low absorb dose of irradiation have the same effect with the Heating absorption without any heating in the whole absorption.

The removal rate of Cd^{2+} in working solution of Section 2.3.2 and 2.3.3 were given in the Fig. 2. Also, the removal rates of Cd^{2+} in Heating experiment and Control experiment, which is 97.01% and 79.82%, respectively, were obtain. From the Fig. 2, we can know that unlike that of Pb^{2+}, the removal rate of Cd^{2+} rise at 2 kGy of absorb dose of irradiation, and decreased at 4 kGy, and finally reach a peak at 6 kGy. The highest removal rate of Cd^{2+} was 88.39%, which was lower than Heating absorption. Moreover, the removal rate of Cd^{2+} was reduced if the addition heating was added after Irradiation absorption. It was obvious that the Irradiation absorption has positive effect on absorption of Cd^{2+}, but did not reach the removal rate of Heating absorption.

3.2. *The removal rate of heavy metal ion in real waste water*

The pH and the concentration of Pb^{2+} and Cd^{2+} in the waste water from different area were given in the Table 1. From the Table 1, we can see that waste water in the selected area was weakly alkaline and the concentrations of Pb^{2+} and Cd^{2+} were low.

Tab. 1. pH and concentration of $Pb^{2+}$and $Cd^{2+}$of real waste water

No.	site	pH	Pb2+concentration ng/mL	Cd2+concentration ng/mL
1	Don River, Baiyangdian	8.56	5.9510	1.8954
2	Burningcars itch, Baiyangdian	8.19	8.1891	2.2114
3	fish Ravine, Baiyangdian	8.29	5.9737	2.5510
4	East Gate, Baiyangdian	7.47	4.2739	1.7482
5	Yuanfeiheyuan, baiyangdian	7.78	6.0000	2.3213
6	Baiyangdian Lake	8.39	8.2449	2.0944
7	Liu Feng Dong Dam (North)	8.70	7.3250	2.4214
8	southwest corner ofLake Avenue Hengshui Lake of the Great Lakes	8.58	8.7997	3.0523
9	106 State Road Hengshui Northeast gas station	8.65	6.0463	3.6643
10	middle North Shore	8.62	7.3378	3.0854
11	East of Water Conservancy Environment Management Office	8.85	9.5237	2.4553
12	Jigme shopping nearby lake	8.76	67.4274	9.0060

The removal rates of Pb^{2+} and Cd^{2+} from the real waste water were shown in the Figs. 3 and 4, respectively. As Figs. 3 and 4 shown, the chestnut shell can absorb Pb^{2+} and Cd^{2+} in Heating absorption, even though the concentrations of Pb^{2+} and Cd^{2+} were low. The highest removal rate of Pb^{2+} from real waste water could reach 83.93%, which was lower than that from simulated waste water. We can know that there were a variety of substances in real waste water, which have an impact on the adsorption of heavy metal ions. From the Figs. 3 and 4, we can know that the removal rates of Pb^{2+} was higher than that of Cd^{2+}. It can deduce that the chestnut shell had selective adsorption for different heavy metal ions.

As it can be seen in Fig. 3 and Fig. 4, the radiation absorption could reduce the concentration of Pb^{2+} and Cd^{2+}, however the removal rates were different in different waste water with different concentration of heavy metal ions. Most of removal rate of Pb^{2+} and Cd^{2+} in Irradiation absorption is larger than that in Heating absorption. The highest removal rate of Pb^{2+} reached 93.35 %, and Cd^{2+} was 76.82%. This may be that the organic compounds in the real waste water were changed after the irradiation, and precipitated with heavy metal ions, which cause the reduce of the concentration of heavy metal ions in water. Addition heating after Irradiation absorption would generally declined the removal rate.

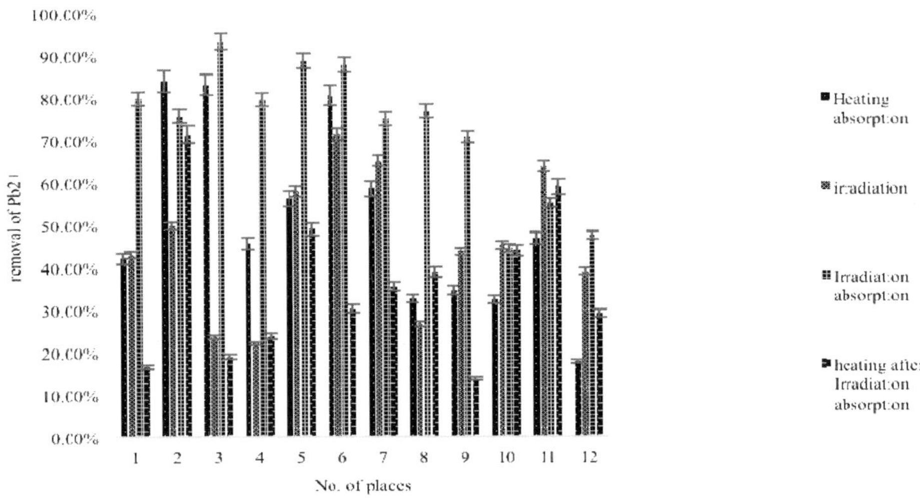

Fig. 3. Removal rate of Pb^{2+} in real waste water

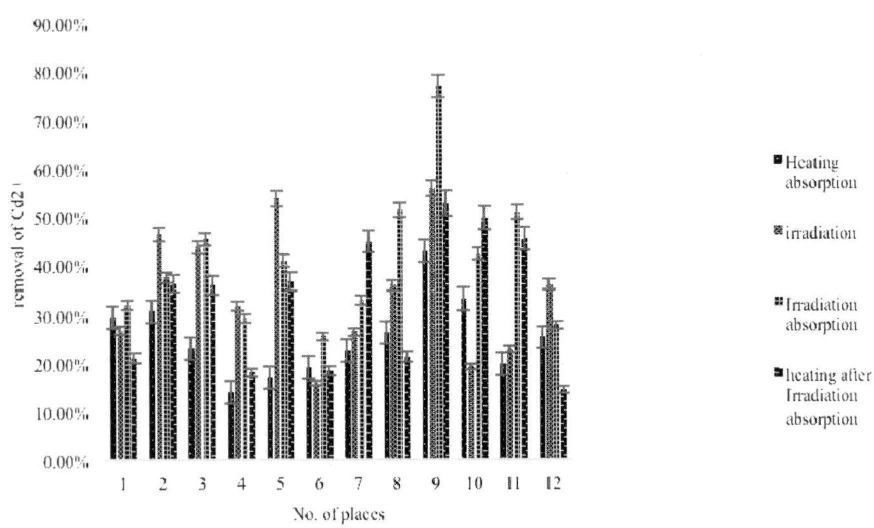

Fig. 4. Removal rate of Cd^{2+} in stimulated water

4. Conclusion

Irradiation can be used in the adsorption of Pb^{2+} onto chestnut shell to replace of heating source. The highest removal rate of Pb^{2+} could up to 96.81%. While the removal rate of Cd^{2+} by Irradiation absorption was lower than Heating absorption in stimulate waste water, which the maximum was 88.39%. In the most of real waste water, the removal rates of both Pb^{2+} and Cd^{2+} were improved. However, the addition heating reduced the removal rate.

Acknowledgement

This work was supported by the Fund for the Natural Science Foundation of Hebei, China (The degradation of phosphate esters organic phosphorous by γ irradiation No.B2012204073) and Science and Technology Support Project of Hebei, China (Research and demonstration of ecological restoration of heavy metal wastewater in typical coastal zone and estuary wetland NO.13273301D; Research on the key technology of pectin extraction and processing of persimmon juice NO.14236810D-3).

References

1. G. Vázquez, E. Fontenla, J. Santos, M.S. Freire, J. González-Álvarez, G. Antorrena. Antioxidant activity and phenolic content of chestnut (Castanea sativa) shell and eucalyptus (Eucalyptus globulus) bark extracts. Industrial Crops and Products, 28, 279(2008).
2. N. Jung-Ran, G. Gil-Tae, K. Yong-Hoon, Y. Keum-Jin, H. Jung-Hwan, L. Hyun-Sun, O. Won-Keun, S. Kyung-Sik, L. Chul-Ho. Antioxidant effects of the chestnut (Castanea crenata) inner shell extract in t-BHP-treated HepG2 cells, and CCl4- and high-fat diet-treated mice. Food and Chemical Toxicology, 48, 3177(2010).
3. G. Vázquez, M. S. Freire, J. Santos, G. Antorrena, J. González-Álvarez. Optimisation of Polyphenols Extraction from Chestnut Shell by Response Surface Methodology. Waste and Biomass Valorization, 1, 219(2010).
4. W. Guohui. Study on adsorptive performance of heavy metal chromium(ε) by chestnut shell. Chinese Journal of Environmental Engineering, 3, 791(2009).
5. Q. Jianhua, L. Zongsuo, D. Xiping, Y. Zengyu. Adsorption of Cu^{2+} on to chestnut (Castanea mollissima) shells: Equilibrium, kinetics and process design. Acta Scientiae Circumstantia, 29, 2141(2009).
6. D. Zhenli, H. Qiaokai, W. Jian. Research on the preparation of chestnut shell active carbon and its absorption process of Cr^{6+}. Hubei agricultural science. 51, 2556 (China. 2012).
7. G. Vázquez, M. Calvo, M. S. Freire, J. González-Alvarez, G. Antorrena. Chestnut shell as heavy metal adsorbent: Optimization study of lead, copper and zinc cations removal, Journal of Hazardous Materials, 172, 1402(2009).
8. G. Vázquez, O. Mosquera, M. S. Freire, G. Antorrena, J. González-Álvarez. Alkaline pre-treatment of waste chestnut shell from a food industry to enhance cadmium, copper, lead and zinc ions removal. Chemical Engineering Journal. 184, 147(2012).
9. D. Özçimen, A. Ersoy-Meriçboyu. Removal of copper from aqueous solutions by adsorption onto chestnut shell and grapeseed activated carbons. Journal of Hazardous Materials. 168, 1118(2009).

10. Z. Renbang, B. Huaying, X. Lingyu. γ-irradiation Degradation of Methamidophos. Chinese Journal of Chemistry, 27, 1749 (China. 2009).
11. Y. Jing, G. Yanpan, L. Weihua, G. Wei, Z. Renbang. The Degradation and Mineralisation of Acephate by Ionisation Irradiation. Environmental Progress & Sustainable Energy. DOI: 10.1002/ep.11983
12. Z. Renbang, B. Huaying, X. Lingyu. γ-irradiation Degradation of Methamidophos. Chinese Journal of Chemistry, 27, 1749 (China. 2009).
13. B.P. Fairand. Radiation sterilization for health care products -X ray, gamma and electron beam, CRC Press, New York (USA. 2002).
14. world health organization, High-dose irradiation: Wholesome-ness of food irradiated with doses above 10 kGy, Report of a Joint FAO/IAEA/WHO Study Group, WHO Technical Reports.
15. R.A. Molins(Ed.), Food irradiation: Principles and applications, John Wiley & Sons, New York (2001).
16. J.G. Drobny. Radiation technology for polymers, CRC Press, New York (2003).

A study on what cause the lack of integrity in China's food companies

Yulan Jin, Yong Meng, Yinchuan Xu and Fan Xie
School of Management, Shanghai University of Science,
Shanghai, 201620, China
E-mail: jinyulan12@163.com

By comparing the different definitions of integrity in western and eastern cultures, this paper analyzes the external and internal causes to the various unethical behaviors of China's food companies such as adulteration, sub-standard quality, counterfeit, sales of outdated food, based on the study of the uniqueness of the food industries as it is a fast-developing sector in China producing the irreplaceable sustenance that are essential to people's lives.

Keywords: Food Company, Corporate Integrity, Dishonesty.

1. Introduction

Food industry is closed related to life qualities and health conditions. Early in the year of 1820, German scholar Frederick revealed some dishonest behaviors by the food manufacturers and gave the possible explanations to such behaviors. Between the year 1851 and 1854, British doctor Hassal sampled about 2500 types of food products and pointed out the significance of building a trust and regulation mechanism in the food industry. MacDonald (1996), after studying the United States' meat processing companies, concluded that large-scale enterprises (over 270 employees) are more suitable for the application of HACCP system than the small-scale enterprises (less than 20 employees), as they will benefit more from keeping themselves honest. Deepananda Herath (2006) researched the Canadian food companies, and his study shows a higher inclination to ethical operations by bigger companies. Some researchers, such as Jayasinghe-Mudalige and Henson (2007), based on their study on the Canadian meat processing companies, believe that market motivation plays a more important role in keeping the food companies honest to their customers [1].

The issue of dishonesty by food companies has been given much attention in the recent years. However, the studies in this area make their analysis mostly from a legal perspective. Our research approach the same issue from a moral standpoint emphasizing the effects of eastern culture ethics have on the honesty of food enterprises' conducts. First, a definition of integrity is given, then, we further examine the dishonest behaviors in China's fast-developing food industry to find out the internal and external causes to the dishonesty of China's food manufacturers.

2. Definition of integrity for food companies

2.1 Differences in definitions of integrity between the East and the West

Both the eastern and the western culture deem integrity as an essential element to the development of individual, business and society. Both agree on the similar ideas that honesty means the respect to truth, the honoring of one's words and the fulfillment of one's commitment. Chinese pay more attention to morality and they think integrity is the foundation for both an individual's development and a nation's prosperity. Whereas in the west, integrity and contracts are the preconditions for interpersonal communications, economical function and the normal development of the society.

However, the eastern and western cultures differ in how they view integrity in many aspects, as is shown in Table 1. From the different perspectives of individual, business and society, eastern and western definitions of integrity have their own meanings ideologically and institutionally. From a social perspective, eastern integrity basically is a type of mutual trust grounded on morals and ethics. Whereas western integrity is a contractual relationship about mutual recognition, trust and keeping one's promise. The eastern society set up an Institutionalized integrity assessment and supervision mechanism, and the western society legislate integrity by building a trust mechanism based on laws and regulations. From a corporate perspective, the eastern culture approves the temperance in pursuing profits; disapproves the promotion of interest by devious means. It calls for people's self-discipline to build the social integrity system. While the western enterprises are bound by contracts to build the trustworthy business relationships with each other. From the individual perspective, eastern integrity is the bedrock of moral principles and a kind of personal trait. However, western integrity put more emphasis on rationality and legality.

Tab. 1. The differences in the definitions of integrity between east and west

		Eastern definition	Western definition
Social perspective	Ideology	Ethical integrity, moral integrity	Contract-binding integrity
	Institution	Institutionalized integrity assessment and supervision mechanism	Legislated credit system
Corporate perspective	Ideology	Moral value before financial value, moral self-discipline	Based on contracts, external-discipline
	Institution	Integrity self-discipline mechanism	Comprehensive contractual system with detailed procedure
Individual perspective	Ideology	Emphasis on personal relationship and emotions, bedrock for one's moralities	Emphasis on rationality and legality
	Institution	——	Personal credit system

Eastern traditional integrity is based on people's self-discipline, while in the west it is more on legal restraint. Dishonest behaviors in China will only lead to public condemnation on a moral basis, but similar behaviors by a westerner may subject the person to not only social isolation but also severe financial and legal punishment.

Therefore in China, traditionally, integrity is more a moral constraint than a legal or financial one.

2.2 Business Integrity defined in this paper

The eastern culture emphasizes the significance of morals and ethic, as a result, "integrity" in the eastern culture refers to a kind of moral restraint at its core. Business integrity relies on the cultivation and development of the right culture and customs of the people working in the business, and consequently the self-discipline of the business itself.

3. Analysis on the characteristics and development of the China's food industry

Food industry is closed related to life qualities and health conditions. It is a sector that takes in more urban and rural labor migration than any other industry. And the food industry is highly dependent on the agricultural industry. China's food industry has been fast developing in recent years, as is shown in the following table.

Tab. 2. Economic indicators of China's food industry growth 2006-2011

Year	2006	2007	2008	2009	2010	2011
Enterprise amount	25298	28381	31649	37245	41867	47406
Growth rate	12.35%	12.19%	11.51%	17.68%	12.41%	13.23%
Sales revenue (Billion Yuan)	1872.764	2484.19	3264.124	3868.895	5816.125	7654.020
Growth rate	24.73%	32.65%	31.40%	18.53%	50.33%	31.6%
Output (Billion Yuan)	2158.695	2864.938	3788.437	4464.53	5933.003	7807.832
Growth rate	23.47%	32.72%	32.23%	17.85%	32.89%	31.60%

Note: Compiled from data published by Bureau of statistics

Edible animal and vegetable are the main ingredients for the China's food industry, along with numerous additive, sweetener and condiments. The supply chain includes food processing companies, food transporters, distributers and other related businesses. China's food industry is a fully competitive, low-entry-barrier, low-margin business, comprising mainly of small and medium enterprises. The 12 main characteristics of the food industry are summarized as (1) perfect competition, low entry barrier, highly competitive; (2) everlasting, big customer base; (3) small product, large market, low unit price, high sales volume; (4) wide variety, large quantity products with shelf lives; (5) repetitive manufacturing pattern; (6) trend toward internet banding and distribution; (7) formulized manufacturing; (8) high level of automation; (9) complex distribution system; (10) flexible marketing strategy, regular price adjustment; (11) fast response to customer needs and speedy delivery; (12) emphasis on quality control.

On one hand, these characteristics may motivate the food companies to relentlessly pursue profit by dishonest means, such as disguising inferior goods as premium, outdated as fresh. On the other hand, we can see a highly repetitive production method with relatively low levels of technology and employees education. Loopholes in the corporate monitoring are more likely to be expected.

The above characteristics of the food industry require the food companies to cultivate a culture that values integrity and build the internal and external food safety monitoring system.

4. The dishonest behaviors of the food companies and their hazards

4.1 Dishonest behaviors

The food companies' unethical behaviors mainly include the followings:

- Adulteration: Adulteration means to add in similar but inferior substances to the food products, and it is hard to discern by the customers.
- Counterfeit and inferior products: Due to asymmetric information, food manufacturers can delude the customers using their advantages with the information they control. Fake or inferior food products may fail to meet the customers' needs or pose a threat to consumers' health.
- Selling outdated products: Food products, unlike other goods, have shelf lives. When the goods are outdated, they may generate hazardous chemicals that cause harm the consumers' health.
- Over-packaging and false advertising: We can see over-packaging especially well on the moon cake market and false advertising on products with special efficacy.

4.2 Hazards

The lack of integrity may result in grave consequences and hazardous effects. Some of the hazards are immediate and some are potential.

(i) Immediate hazards

Accidents caused by dishonesty of food enterprises pose threats to consumers' lives. According to "Report on food poisonings incidents 2008-2012", 174 food poisoning incidents occurred in 2012, with 6685 victims and 146 deaths. Comparing to 2011, the happenings of food safety accidents and the victims count drop 7.9% and 19.7% respectively, but the death toll rose 6.6%.

Dishonesty of food companies lead to a decrease in market share and inefficiency of the company. According to the 2011 annual reports posted by meat processing company Shuang Hui, it manufactured 766.6 thousand tons of meat products, 23.94% less than the previous year. 2480.5 Thousand pigs were processed in the company, 29.72% less than the last year. Sales revenue and operating income are 37.615 billion Yuan and 699.73 million respectively. The Former increased 3.59% than last year and the latter declined 60.85. The net profit also dropped 51.27% at 564.89 million Yuan. The drastic drop in profit as is shown by the above figures is a result of the March "lean meat powder" incident.

Also, dishonesty of food companies have a detrimental effects on China's national image and the food trade. According to the statistics released by China's Customs, after the "San Lu" Melamine incident, China imported 35.1 more tons of dairy products, a

17.4% increase from the last year, and exported 12.1 less tons of dairy products, 10.4 less than last year.

(ii) Potential hazards

The lack of integrity of food enterprises severely lower the customers' trust. In a questionnaire investigation, Li Hong (2010) [2] examined the influence of food safety accidents on consumers. Her research finds out a sharp decline in customers' faith in food companies. 21.5% of the surveyed express a complete lack of confidence in food safety. And many people chose to turn away from foreign fast food brands such as MacDonald's and KFC after these enterprises were revealed to have used outdated meat in their products.

5. Causes to the dishonesty

5.1 External causes

5.1.1 Lack of social integrity environment

China has experienced fundamental economic and political reform and the fastest growth in the last 30 years, turning from a government planned economy to a market economy, followed by profound change in its social ideology. The social moral and trust mechanism of the past could not adapt to the drastic change. As a result, money worshiping became more dominant in the main-stream culture. Although the traditional Chinese ethics still is influential on many people and enterprises, by and large, there is an increasing lack of integrity in the Chinese society.

Compared to a lot of the developed countries, China is a much bigger food market with more customers and enterprises, but with fewer and less rigorous rules and regulations to restrain the immoral behaviors by the food companies. The markets also have higher tolerance to such behaviors than other countries. For example, Shanghai's food supplier Fu Xi used to sell their higher-quality material to their Japanese clients, and the inferior or outdated ones in the Chinese market.

5.1.2 Underdeveloped legal system

(i) Absence of integrity related legislatures

The integrity related legislature stops at the institutionalized credit assessment and monitoring mechanism, and they are more of a moral requirement to the food companies. What is needed now for the china's food industry is a social record system keeping track of the dishonest behaviors of food companies and a mature legal system to punish such behaviors. Combined they will increase the cost for dishonesty and eventually increase the social responsibility awareness among the food enterprises and cultivate a culture that value business integrity.

A series of bills concerning business integrity and honesty was passed between 1860s – 1880s in the United States, and thus forming a sophisticated legal system governing this area. By contrast, China's legal system only lay out some of the principles about business integrity at legislations such as "General rules of the civil law", "Contract law" and "Anti-monopoly law". The underdevelopment in legislations make it impossible to administer punishments on businesses for their dishonest conducts [3].

(ii) Inadequate legal punishment and low legal cost

In the recent years, the government has attached more importance to the business integrity problem, and sped up the related legislation to impose harsher punishment on dishonest companies. Many new laws and regulations have been passed concerning the food safety issue. Nonetheless, tempted with a high return, many food manufacturers maintain their unethical operations, because the risks of getting caught and the legal cost thereof is still relatively small compared to the enormous profits they can get.

Using outdated and reused ingredients are illegal in China, and will be subject to a penalty of up to ten times the value of the goods and the company's business licensed could be revoked for such acts. In accordance with "Food Security Law", consumers can file for a compensation 10 times the value of unsafe food products they buy, if such unsafety is confirmed by the food quality administration. However, there is no food company that can't bear such penalty. Even when their licensed are revoked, they can just found a new firm.

Death penalty could be the highest punishment to crimes of food safety in China. In reality, however, many food safety related illegal conducts are inadequately punished. For example, the president of "San Lu" was sentenced with life imprisonment for the melamine found in the powered milk the company manufactured, but the sentence was actually lowered to 17 years eventually after the event quieted down. Also, many government officials involved with the crime were reinstated into other government posts after they got demoted or resigned.

(iii) Favorable treatment to large scale food enterprises may cause food safety issues

According to "Famous brand products supervisory rules", any brand recognized by the Administration of Quality Supervision and Inspection as Famous Brand will be exempt from any inspections by the government. This regulation mean to encourage and motivate those celebrated companies for what they have done to the economic growth. In fact, the special treatments they get from the government only cause their negligence to their own internal supervision.

5.1.3 Absence of government regulation

At present, there is yet no unified official standards governing the food products quality. From central government to local governments, and among the different government agencies, the standards they issued are overlapping and contradictory. This renders the enforcement of these standards very hard. For the time being, there are state, industrial, local and corporate quality standards. Among them, 1070 food industry state standards, 1164 industrial standards and 578 export food standards [4].

Government departments such as health, quality supervision, agriculture, commerce, food and drugs all have regulatory authority over food companies. As a result, two or more authorities will regulate the same issue in a food industry supply chain.

For medium and small food enterprises, the government would choose to punish their illegal operations rather than take precautions to prevent them from happening. It is usually after some food safety accident happens, the regulatory department of the government then take measures to remedy, fining as the most important one of those measures. Some food companies with a famous brand contribute a lot to the local economic growth and the tax revenue of the local government. Then these companies may be treated favorably by the government, leading to an absence of supervision. When the "San Lu" incident just occurred, local government was well aware of it. But it chose to not report it to the central government and other related authorities, even though it already had the detailed reports on the poisoned powdered milk [5].

In Shanghai's "Fu Xi Incident", as the meat supplier to fast food chains like MacDonald's and Pizza Hut, Fu Xi has been hiding its supply of outdated meat for a long time. It's the undercover journalist who unearthed such crime, not the regulatory departments of the government. Because in the eyes of the government, large food enterprises are less prone to irregularities in food safety, thus they monitor the large companies less rigorously than small plants. This may encourage the large companies to lower their own safety or quality standards.

5.2 Food enterprise internal reason

In the food business, the food manufacture products to make a profit, but the consumers of the products fully assume the safety risk. Business integrity of food company comes with a cost, shared by the company and the consumer and even the society. A lack of integrity may lower the cost to maximize the profit for the company, but the customers will face the health hazards and the whole food industry will suffer the loss when the consumers lose their faith. With the temptation of high return, some companies choose to pursue profit by dishonest means. Therefore, the main reason for the lack of integrity lies with the enterprise itself.

5.2.1 Staff's lack of integrity

Motives determine how people behave. The continued presence of uncertainty, limited rationality and opportunism in people's behaviors often leads to the employees' short-sighted decision especially when the return is uncertain and unstable. Moreover, since the information and knowledge one can obtain is limited, we can only expect so much rationality from the staff. In a socialist market economy that is constantly evolving, the staff of an enterprise (managers and other employees) will ignore ethics when they try to maximize their personal gains. Between 2009 to 2013, 2173 counts of food safety accidents came to public in China, 90% of which are caused by staff of the company.

5.2.2 Loopholes in the management system

The right management system can correct people's behavior by punishing their dishonest behaviors. Among the almost 50 thousand food manufacturers in China, 80% are small and medium sized. Most of these small plants have poor working conditions and low technological level. The loopholes in their quality management systems cause a lot of problems to the whole food industry supply chains and is the obstacle to the quality improvement of the whole industry.

Bigger companies also have their own problems, especially in the way they choose their suppliers, which largely determine the qualities of the final products. The faulty mechanism in this area results in a weak link between the food companies and their suppliers. The suppliers' more opportunist behaviors are the direct cause to numerous food safety accidents, such as the 2012 "Preserved fruit Incident". When there is public relation crisis caused by food safety, management are often flustered and fail to respond in the right way. Some try to cover the truth, end up losing the trust of their customers.

6. Conclusion

Integrity is a type of competitiveness and intangible asset, and a precondition for a food company's sustainable growth. China's food industry has been on the rise recently and contributed a lot to the economic growth and improvement of life qualities. Nevertheless, the frequent occurrence of food safety accidents make people reflect more on this issue. Dishonesty of food companies are not only a barrier to the future growth of food enterprises, it also becomes a bottleneck to impede China's economic growth and is hurting China's national image.

Dishonest behaviors of food companies are closely related to a lack of related legislation, underdeveloped legal system, and an asymmetric information between the enterprises and the consumers. However, at the end of the day, the food company is the one responsible for its own behaviors. On one hand, staff neglect moralities in pursuit of profit or personal gains, trying to find loopholes in the system for their opportunist acts. On the other hand, faulty quality monitoring system and poor management also bring forth a lot of problems to the whole industry supply chain.

In conclusion, we believe that in order to build a trust mechanism between the food companies and the customers. It is essential to have a corporate culture that values integrity. The government should tighten its supervision over food industries by speeding up the food safety related legislations. Thus, both the internal and external binding forces will keep the food companies in the right path to a sustainable growth.

References

1. WANG CHANG-wei, GU HAI-ying. Industry Environment, Supervision Intensity and the Choice for Credit of Food Enterprises in China: Analysis Based on Incentive Compatibility Constraint Conditions [J]. *Journal of Business Economics*, 2013, 08:18-25.
2. LIN YUN-yue. Legal system of credit management in the United States [J]. *WORLD ECONOMY,* 2000(4) : 62-67.
3. ZHOU CAI-qiong. Food standards and regulations [M]. *Beijing:China Agricultural University press*, 2009:187-188.

4. SONG YI-wen, Establishment of food safety supervision system for small food production enterprises [D]. *Shandong University*, 2012.46-47.
5. LIAO YONG-gang. Enlightenment of German social credit system construction to China [J]. *Qinghai Finance*, 2009(4) :52-54.

Practice and exploration of teaching reform of the pharmaceutical microbiology in pharmaceutical education

Xiao Han, Yan Zhuang, Ke Pan, Mengchuan Zhang, Liping An, Guangyu Xu, Yingnan Zhang*

Department of Microbial and Biochemical Pharmacy, College of Pharmaceutical Science, Beihua University, Jilin, Jilin, China

E-mail: hanxiaorumeng@126.com(H.X.); 648473164@qq.com(Z.Y.N.)

**corresponding author: Zhang Yingnan*

Pharmaceutical microbiology was increasingly important in the pharmaceutical teaching field, which has become an important basic course in pharmaceutical education. The practice and exploration of Pharmaceutical microbiology teaching reform has been carried out in our college for many years. Summarizing and exploring pharmacy microbiology teaching content and methods from many aspects, initially established a suitable professional pharmaceutical microbiology teaching system, it get the students' recognition and praise through years of practice.

Keywords: Pharmaceutical microbiology; Teaching content, Skill training.

1. Optimization of Teaching Content

Pharmaceutical microbiology is not only a basic course, also the main courses of pharmacy. The course objective is to cultivate application-oriented and practical talents for workshop, plant quality inspection center, drug sector engaged in pharmaceutical quality control, drug testing, drug supervision and management. Pharmaceutical microbiology is that an important professional basic course. It is not only the basis for subsequent courses of study, the basic knowledge and skills in itself is an important part of the professional competence. Inoculation technology, disinfection and sterilization technology, environmental microbial control and inspection, pharmaceutical microbiology inspection techniques, and titer determination methods are directly used in pharmaceutical production and testing processes. However, with the adjustment of professional overall course structure, the total hours of this course declining, and each is different. So, it is restrict the curriculum to a large extent. How to coordinate the relationship between the class and the teaching content, ensuring not only teach the necessary theoretical knowledge, but also to strengthen skills training to meet the needs of professional personnel training in a limited time, our school is mainly explore as the following aspects.

1.1 Textbook-based, to broaden the teaching content

Write or use high-quality materials is the premise and key to ensure the quality of teaching. Now, most pharmacy colleges use the People's Health Publishing House textbooks as Pharmaceutical Microbiology materials. However, the development of modern microbiology is rapid by leaps and bounds with each passing day. The speed of its theoretical and technological development is far exceed the speed of textbooks updating. Coupled with the limitations of people's understanding of the life, knowledge perhaps will be update soon,even now that the very right knowledge. Redesigne teaching programs and prepare lectures personally on the basis of the original textbooks make teaching content no longer limited to the materials, but rely on textbooks. Both concerned with the basics, and also fit the frontier development extends to extracurricular broaden learning content. Students could keep abreast of trends in the development of microbiology disciplines, capturing the latest information, and follow the forefront of developments[1]. For example, to find a lot materials about EHEC, superbugs, avian flu, SARS and other related materials through the network when we explaine the time of prokaryotes and viruses, students can try to abstract concepts and rich picture, closely linked with the real life together. It's not only improve students' participation and initiative, but also increase students' grasp of the frontiers of knowledge.

After years of practice, personally prepare handouts has following advantages: content updates quickly, follows the developments of microbiology, allows students to appreciate the changes and charms of the world's cutting-edge knowledge[2]; optimize the teaching content, through repeated studies, make a reasonable choice to avoid the repetition of teaching content in the Cell Biology and Biochemistry course, while alleviating the limited number of hours and the ever-increasing amount of information teaching contradictory; convenient for students to use in class, announced teaching contents and plans to students before class or advance on the course website, to facilitate students to print additional notes, so that students have a full understanding and preparation for teaching contents, students can take the initiative to reorganize knowledge, thinking and asking questions; amplification of teaching information, so that teaching content is no longer confined to in textbooks, extracurricular extend to increase the accumulation of knowledge.

1.2 The implementation of the module teaching

The microorganism teaching professional courses are first to establish a complete theoretical system, and then arrange the experiment. Intersperse with experimental arrangement based on theory, the teaching process has always been the mainly theoretical and experiment-based, giving even increase the proportion of experimental class, always not able to escape the guidance practice teaching under complete discipline system[3]. This teaching model has been completely unable to meet the demands of pharmacy education practice skills of sudden facing with pharmacy students' short time in school, more courses and hours reduced. How to highlight the advantages of pharmacy education in teaching, our school try to find a way out from the implementation of the module teaching.

Module teaching can completely broken the systematic and integrity of discipline system, divide into a plurality of modules using practical technology as the unit, use skills as skeleton and theoretical knowledge as a connection point in teaching, in order to set up new skills framework microorganisms knowledge. Theory is service for skills to the degree of necessary and enough. In teaching, teach the theoretical knowledge combined with skills which is useful in modules experimental techniques, and less than, students study self[4]. According to the this divide Pharmaceutical Microbiology curriculum into eight modules: Microbial morphology and identification techniques, Disinfection and sterilization technology, Inoculation with Isolation Technology, Microbial distribution measurement technology, Strain selection and preservation techniques, Antibacterial Drug Test, Drug microbiological examination techniques, Serological test technology.

2. Teaching method

2.1 Examples of teaching, theory and practice

Microbiology is an experimental science, many basic theory and research methods could guide pharmaceutical research directly, explain the theory of knowledge is sufficient for the degree of use, increase the use of modern microbiology technology in drug research and production, to increase student learning motivation and cultivate students' creative ability. Combine with clinical or marketed successfully biologic drugs, use of examples to guide students to learn, ensure students have a solid theoretical foundation, while master basic strategy of microbiology theory to guide drug development. For example, As explained, teacher details the anticancer mechanism of classic anticancer drug cisplatin which is major targets of DNA replication enzymes, interfere with replication of DNA, the cell can not replicate and kill cancer cells when explain DNA replication process; so that students can quickly mastere the theory and process of cloning, expression, protein purification[5]. Teaching by example can guide students combine pharmacology, medicinal chemistry, pharmaceutics and other basic theoretical knowledge which already studyed with microbiology, design and development of innovative new drugs, cultivates students' groundbreaking thinking which play a very good role in the future of the students engaged in the develop of biotechnology new drugs.

2.2 Carrying out heuristic teaching, guiding students to think scientific

In traditional class, teachers generally play as the subject and the students play as object, one-man one-way teaching model is not conducive to the cultivation of students' thinking ability and interest in learning[6]. Currently, the constructivist educational model and student of the rise of new educational theory suggests that teachers as the teaching activity organizers and instructors should not just act as the role of imparting knowledge, students should be perceived as a subject, they should take seminar-mode teaching , such as heuristic teaching and two-way interactive teaching mode. Teachers should post teaching content and handouts to students on the website in advance, so that students can have a full understanding and prcparation, students with questions in class can take the initiative to think, ask questions, are in exploring of actively learning process. Taking language arts charm into lecture course, to guide students to analysis when they learn.

When you introduce the genetic engineering, introduce the "Human Genome Project" firstly, the extent of development now, what achievements have achieved, what achievement begin to apply to? What are the genetic engineering drugs on the market? Do GM foods are health or harmful to human? Inspiring the curiosity and thirst of the students for knowledge to what they learn, so that they will take the initiative to think. In addition, increasing the time for questions and discussion in the class about the course content, encouraging the students to ask and ask cleverly, teaching the students methods of asking, and actively creating a dynamic class atmosphere.

2.3 Integrating research ideas to promote learning

Microbiology is an experimental subject, the theory can guide specific research work, and the research work can deepen the understanding of the theory[3]. We found that the students pay the most attention when they teach classical microbiology experiments or combine the theory with their research work. Since that the microbiology speaker teachers of the College of Pharmacy in Hebei University have research and practical experience for many years, it is easy to combine their research projects to enrich the content of their classroom teaching. Focus on integrating the research concepts into teaching, explaining basic theory with research work, while teaching scientific thinking and reasoning; encourage students to observe related experimental group, discuss problems and even participate in research. Educational circles generally believe that college students carry out research and training program play an important role in the reformation of undergraduate teaching mode, Hebei University specially set up students' scientific and technological innovation projects, providing funds to support undergraduate research in extracurricular activities. Students are free to choose the subject or they can take part in the teachers' research projects directly for scientific research, to ensure the students' interest, the premise of improving the teaching quality is giving ithem greater freedom. In comparison we found that students who participate in research activities, their theory course test scores, overall quality, practical skills, writing papers, etc. have a more substantial increase, reaching the purposes of combining the basics of teaching with experimental practice, promoting research and teaching mutually.

2.4 Diversity of teaching methods

The basic theory and concepts of microbiology are abstract, experimental means are diverse, experimental techniques is more difficult to grasp, using traditional teaching methods is difficult to explain thoroughly. So we need to optimize a variety of teaching methods, make full use of modern teaching methods, take many different forms to teach[7]. Teachers collect a large number of online resources, produced a rich multimedia courseware, using computer-aided instruction (CAI), to performance what they talk about vivid, intuitive and visual, from deep to shallow, from difficult to easy, intuitive experimental techniques can half the work with double results to make students master the difficulties. In particular, to understand the basic structure of biological macromolecules, replication and transcription of various protein factors, interactions between RNA polymerase gene expression and the regulation of biological macromolecules such as DNA-DNA, DNA-protein, protein-protein and so on, it's very complicated. Using multimedia teaching to make the complex process at a glance. Moreover, some students don't understand can copy the multimedia courseware and observe repeatedly until they understand. In addition, teachers should use teaching aids or material object in teaching as far as possible, such as when to explain the DNA structure, use two strands of rope together to simulate DNA, the two strands are equivalent to two chains of DNA double helix. Rotating ropes in different ways so that students can understand the concepts of DNA helix form, supercoiled, negatively supercoiled better. Another example, in the teaching of the principle of polymerase chain reaction (PCR), move the PCR instrument to the classroom; in the learning of blue-white screening, move the blue-white tablet to the classroom, teaching on site, so that the students can understand the theory more sentimentally.

2.5 Carrying out the construction of network course

Network teaching play an important role in modern university education, compared to traditional teaching, the resources and the content of network teaching are rich; the content update fast; it can be used repeatedly; and it can be even downloaded and saved, making it easy to learn; it can find content quickly and easily, improving the learning efficiency. With the popularity of the network in universities, the basic Internet material conditions of the institutions are equipped, but the application in teaching is not very popular. The network teaching mainly includes the remote network teaching method, the on-line teaching method, the situational teaching method, the exquisite curriculum website method and so on. Exploring online teaching, combined with traditional teaching, primarily as a complement to traditional teaching. Because of the limit of conditions, mainly take exquisite curriculum website to organize teaching. Using the resources of traditional teaching to digitize, and build the network teaching platform according to the teaching needs, providing teaching content in class, taking full advantage of the shared network resource in addition. Their role should be changed promptly, to update the teaching content timely, adjust the arrangement of curriculum, guide students to learn to use the network resources to establish a network learning mode for their own, the teaching role is changed from dominated position to guidance position. Overall, it proves through practice, via the network teaching platform assists classroom teaching to enhance the effectiveness.

In summary, summarizing and exploring pharmacy microbiology teaching content and methods from many aspects, initially established a suitable professional pharmaceutical microbiology teaching system, it get the students' recognition and praise through years of practice. However, microbiology teaching is a continuous exploration and improvement process, will further deepen the reform, the introduction of bilingual education, special education, discussion of teaching and other teaching methods, will constantly optimize the teaching content, explore a better education system and improve the quality of teaching.

Acknowledgments

This work was funded by project 31201061 supported by the National Natural Science Foundation of China, project 2015144 supported by Education Department of Jilin Province, and project 2013625030 supported by Jilin City Sci-tech Bureau (China).

References

1. X.L. Qin, J.Y. Sun. Research and Practice on the Teaching Reform of Microbiology Courses. Journal of Jilin Agricultural Science and Technology College. (Jilin, China, 2011)
2. Y.T. Zhong, L.L. Ma, G.L. Gao. Research and Practice on Teaching Reform of Medical Microbiology Experiment. Jounal. microbiol. china. (Ganzhou, China, 2007)
3. B. Yu. Practice of and Research on Microbiology Teaching Reform for the Speciality of Biology Technology. Journal of Jiangxi Institute of Education. (Nanchang, China, 2000)
4. H. Liu. The Teaching Reform of Pharmacetutical Microbiology of Higher Vocational Pharmacetutical. China Educetlon Innovation Herald. (Haerbin, China, 2013)
5. H. Li, G. Yan, W.F. Dou. Teaching Reform and Practice of M icrobiology in the Biopharmaceutical Engineering. Guangzhou Chemical Industry. (Jiangsu, China, 2012)
6. C.Y. Sun, M.H. Ye. Teaching Reform and Exploration of Microbiology of Higher drug sub-professional. China EduOation Innovation Herald. (Guangdong, China, 2008)
7. H.Y. Zhao. Teaching Reform of Microbiology Independently Designed Experiment. China Journal Modern Drug Apply. (Haerbin, China, 2009)

The important role of the case-based learning in medicinal chemistry for training qualified pharmacy talents

Jingbo Sun, Yujie Ye, Shuo Yang, Zhenyao Han, Jianlei Yan, Cangjian Jia, Xiao Meng, Ke Pan*

Pharmaceutical College of Beihua University,
Jilin, 132013, China
**E-mail: sjb781219@163.com*
Corresponding author: Pan Ke

Medicinal chemistry is a very important subject in the pharmaceutical field, and it is also the necessary special knowledge and skill of the pharmacy job holders. Case-based learning (CBL) is an instructional design model that is a variant of project-oriented learning. It is the most appropriate teaching methodologies for training qualified pharmacy talents. Reasonable case–based learning can fully inspire the abilities and desires of the students' to analyze and solve problems. Students can broad their specialist knowledge, expand their field of vision during the process.

Keywords: medicinal chemistry; case-based learning; practicing qualification

1. Introduction

Pharmacy is a health care industry which connects life science and chemical science, and the pharmaceutical practitioners take an important responsibility of safety, reasonable, effective use of drug, and also bear the important mission for human beings fighting against disease. Therefore, the dissemination and study of the relevant professional knowledge are very important. The knowledge and technology of pharmacy includes pharmaceutical research and development, drug production, drug quality control, clinical application, supervision and management. And "medicinal chemistry" played an important position and role in the professional knowledge. In the construction of medicinal chemistry, the teacher should prepare earnestly and attach important teaching methods, teaching tools and experiment teaching so as to develop lots of excellent students. Among of these, only by adopting appropriate teaching method - case teaching, can we spread the knowledge of medicine to students well.

2. The importance of medicinal chemistry course

2.1. *Required courses of related pharmaceutical specialty*

"Medicinal chemistry" is a comprehensive discipline which found and invented new drugs, synthesized chemical drugs, demonstrated chemical characteristics of drugs, studied the interactions between drug molecules and the body cells (biological macromolecules). It is a professional course for pharmacy related majors (such as pharmaceutical, clinical pharmacy, etc.)[1] and the research content involves the discovery, development and identification of the new drugs, as well as to explain the mechanism of drugs action and bioactive compounds in molecular level. It can provide the necessary chemical basis and conditions for the drug analysis, qualitative research and quantitative research, impurity inspection, quality standard study of the subject, the exposition of "Pharmacology" mechanism, structure-effect relationship research, determination of drug form of "pharmaceutics", the choice of solvent and accessories, also for the “pharmacokinetics” and the drug analysis in vivo. Therefore, “Medicinal chemistry” is the bridge between life science and chemistry discipline, it is a key subject in the pharmaceutical field.

2.2. *The important role of "Medicinal chemistry" in the practice of exam*

"Medicinal chemistry" is a comprehensive discipline which found and invented new Pharmaceutical professionals in the vast majority of the employment need to have a certain qualification. They also need to take the appropriate qualification examination, such as medicine, pharmacists, licensed pharmacist qualification examination. "Medicinal chemistry" has a certain score in all kinds of qualification examination listed above.

“Medicinal chemistry” is a compulsory subject in Western medicine licensed pharmacist qualification examination requirements. Along with the reform of the national licensed pharmacist examination system, the examination syllabus of licensed pharmacists also have different requirements for this subject. 2015 is a major change in the examination of licensed pharmacists in our country, the content and the scope of the exam are very different from the previous ones[2]. Version of the syllabus for each subject content have restructured, among which the pharmacy specialized knowledge "(with the original professional knowledge of pharmacy and pharmaceutical professional knowledge of content: Pharmacy, pharmaceutical chemistry, pharmacology, pharmaceutical analysis)[3], the pharmacy professional knowledge" (for new subjects and included clinical drug treatment and clinical pharmacology)[4] and the pharmacy’s comprehensive knowledge and skills" (making large adjustments in the original content) [5]. Three subjects have great changes, adding more than 73%. At the same time, the new case selection question has greatly increased the difficulty of examination than before.

In view of the above reasons, the teaching of medicinal chemistry should updated continuously to meet the needs of the society and meet the needs of the students after graduation.

3. Advantages and features of Case-based learning

Case-based learning (CBL) is an instructional design model that is a variant of project-oriented learning. It was first proposed by the Harvard University in early twentieth Century[6]. The case teaching method, which is based on the simulation of some real scenes, is to make the students in the teaching method. Case teaching promote students' independent thinking and initiative them to analyze and solve problems, case teaching also do not need students to learn by rote, but to think in the understanding, to explore the knowledge, more importantly, case teaching is a social practice most time-saving, labor-saving. In the practice process, students can truly understand and appreciate the events, and put forward constructive opinions and suggestions for solving problem after scientific and rational use of knowledge. Therefore, the case teaching is very suitable for the current situation in our country in the teaching of "medicinal chemistry". If the teachers can foster their students a long term interest in the medicinal chemistry, then its' the whole "teach a man to fish" principle.

4. Application of case-based learning in medicinal chemistry

After graduation, a considerable portion of students who have graduated from pharmacy are engaged in the work in hospital pharmacy, drug marketing and drug retail. These people should have the qualification certificate, especially the qualification certificate of licensed pharmacist. And the 2015 edition of the new practice medicine examination outline of the new content, clinical drug use, "medicinal chemistry", the new "clinical medicine treatment" and "Clinical Pharmacology" are closely related to each other. These require us to introduce the knowledge of practical application in the teaching process of medicinal chemistry, highlighting the importance of case teaching in the teaching of medical chemistry.

According to planning for case-based learning, the format of a case often influences how to use it with students. Examples of cases with commonly encountered formats are provided with a brief description and likely implementation strategies. The choice of case teaching should be adapted to the purpose of teaching, which is practical, factually-based, complex problems written to stimulate classroom discussion and collaborative analysis, involves the interactive, student-centered exploration of realistic and specific situations. Cases have traditionally been used to teach decision making skills in professional education. More recently, cases are being used for learning medical science in problem based learning. The medical school use of cases differs from that in other professional schools in that problem based learning focuses on medical subject matter content more so than on decision-making. Different topics of the theoretical background are different, in view of the new form of the target, we should set up the case from the following several aspects to achieve a comprehensive knowledge:

4.1. *Drug guidance case for special masses*

The use of the drugs should be different according to the sex, age, physical status and so on of the patients. Such as the elderly or infants, the physiological changes of the body are significantly different from normal adults, the drug absorption, distribution, metabolism and excretion rate is different, it will affect the treatment effect and adverse reactions. Therefore, program attention should be paid more to such special group of people. Such as patients Lee, male, 76, fever because of cold, headache. Doctor Zhang made the prescription: Park (two tablets each time, three times each day, every tablet contains 325mg of acetaminophen), when necessary, take 0.3g paracetamol. As a clinical pharmacist, please review if the drug regimen is reasonable[7]. This is a typical irrational phenomenon for repeat of drug use, for the elderly, their metabolism is slow, repeated drug use is easy to cause liver damage in more than a conventional dose.

4.2. *Case of drug compatibility taboo*

There are many patients who need to use combination of multiple drugs at the same time, which requires the selected drugs can not produce obvious adverse reactions or side effects and enhance treatment effect at the same time. This requires a reasonable determination of the program. This problem should be noticed also in the validation of the doctor's treatment plan's reasonableness. Such as theophylline is metabolized by hepatic drug enzymes, and ciprofloxacin, erythromycin, phenytoin and other drugs will interfere with theophylline metabolism in the liver, which increase the blood concentration of theophylline and increase the toxicity at the same time. Therefore, when theophylline and these drugs are in combination use, we must balance the pros and cons. And when theophylline and corticosteroids are used in the treatment of moderate asthma, they played the advantages of combination use[8]. Similarly, ciprofloxacin combined with azithromycin also achieved good efficacy in the treatment of Pseudomonas aeruginosa resistant Pseudomonas aeruginosa infection[9]. All the pharmacy staffs have to remember the interactions of the common medicines and offer the patients and doctors a timely reminder that there are potential dangerous factors when these medicines are used together.

4.3. *Guidance case of special drug use*

Now, there are lots of drugs in clinical treatment, such as digoxin has good effect by strengthening myocardial contractility, slowing down heart rate, affecting myocardial electrophysiological characteristics and for the treatment of chronic cardiac insufficiency and some arrhythmia. However, the therapeutic index is low, safe range is narrow, such as digoxin and other cardiac glycosidic drugs, they easily lead to drug poisoning[10]. For this kind of medicine, it is very important to determine the scientific and rational drug delivery plan for ensuring the safety of the patients and the treatment effect.

5. Conclusion

To sum up, in order to cultivate pharmaceutical professional talents in the new situation, in order to make students grasp the knowledge of medicinal Chemistry, integrate chemical knowledge into other specialized courses and apply in future work well, we should actively carry out teaching reform, the use of newer, better teaching methods and means to enhance teaching level and students' comprehensive quality to make due contribution for the development of China's pharmaceutical industry.

References

1. You Qidong. Pharmaceutical Chemistry (Seventh Edition) [M]. 2014, people's medical publishing press.
2. National professional pharmacist qualification examination syllabus of the State Food and drug supervision and Management Bureau of licensed pharmacist

qualification certification center organization. (7th Edition) [M]. 2015, Chinese medical science and technology press.

3. Professional knowledge of Pharmacy (I) State Food and drug supervision and Management Bureau of licensed pharmacist qualification certification center to organize writing. (7th Edition) [M]. 2015, Chinese medical science and Technology Press.
4. The professional knowledge of Pharmacy (II) State Food and drug supervision and Management Bureau of licensed pharmacist qualification certification center to organize writing. (7th Edition) [M]. 2015, Chinese medical science and Technology Press.
5. Comprehensive pharmaceutical knowledge and skills (Seventh Edition] State Food and drug supervision and Management Bureau of licensed pharmacist qualification certification center to organize writing. [M]. 2015, Chinese medicine science and Technology Press.
6. Shi Meilan. Experience of Harvard Case Study of the National Education Administration College of education, 2005, 6 (2): 84-86.
7. Wang Lin, Xiao Huai. Case teaching in medicinal chemistry teaching [J]. Chinese medicine, 20, 2011 (24): 18-19.
8. Zhu Wenjun. Analysis of the clinical efficacy of theophylline combined with corticosteroid in treatment of moderate asthma [J]. Shenzhen Journal of Chinese and Western medicine, 2015, 25 (12): 75-76.
9. Liu qiming, Zhang zhi, Liao Hua. Efficacy of ciprofloxacin and azithromycin in the treatment of carbapenems carbapenem resistant Pseudomonas aeruginosa infection [J]. Journal of Chinese Pharmaceutical Sciences 2015, 17 (6): 591-593.
10. Li Hualiang, Yan Qiulan, Yi Linxin, et al. Analysis of risk factors for the elderly patients with digoxin poisoning [J]. Chinese prescription drugs, 2015, 13 (7): 18-19.

Session 2: Agriculture

Dynamic variation of groundwater evaporation and soil temperature under plastic mulch with openings

X. G. Xing[1], X. Y. Ma[1,*] and W. J. Shi[2]

[1]*College of Water Resources and Architectural Engineering, Northwest A&F University, Yangling, Shaanxi, 712100, China*

[2]*Institute of Water Resources and Hydro-electric Engineering, Xi'an University of Technology, Xi'an, Shaanxi, 710048, China*

E-mail: xg_xing@yeah.net

Groundwater evaporation (GE) and soil temperature (ST) are important for field crops. The characteristics of GE and ST using non-weighing lysimeters were thus studied. The study consisted of five treatments: plastic mulch with a water table at 1.5 m but with five percentages of open areas (POs), that is 0.78%, 2.50%, 5.00%, 10.00% and 100% (CK). And the ST in the 5, 10, 15 and 20 cm layers were monitored. The results indicated that (1) The salt crusts formed around punched holes could inhibit GE and soil water lost, leading to a non-linear relationship between accumulated GE and opening area. The accumulated GE reduced by 73.77%, 69.07%, 60.17% and 66.63% for the design with PO of 0.78%, 2.50%, 5.00% and 10.00%, respectively, comparing with CK. (2) Modified Averiyanov and Ye equations had a high accuracy in calculating GE under plastic mulch with PO less than 10%. (3) In summer of experimental area, GE was larger at night than in daytime, and its major consumption periods were around 2:00~10:00 and 18:00~22:00. Further analyses showed that GE was significantly affected by topsoil temperature. This meant that it was restrained when ST decreased with increasing depth; conversely, it was enhanced when ST increased with increasing depth.

Keywords: Groundwater Evaporation; Soil Temperature; Plastic-film Mulch; Opening Ratio.

1. Introduction

Salinity is a serious and chronic problem, significantly affecting agricultural productivity in arid regions, such as Xinjiang. In such region, changes in the phreatic water are largely determined by the supply of water to the upper soil layers and little by percolation to lower layers [1]. Consequently, the soluble salts in the groundwater or saline soil easily move upwards with water flow under strong evaporation, ultimately accumulate on the surface, and are likely to salinize the soil. Thus, an understanding is required of GE in Xinjiang region.

Mulching with plastic film tends to conserve water, inhibit salinization, and improve the physical condition of the soil. Plastic-film mulching has thus become popular in arid Xinjiang [2]. Plastic mulch, however, requires holes for seeding or irrigation. Different areas of plastic-mulch openings may thus differentially influence water flow and ST [3]. On the other hand, the effects of plastic film sheet should be taken into account when computing GE under plastic-film mulching condition, which is different from that under

* Corresponding author: X.Y. MA *E-mail address: xiaoyima@vip.sina.com*

bare soil condition. Averiyanov and Ye equations are widely used to calculate GE under bare soil, and based on which, the influence coefficient of plastic mulch is introduced, so the modified Averiyanov and Ye equations could be established.

The objectives of this study were (1) to identify the characteristics of daily GE and ST under plastic mulch with different POs, (2) to present new, simple and acceptable models for simulating GE under plastic mulch with openings, (3) to quantitatively analyze day- and nighttime GE.

2. Martials and Methods

2.1. *Study area and lysimeters*

The field experiment was conducted at the Bazhou Irrigation Experimental Station (41°36′N, 86°12′E and 950 m a.s.l.) in town of Xinir in the city of Korla and near the Peacock River, in Xinjiang of China. The study area has a typical temperate continental desert climate with a mean annual temperature of 11.5 °C and a mean of 3036 hours of sunshine per year. The mean annual precipitation and potential evaporation are approximately 58.6 and 2788 mm, respectively.

The lysimeter system was in the middle of a cotton field. This system consisted of steel soil containers, with a height of 1.5 m, and of rectangular soil containers, with lengths and widths of 1.5 and 1.0 m, respectively. The entire system comprised these soil containers, a water supply, and monitoring equipment. The monitoring system was installed mainly in an underground compartment, and the surface of each soil container was flush with the ground surface. Each soil container had a 0.3-m base, containing a filter layer of gravel and sand. A graduated Mariotte bottle connected to the container maintained a constant water table and monitored the amount of water (groundwater) supplied to each container. The experimental soil was collected from a nearby field, air-dried, sieved through a 2-mm screen, and then compacted into the lysimeters. All containers were equally irrigated before the trial to ensure identical initial conditions. The experimental soil was silt loam.

2.2. *Treatments and observations*

PO was calculated as the area of punched holes divided by the area of unbroken plastic film. There were five treatments in total based on five POs each having three replicates. The POs were set at 0.78%, 2.50%, 5.00%, 10.00% and 100% (i.e., bare soil, CK). Variable numbers of 9-cm^2 holes were cut evenly distributed on sheets of plastic mulch. The groundwater table depths were set at 1.5 m based on actual changes of groundwater table in the local area. Rectangular geothermometers were buried in all treatments at depths of 5, 10, 15 and 20 cm layer to monitor ST. The salts in soil or groundwater can move upwards with upflow under evaporation, resulting in the accumulating of salts in upper soil layers [4]. Water from a local crop-land channel was thus used as the source of groundwater to lessen the effects of salt transfer to the surface. This water had total dissolved solids of 0.74 g L^{-1}, which could be viewed as freshwater.

Observations commenced in July and continued to September, a typical high temperature period with no irrigation events or rainfalls. Daily evaporation was calculated from the amount of water lost from the Mariotte bottle recorded at 8:00 and 20:00 each day. The water surface evaporation, representing the daily intensity of atmospheric evaporation, was measured by a volatiliser using the weighing method. ST in each layer was obtained from the geothermometer recorded at 8:00, 14:00 and 20:00 each day. Five sunny days were selected for observations of diurnal variation. In such five days, 7/31, 8/7, 8/15, 8/24 and 8/28 (Month/Day), the GE and ST were observed at 2-h intervals from 8:00 to 8:00 of next day. Day- and nighttime were defined as 8:00~20:00 and 20:00~8:00, respectively.

2.3. *Models*

Averiyanov and Ye equations (Eq. 1~2) are the most widely used among many empirical models in computing GE under bare soil condition. However, as for plastic-film covering, the influence of plastic film sheet on GE should also be taken into account. So, the influence coefficient of plastic mulch was introduced (Eq. 3).

$$E = E_0 \left(1 - H/H_{\max}\right)^n \tag{1}$$

$$E = E_0 e^{-\alpha H} \tag{2}$$

$$C = E_{mulch} / E_{bare} \tag{3}$$

Where E is GE (mm d^{-1}); E_0 is water surface evaporation, or soil surface evaporation with groundwater depth of zero (mm d^{-1}); H_{max} is extinction depth of GE (m); H is actual groundwater table depth (m); n is a coefficient related to soil texture; α is attenuation coefficient; C is influence coefficient of plastic mulch; E_{mulch} is steady GE intensity under plastic covering condition (mm d^{-1}); E_{bare} is steady GE intensity under bare soil condition (mm d^{-1}). Modified Averiyanov and Ye equations were then established (Eq. 4 and Eq. 5).

$$E' = CE = CE_0 \left(1 - H/H_{\max}\right)^n \tag{4}$$

$$E' = CE = CE_0 e^{-\alpha H} \tag{5}$$

Where E' is GE after modification (mm d^{-1}).

3. Results

3.1. *Dynamic variation of GE*

Fig. 1 showed that the inhibitory effect of mulch on GE became stronger with evaporating time. Accumulated GE increased with increasing PO during the first 20 days, and the accumulated GE for different PO treatments showed the trend 10.00% > 5.00% > 2.50% > 0.78%. However, daily evaporation intensity for each treatment experienced some differences thereafter till off-test. At the end of experiment, the accumulated GE for different POs of 0.78%, 2.50%, 5.00% and 10.00% decreased by 73.77%, 69.07%, 60.17% and 66.63%, respectively, comparing with CK. This meant that the accumulated GE for different POs was ranked in the order of 5.00% > 10.00% > 2.50% > 0.78%. This is because during the first 20 days, the inhibitory effect of plastic mulch on water loss weakens with PO increasing from 0.78% to 100%, as expected. However, from then (about mid-August) on, air temperature continuously increases, finally resulting in much stronger atmospheric evaporation capacity than soil water-carrying capacity. Much soluble salt moves upwards with upward flow under strong evaporation, gradually accumulates exactly at punched holes after rapid water loss, and finally forms salt crusts. And this salt crust could in turn prevent water loss to some extent, which leads to less accumulated GE for 10.00% treatment than that for 5.00% treatment. Nevertheless, as for POs of 0.78%, 2.50% and 5.00%, the inhibitory effect of salt crusts on water loss is always weaker than evaporation intensity, though some salts also accumulated on surface. And under this circumstance, water transport is mainly affected by air evaporation, resulting in increasing accumulated GE with increasing PO from 0.78% to 5.00%.

In conclusion, as for GE under surface condition of plastic mulch with openings, especially for saline soil, atmospheric evaporation intensity may not the most important factor affecting soil water transport. And we should consider whether there is salt crust accumulating at punched holes, which is one of the significant differences in characteristics of GE under saline soil and non-saline soil, requiring further studies. And this also has been verified in some researches [5-7].

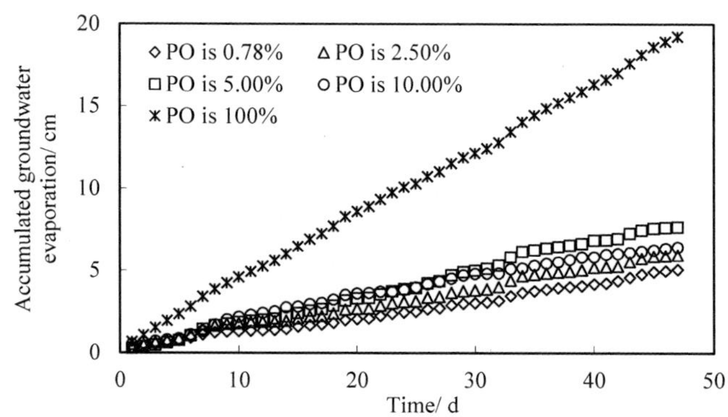

Fig. 1. Relationship between accumulated groundwater evaporation and time under different percentages of open area (POs).

3.2. *Calculation of GE*

The GE could be calculated by Eqs. 4~5, however, each model has to be calibrated twice based on the two meanings of parameter E_0 during GE calculation due to significant effect of E_0 on GE [8]. In this case, parameters n and α in Eqs. 4~5 were fitted based on mean daily GE for CK treatment. The results showed that for Averiyanov equation, n

equaled 1.53 when E_0 was selected as water surface evaporation, while it equaled -2.40 when E_0 was selected as soil surface evaporation with groundwater depth of zero, which was not reasonable. For Ye equation, α equaled 0.49 when E_0 was selected as water surface evaporation, while it equaled -0.76 when E_0 was selected as soil surface evaporation with groundwater depth of zero, which was not reasonable. In all, parameter E_0 should be considered as water surface evaporation when using these two equations.

Both modified Averiyanov and Ye equations had high simulation accuracy, with low relative error less than 10%. However, the simulation accuracy markedly decreased, with high relative error of about 60% when PO increased to 10%. This may be mainly attributed to the fact that the salt crusts at punched holes would significantly affect water loss. This indicated that when PO was beyond a limitation, simulation accuracy was likely to decrease, though influence coefficient of plastic mulch was introduced, especially for saline soil. And meanwhile, it also demonstrated that both modified Averiyanov and Ye equations could be well applied to GE calculation under plastic mulch with openings. However, the accuracy of calculated results obtained from these two equations may decrease when soil salt content was high and PO was large.

3.3. *Dynamic variation of ST*

3.3.1. *ST in different layers*

ST in each layer for all treatments showed similar trend. So, 5.00% treatment was taken as an example to analyze ST at 8:00, 14:00 and 20:00 (Fig. 2), and to further analyze mean ST and daily variation of ST from 0:00 to 22:00 (Figs. 3a, b).

At 8:00, ST increased with depth and was higher than air temperature (Fig. 2a), because air humidity was high and air temperature was low in the morning. And this resulted in lower temperature in shallow layers due to high level of synchrony in temperature between shallow layer and air. At 14:00, air temperature obviously increased, leading to increasing ST, which showed that ST gradually decreased as depth increasing (Fig. 2b). Furthermore, the temperature in the 5, 10, 15 and 20 cm layers increased by 17.39°C, 9.33°C, 4.17°C and 1.53°C, respectively, comparing with that at 8:00 (Fig. 3a). This indicated that the range of ST weakened with depth. This was mainly due to the fact that air temperature is higher at noon than in the morning, leading to a heat exchange at soil-atmosphere interface, and the ST in shallow layers grows faster than in deep layers. At 20:00, firstly, ST decreased as depth increasing, with small temperature difference (Fig. 2c). Secondly, the ST in deep layers decreased with very small extent because of insensitive response of deep-soil temperature to air temperature. However, topsoil temperature obviously decreased by 4.94°C due to the direct effect of air temperature on surface soil, while the temperature in other soil layers slightly increased (Fig. 3a). The highest ST in the 5 cm layer reached 40.45°C, resulting in heat transfer driven by large temperature gradient, which leads to slight increasing temperature in deep layers. This indicated that although air temperature decreased from morning to night, it still hardly affected deep-soil temperature. Furthermore, ST in shallow layers had more significant influences than air temperature on deep-soil temperature.

As for diurnal variation of ST on 7/31, 8/7, 8/15, 8/24 and 8/28 (Fig. 3b), average ST in the 5 and 10 cm layers increased by 20°C and 14°C from 8:00 to 16:00, peaking at about 43°C and 37°C, respectively. However, the average ST began to decrease thereafter, synchronizing with air temperature. The ST at 20:00 in the 15 and 20 cm layers peaked at about 34°C and 33°C, respectively, attributing to the delay between deep-soil temperature and air temperature. Besides, before 12:00, ST increased with increasing depth; from 12:00 to 22:00, although ST experienced rising and dropping along with air temperature, it always decreased with increasing depth, attributing to insensitive response and of deep-soil temperature to air temperature; after 22:00, variation tendency of ST was the same as that before 12:00.

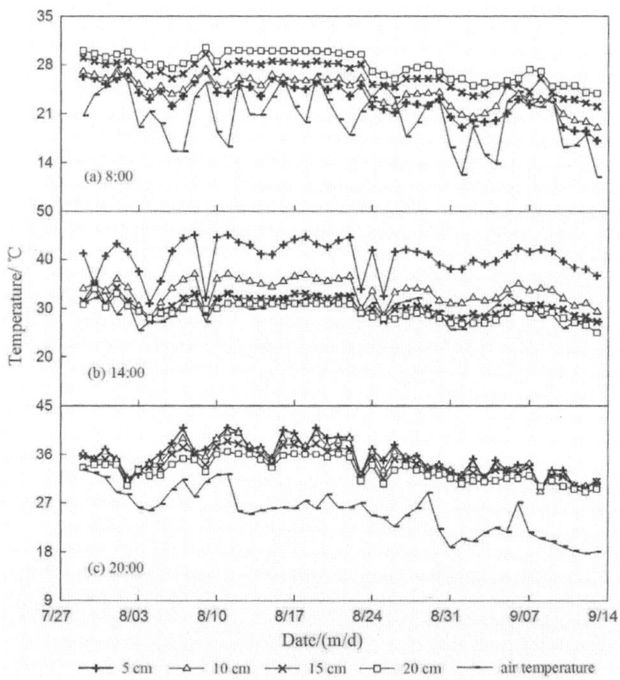

Fig. 2. Soil temperature in the 5-20 cm layers for 5.00% treatment.

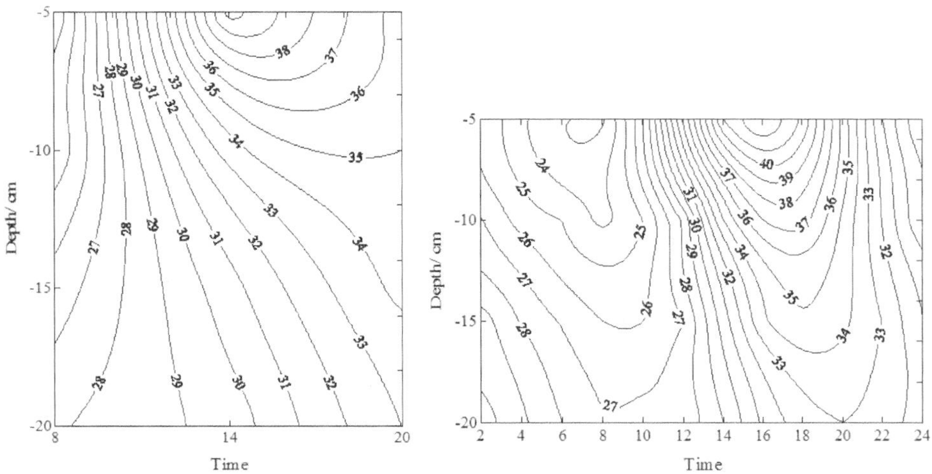

Fig. 3. Average soil temperature (a, left) and daily variation of soil temperature (b, right) in the 5-20 cm layers for 5.00% treatment.

3.3.2. *ST for different POs*

Similar with the above, the ST in the 5 cm layer (*T5*) for 0.78%, 5.00% and 100% (CK) treatments were taken as examples (Fig. 4).

Fig. 4 showed that ST had a stronger fluctuation at 14:00 and 20:00 than at 8:00. Furthermore, the *T5* at 8:00 for different treatments was ranked in the order of 100% < 5.00% < 0.78%, reaching 22.60°C, 23.06°C and 24.21°C, respectively. The order in *T5* at 14:00 was then 100% < 0.78% < 5.00%, reaching 37.67°C, 38.80°C and 40.45°C, respectively. The order in *T5* at 20:00 was again the same as that at 8:00, with temperature of 35.51°C, 35.88°C and 36.55°C, respectively. This was largely attributed to the fact that the ST under plastic mulching is higher than CK when air temperature is relative low at 8:00 and 20:00; however, when air temperature is relative high at 14:00, although ST under plastic mulching is higher, ST variations of different POs do not follow regular trend, which is described that ST increases with increasing covering area (i.e., decreasing PO). This indicated that the mechanism of plastic mulch varied with air temperature. And this meant that as for low air temperature, plastic film cut off heat exchange between soil and atmosphere, and this could prevent heat loss, resulting in increasing ST; as for high air temperature, plastic film may lower topsoil temperature to some extent, narrowing ST difference.

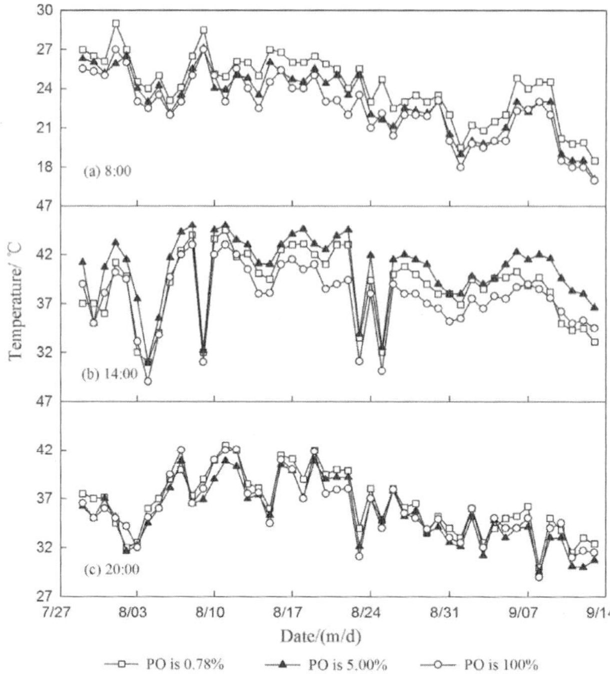

Fig. 4. Soil temperature in the 5 cm layer of different percentages of open area (POs).

3.4. *Relationship between GE and ST*

There was a closed relationship between GE and ST (Fig. 5). GE obviously decreased with increasing *T5* from 10:00 to 16:00 but increased with decreasing *T5* from 16:00 to 6:00, seeing a valley at 2:00. The GE for 0.78% and 5.00% treatments was much smaller than CK due to plastic film. Furthermore, Fig. 5 showed that major consumption periods of GE were 2:00~10:00 and 18:00~22:00. This meant that GE was larger at night than in daytime, and these results were consistent with those by Li [9] and Xia [10]. The GE during around 12:00~18:00 was nearly zero, largely attributing to the highest *T5*.

Fig. 6 showed that water surface evaporation during 12:00~18:00 accounted for larger proportion of the total amount due to very high air temperature, but GE during this period was nearly zero, just when ST decreased with increasing depth (i.e., $T5_{5cm} > T5_{10cm} > T5_{15cm} > T5_{20cm}$). This indicated that strong atmospheric evaporation intensity was likely to weaken GE. Water surface evaporation during 20:00~8:00 obviously decreased due to weak atmospheric evaporation intensity, but GE increased, just when ST increased with increasing depth (i.e., $T5_{5cm} < T5_{10cm} < T5_{15cm} < T5_{20cm}$). This indicated that relative weaker atmospheric evaporation intensity may promote GE.

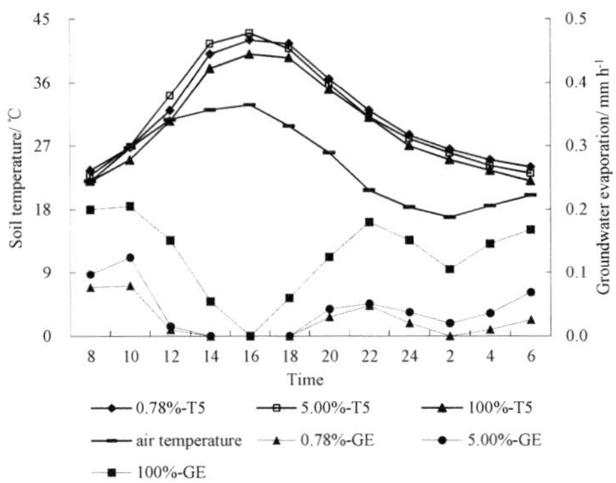

Fig.5. Daily variation (8/28) of soil temperature in the 5 cm layer (*T5*) and of groundwater evaporation (GE) of different percentages of open area.

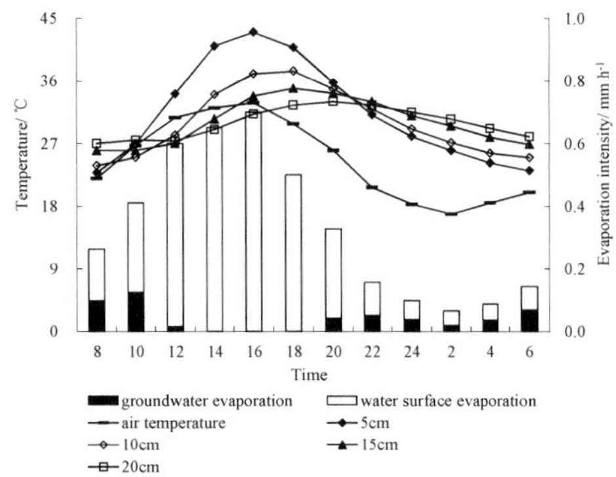

Fig.6. Daily variation (8/28) of soil temperature in the 5-20 cm layers and of evaporation intensity for 5.00% treatment.

4. Discussion

In this field experiment, a modified mulching pattern was introduced based on a traditional plastic-film mulching, namely, punching some holes on plastic films. However, in this study for saline soil, salt crusts may form at punched holes under strong evaporation in Xinjiang region, which had inhibitory effects on water loss (Fig. 1). Also, the salt crusts may promote heat preservation. Issues about mechanism of salt crusts need to be further studied.

The experiment showed that GE was larger at night than in daytime (Figs. 5~6), which deviated from the conventional understanding that higher daytime temperatures would increase the evaporation intensity. Our results may be due to one or more of three mechanisms. (1) To some extent, there is a delay between water supply and atmospheric evaporative demand. (2) As solar radiation increases after midday, the water-carrying capacity of the soil decreases due to the lower soil-water content in topsoil layers. The evaporative surface may become slightly deeper due to the disrupted capillary action in the upper soil layer. And this deeper evaporative surface would lead to less evaporation of groundwater [11]. As solar radiation decreases near 18:00, the water-carrying capacity of the soil exceeds the atmospheric evaporative demand, which would increase the soil-water content in the surface layers. The evaporative surface, however, would return closer to the soil surface when the disrupted capillary action recovered [12]. (3) The surface tension at the interface of the soil water and the atmosphere increases at night under the influence of the lower temperatures of the surface soil, so the soil-water suction increases and the soil-water potential decreases. This mechanism may lead to the flow of soil water from low-suction sites to high-suction sites, namely, from warm to cold sites (i.e., from underground to the surface). These mechanisms suggest that groundwater flows upwards not only in daytime with high temperatures but also at night, which was even stronger in our experiment. Our results could thus benefit the reasonable scheduling of irrigation.

5. Conclusion

These field experimental results demonstrated that soluble salts in saline soil could move to the surface and form salt crusts, which in turn prevented further water loss. Modified Averiyanov and Ye equations were recommended to calculate GE under plastic mulch condition with PO less than 10%. Parameter E_0 in these two models should be viewed as water surface evaporation. GE was larger at night than in daytime during the summer of Xinjiang, and as for plastic-film-covering, its major consumption periods were around 2:00~10:00 and 18:00~22:00. GE was significantly influenced by the ST in shallow layers, and it may be restrained when ST decreased with increasing depth, but it may be enhanced when ST increased with increasing depth. ST in shallow layers was sensitive to air temperature, and it had more obvious effects than air temperature on deep-soil temperature. Plastic-film mulch could improve ST, creating good water-heat environment for crop growth, but its mechanism of action was varied with air temperature.

Acknowledgements

This work was financially supported by National Sci-Technology Supporting Program for "Twelfth Five" Plan of China (grant no. 2012BAD08B01), Special Scientific Research Fund of Public Welfare Profession of MWR, China (grant no. 201301016) and National Natural Science Foundation of China (grant no. 51379173, 51279167).

References

1. Y. Liu, T. Yamanaka, X. Zhou, F. Tian, and W. Ma, Combined use of tracer approach and numerical simulation to estimate groundwater recharge in an alluvial aquifer system: A case study of nasunogahara area, central Japan, *J. Hydrol.* **519**, 833-847 (2014).
2. Q. Zhang, S. Wang, L. Li, M. Inoue, J. Xiang, G. Qiu and W. Jin, Effects of mulching and sub-surface irrigation on vine growth, berry sugar content and water use of grapevines, *Agr. Water Manage.* **143**, 1-8 (2014).
3. D. Yang, Z. Bian, K. Zhang and S. Lei, Study on modeling soil moisture responses under steady-state evaporation conditions from a shallow water table, *J. Sichuan Univ.* **45**, 13-18 (2013).
4. M. Zammouri, T. Siegfried, T. El-Fahem, S. Kriâa and W. Kinzelbach, Salinization of groundwater in the nefzawa oases region, Tunisia: Results of a regional-scale hydrogeologic approach, *Hydrogeol. J.* **15**, 1357-1375 (2007).
5. X. Xing, W. Shi, and Q. Wang, Analysis on laws of soil water and salt distribution characteristics and cotton emergence rate in seeding stage under different opening ratios, *J. Xi'an Tech. Univ.* **30**, 96-101, 107 (2014).
6. W. Shi, X. Xing, Z. Zhang, and Q. Wang, Groundwater evaporation from saline soil under plastic mulch with different percentage of open area, *J. Food Agric. Environ.* **11**, 1268-1271 (2013).

7. O. Grünberger, P. Macaigne, J.L. Michelot, C. Hartmann, and S. Sukchan, Salt crust development in paddy fields owing to soil evaporation and drainage: Contribution of chloride and deuterium profile analysis, *J. Hydrol.* **348**, 110-123 (2008).
8. Q. Fu, J. Zhang, and Q. Wang, Impact of E_0 value on calculation accuracy of phreatic evaporation empirical formulae, *Arid Land Geog.* **30**, 820-825 (2007).
9. X. Li, J. Zhou, M. Jin, and Y. Liu, Experiment on evaporation of high-TDS phreatic water in arid area, *J. Water Res. Eng.* **23**, 6-10 (2012).
10. Z. Xia, B. Guo, and H. Jiang, Effects of temperature variation on soil water movement and water exchange between soil water and phreatic water, *Adv. Water Sci.* **6**, 15-22 (1995).
11. Y. Ma, and X. Li, Water accumulation in soil by gravel and sand mulches: Influence of textural composition and thickness of mulch layers, *J. Arid Environ.* **75**, 432-437 (2011).
12. H. Li, W. Wang, and B. Liu, The daily evaporation characteristics of deeply buried phreatic water in an extremely arid region, *J. Hydrol.* **514**, 172-179 (2014).

A measurement study on the residents' attitude towards ecological compensation in tourism destination: A case in Li river basin

D.D. Lu[1,a], Y.D. Zhong[2,b,*], S.Y. Chen[3,c]

[1]*Tourism College, Central South University of Forestry and Technology, Changsha Hunan, China*
College of Business Administration, Guangxi University of Finance and Economics, Nanning Guangxi, China
[2]*Tourism College, Central South University of Forestry and Technology, Changsha Hunan, China*
[3]*College of Business Administration, Guangxi University of Finance and Economics, Nanning Guangxi, China*
E-mail: [a]ludandan163@163.com, [b]yongde65@yahoo.com.cn, [c]thecsy@126.com
[*]*Corresponding author*

The dimensions, contents and influencing factors of the residents' attitudes are constructed as the starting point and foundation of the ecological compensation attitude study. This paper chooses Li River Basin as the research object, and measuring the attitude of local residents as the starting point, by using factor analysis to analyze the dimensions, contents and influencing factors of the residents' attitudes towards ecological compensation. The result shows that the dimensions of Li River Basin residents' attitude towards ecological compensation can be divided into three areas: "compensation cognition", "environmental benefits", and "economic interests"; Li River Basin residents strongly hope to get ecological compensation, meanwhile, hold neutral attitude towards the environmental benefits and economic interests brought by Tourism. There are significant differences between the residents' attitudes towards ecological compensation in different demographic groups of Li River Basin.

Keywords: Ecological compensation, Attitude measurement, Local residents, Li River Basin.

1. Research Background

As a new environmental management system, the study on ecological compensation has been a hot issue in multiple disciplines and fields both at home and abroad. Good ecological environment is the foundation for Li River tourism development. In order to protect the ecological environment and construction, local residents along the Li River lost too many opportunities to develop Li River tourism, which kept them in a lower standard of living. The implementation of ecological compensation to the upper water conservation forest and the coastal residents in the middle of the Li River guarantees Li River tourism in sustainable, stable and healthy development. Therefore, based on the concept of ecological compensation and the factor analysis, an empirical research on local residents' perception of ecological compensation is normatively conducted in this paper.

2. Overview of Li River Basin

Situated in the northeast of the Guangxi Zhuang Autonomous Region, 24°18′N~25°41′N latitude, 109°45′E~110°40′E longitude, Li River Basin covers an area of 5,660 km^2, averaging about 34km in width and 115km in length. The research focuses on Wugui river in Xing'an upstream to the Li River in Yangshuo downstream. The area is abundant in world-class tourist resources, among which the Li River (one of the national 5A grade scenic spots) is becoming an important tourist attraction in China.

3. Research Design

3.1. *Questionnaire Design*

Residents' Attitude questionnaire uses the five-point Likert scale method. Taking into account of tourism, measuring dimension of local residents' perception of ecological compensation mainly bases upon the following four aspects: (1) benefit; (2) cost; (3) perception of social status; (4) compensation cognition and behavior tendency. Questionnaire design and sample survey were carried out in preparatory work for research. The author asked 30 specialists and students from Tourism Management major to participate in the sample survey and make appropriate adjustments before contributing the questionnaire to local residents.

3.2. *Formal Survey and Sample Recovery*

The study on measuring the local residents' attitude towards ecological compensation in Li River Basin consists of two parts: demographic information and relevant options to ecological compensation. The Likert scale was employed in the questionnaire by ranking the responses accordingly: strongly agree=5 points, strongly disagree=1 point. From May 1, 2015 to May 9, 2015, 230 formal questionnaires were distributed by project group members at random to households and immediately recovered in Xing'an, Lingqu, Guilin urban district, Yangdi, Xingping and Yangshuo. Finally, a total of 200 responses as valid was received out of recovery of 225 questionnaires, with a valid rate of 88.89%.

3.3. *Data and Analysis*

3.3.1. *Sample Data*

Article uses SPSS 19.0 software to analyze the data. In the selected sample, 54% of the respondents are male, 46% are female; most of the respondents are 19-29 years old, amounting to 61.5%; most of residents have junior high school diploma or above; the overall monthly income of residents in Li River Basin is low, 45% less than 1,000 yuan, 44.5% ranging from 1,000 to 2,000 yuan and 10.5% more than 2,000 yuan. As the largest proportion in household income, agriculture accounts for 33.5%, followed by tourism and industry, holding 23.5% and 18.5%, others far less than 30%.

3.3.2. *Dimension Analysis of the Local Residents' Attitude Towards Ecological Compensation*

From Table 1 we could see that 14 items were fell into the corresponding common factors. The accumulative variance of the first three factors was 64.351%, which proved that the first three common factors basically contained the key information of variables, therefore, the first three were chosen as main factors. Meanwhile, according to the property of each item, they could be respectively named "compensation cognition factor", "environmental benefits factor" and "economic interests factor", which became the most important influential factors arranged in descending order on local residents' attitude to ecological compensation.

Tab. 1. Factor analysis of residents' attitudes towards ecological compensation in Li River Basin

Attitude dimension	Item	Factor Loadings			Eigen values	Variance Explained
Compensation cognition	20	0.844	0.065	0.095	5.399	38.561%
	19	0.820	0.216	0.019		
	21	0.776	0.004	0.250		
	22	0.765	-0.026	0.168		
	18	0.501	0.294	0.482		
Environmental benefits	14	-0.025	0.757	0.245	2.306	16.474%
	13	-0.021	0.752	0.152		
	17	0.261	0.709	0.192		
	12	-0.141	0.682	0.222		
	15	0.363	0.647	0.034		
	16	0.254	0.502	0.280		
Environmental benefits	7	0.184	0.191	0.892	1.304	9.316%
	6	0.158	0.304	0.821		
	8	0.135	0.262	0.812		

3.3.3. *Content Analysis of the Local Residents' Attitude Towards Ecological Compensation*

From statistical data of the questionnaire we could assess the local residents' attitude towards ecological compensation in Li River Basin. They kept a positive attitude to compensation while perceived neutrally to the environmental and economic effects caused by tourism. However, the residents had an awareness of how important the value of ecosystem services in Li River Basin was and had a profound understanding of the relationship between tourism development and Li River's protection. They realized that good environment could help tourism to make higher profits. Hence, in order to protect the ecological environment and construction along Li River Basin, they should gain ecological compensation.

Tab. 2. Perception analysis of residents' attitudes to ecological compensation in Li River basin

Factor & Item	Mean	Std. Deviation
Compensation cognition	**3.590**	
The more compensations the better;	3.46	0.981
Making a contribution to protection of Li River should be compensated;	3.41	0.968
Income of tourism should be one of sources of compensatory funds;	3.63	0.948
It is necessary to turn this compensation into laws and regulations;	3.58	0.927
If we don't protect well, economic benefits of tourism will be minimized;	3.87	1.095
Environmental benefits	**3.162**	
Tourism destroys the natural environment and landscape quality;	2.96	1.002
The development of tourism disturbs local residents' daily lives;	3.03	1.051
The local natural envirnment is maintained very well;	3.40	1.056
Tourism activities will cause conflicts between local residents and tourists;	2.86	1.077
Residents' living has a strong dependence on the resources of the Li River;	3.56	1.115
Local sanitary environment is good and looks well;	3.32	1.096
Economic interests	**3.210**	
Tourism development has increased personal income;	3.17	1.085
Tourism development make it easier for one to find a job;	3.29	1.176
Tourism development contributes to local living standard.	3.17	1.016

Note: 1=Strongly disagree, 2=Disagree, 3=Neither agree nor disagree, 4=Agree, 5=Strongly agree

3.3.4. *Different Attitudes and Discriminate Analysis of Individual Variable about Ecological Compensation in Li River Basin*

The heterogeneity factors in this paper include gender, age, qualification, income and sources of household income. MANOVA was employed to analyze different attitudes and discriminating indicators of residents about ecological compensation in Li River Basin. Different Attitudes of Residents to Ecological Compensation with Different Ages MANOVA Model: Wilks' Lambda=0.726, F=5.489, P=0.000<0.001. Different Attitudes of Residents to Ecological Compensation with Different Qualifications MANOVA Model: Wilks' Lambda=0.757, F=6.358, P=0.000<0.001. Different Attitudes of Residents to Ecological Compensation with Different Income MANOVA Model: Wilks' Lambda=0.719, F=5.643, P<0.001. Different Attitudes of Residents to Ecological Compensation with Different Sources of Household Income MANOVA Model: Wilks' Lambda=0.580, F=9.731, P<0.001.

Tab. 3. Differentiation characterization of residents' attitudes to ecological compensation

Discriminating indicator		Differentiation characterization
Primary	Secondary	
	gender	No effect on residents' attitude to ecological compensation.
	age	Over 30 years old residents strongly hoped to get ecological compensation; residents aged between 30 and 60 were more conscious of economic and environmental interests brought by tourism, while young people kept a rather neutral attitude towards ecological compensation.
	educational status	There were significant differences between educational status and residents' attitude to ecological compenation. On the one

Residents' socio-demographic characteristics		hand, the higher degree they gained, the stronger will to obtain the compensation. On the other hand, those who held junior middle school diploma or below were extremely intense about environmental benefits. Meanwhile, residents that were highly educated had a strong awareness of economic interests. Those had indicated that the lower the educational status was, the stronger dependence the residents had upon mountain, forest and water resources in Li River Basin.
	income	In terms of residents with different income, their attitudes to ecological compensation were presented in compensation cognition and economic interests. Both high and low income groups strongly hoped to get ecological compensation. Moreover, the higher income they received, the stronger consciousness they could have.
	sources of household income	The residents, who were engaged in the service industry specially the tourism, held a rather positive attitude to dimensions of compensation cognition, environmental benefits and economic interests, whereas those whose family income had dominated by agriculture and industry maintained quite neutrality.

4. Conclusion

The local residents' attitude towards ecological compensation in Li River Basin can be divided into three dimensions: "compensation cognition", "environmental benefits" and "economic interests". The residents strongly hope to get ecological compensation, meanwhile, hold a neutral attitude towards the environmental benefits and economic interests brought by tourism. Demographic characteristics, such as age, income, educational status and the source of household income significantly lead to difference of residents' attitude to ecological compensation, among which, ecological compensation and economic interests become the most essential part. In conclusion, those above illustrate that the local residents have achieved a certain level of understanding on ecological compensation; in addition, the fundamental driving force for economic development of Li River Basin comes from tourism, therefore, the residents attach great importance to economic benefits in hope of getting rid of poverty and becoming prosperous; meanwhile, residents' environmental awareness need to be further improved in the future.

Acknowledgement

This research was financially supported by the National Science Foundation (71163004; 41461110).

Appendix

Residents' Attitude Towards Ecological Compensation Questionnaire

Basic Information

1. Gender:□A. male □B. female
2. Income:□A. less than 1000yuan □B. 1000-2000yuan □C. 2000-3500yuan □D. 3500-5000yuan □E. more than 5000yuan
3. Age:□A. under 18years old □B. 19-29 years old □C. 30-45 years old □D. 46-60 years old □E. over 60 years old
4. Educational Status:□A. below junior middle school □B. senior middle school □C. Vocational secondary school& three-year college □D. bachelor degree or above
5. Sources of Household Income:□A. agriculture □B. industry □C. tourism □D. other service industries □E. others

Residents' Attitude Towards Ecological Compensation

(Five-point Likert Scale Method, Note:1=Strongly disagree, 2=Disagree, 3=Neither agree nor disagree, 4=Agree, 5=Strongly agree)

		1	2	3	4	5
6.	Tourism development make it easier for one to find a job;					
7.	Tourism development has increased personal income;					
8.	Tourism development contributes to local living standard;					
9.	Local utilities, communication, transportation have been improved;					
10.	Tourism development to improve the appearance and health of local;					
11.	Local price increasing, cost of living increasing					
12.	Tourism activities will cause conflicts between local residents and tourists;					
13.	The development of tourism disturbs local residents' daily lives;					
14.	Tourism destroys the natural environment and landscape quality;					
15.	Residents' living has a strong dependence on the resources of the Li River;					
16.	Local sanitary environment is good and looks well					
17.	The local natural environment is maintained very well;					
18.	If we don't protect well, economic benefits of tourism will be minimized;					
19.	Making a contribution to protection of Li River should be compensated					
20.	The more compensations the better;					
21.	Income of tourism should be one of sources of compensatory funds;					
22.	It is necessary to turn this compensation into laws and regulations;					

References

1. S. Richard. Host perceptions of tourism: A review of the research[J]. Tourism Management, 2014, 42(6):37-49.
2. V.S. Alfonso, O.V. Patricia, C.M. Júlio, et al. Residents' attitude and level of destination development: An international comparison[J]. Tourism Management, 2015, 48(6):199-210.
3. A.G. Fernando, B.V. Antonia, C.M. Rafael. Resident's attitudes towards the impacts of tourism[J]. Tourism Management Perspectives, 2015, 13(1):33-40.
4. R. Nunkoo, D. Gursoy. Residents' support for tourism: an identity perspective[J]. Annals of Tourism Research, 2012, 39(1):243-268.
5. Y.Z. Zhao, D.H. Li, M.L. Huang. On the residents' perception and attitudes towards tourism development in tourist destinations overseas: A review[J]. Tourism Tribune, 2005, 20(4):85-92.
6. B. Florence, S.G. Rudolf, J.C. José. Valuation of tropical forest services and mechanisms to finance their conservation and sustainable use: A case study of Tapantí National Park, Costa Rica[J]. Forest Policy and Economics, 2009, 11(3): 174-183.
7. S.V. Lankford, D.R. Howard. Developing a tourism impact attitude scale[J]. Annals of Tourism Research, 1994, 21(1):121-139.
8. D. Gursoy, C. Jurowski, M. Uysal. Resident attitudes: A structural modeling approach[J]. Annals of Tourism Research, 2002, 29(1): 79-105.
9. V. Teye, E. Sirakaya, S.F. Sönmez. Residents' attitudes toward tourism development[J]. Annals of Tourism Research, 2002, 29(3): 668-688.
10. M.L. Wu. Statistical analysis questionnaire -SPSS practical operation and application [M] Chongqing: Chongqing University Press, 2012:184-192.

Research on China's knowledge sharing system: Agricultural knowledge sharing

Xi Wang*, Liliana Mitkova
Central University of Finance and Economics
39, South College Road, Haidian District, 100081, Beijing, China
Institut de Recherche en Gestion, University of Paris Est Marne la Vallée (UPEM)
Bât. BDE – Rue Gallilée – 77420 Champs-sur-Marne - France
miracle_wangxi@163.com, Liliana.Mitkova@u-pem.fr
**Corresponding author*

Abstract: Knowledge sharing is one of the major driving forces to success and in the future, the development of agriculture depends to a great extent on the opening degree of knowledge sharing. This essay emphasizes on studying from the system and the agricultural practice. On the matter of the system, we study the series of policies the government takes to support the knowledge sharing system based on the opening of industry, university and research. On the matter of agricultural practice, we have analysed the composition and the future developing trends of the knowledge sharing system in the present development of agriculture in China.

Keywords: Knowledge Sharing, Agricultural, Chinese Innovation Model.

1. Introduction

In the knowledge management, knowledge sharing is analyzed by various approaches that refer to how organizations create, retain, and share knowledge (Bogers, 2012). Recently, the studies of knowledge sharing, which is the means by which an organization obtains access to its own and other organizations' knowledge, has emerged as a key research area from a broad and deep field of study on technology transfer and innovation, and more recently from the field of open innovation management (Cummings, 2003). Cummings (2003) suggests three types of researches concerning the knowledge-sharing activities. First, the analyses of the form and the location of the knowledge investigate how they affect the sharing processes. Second, the studies concerning the types of rules of engagement and managerial practices adopted by the parties evaluate in that they can shape both the flows of resources and knowledge between the parties. Third, the researches of specific knowledge-sharing activities are important in that they are the means through which the parties seek to facilitate knowledge sharing. However, despite these studies, how to effectively manage knowledge sharing in open innovation perspective is not yet fully understood (Enkel et al., 2009). The main purpose of this article is focused on the second approach, offering an overview of the different measures and tools implement in the China innovation system in order to support the knowledge sharing.

A company's capacity to make use of knowledge sharing thus to adapt the open innovation model has been viewed as a source of sustainable competitive advantage of the Chinese firms and their successful integration in the international technological landscape. China is determinate to enhance the global competitiveness by its firms and to be defined as an "innovation-oriented" country by 2020 and a "leading science power" by 2050 (Chen and Li, 2011). In general, the innovation policy has been promoting by strengthen the internal R&D capacities and recently by adopting more open approach toward the international technologies in order to catching-up and improve the internal innovative performance (Chen and Qu, 2003). The recent literature debates the relative importance of different tools of the Chinese national policies as a major determinant of Chinese innovation system improvement (Hung, 2009; Liu, 2010). The academic works put the accent on the external factors to implement the open model in China (Savitskaya et al., 2010), especially the development of the intellectual property system and government policy (Deng, 2009).

The paper is organized in two parts. The first, offers the theoretical background of knowledge sharing into open perspective. The second part proposes the overview of Chinese agricultural knowledge sharing framework.

2. Theoretical Background

Knowledge Sharing, working as a major supporting process for knowledge management and open innovation (Llopis-córcoles, 2011), is an activity through which knowledge (i.e., information, skills, or expertise) is exchanged among friends, families, communities, or organizations (Bukowitz et al., 1999). It is a two-way or dual process, including both the supply of new knowledge and the demand for new knowledge, enquiring and contributing to knowledge through activities such as learning-by-observation, listening and asking, sharing ideas, giving advice, recognizing cues, and adopting patterns of behavior (Bosua and Scheepers, 2007). Authors offer others definition of the knowledge sharing as the means by which an organization obtains access to its own and other organizations' knowledge from a more self-centered or inbound view, a key research sector of technology transfer, innovation and strategic management (Cummings, 2003). Or it refers to the provision of task information and know-how to help others and to collaborate to solve problems, develop new ideas, or implement policies or procedures (Wang and Noe, 2010).

The practices and initiatives of knowledge sharing, in terms of organizational and individual learning, often form a key component of knowledge management programs (Alavi and Leidner, 2001; Riege, 2005). According to Nonaka and Takeuchi (1995), knowledge sharing also acts a critical stage in knowledge transfer process. Some see knowledge management and knowledge transfer as processes that undertake largely for the purpose of creating a knowledge sharing culture, fostering collaboration and communication, and so in turn enhancing organizational innovation (Liebowitz, 2002). Therefore, the coordination of the process within the organization became the key success factor for the company.

2.1 Coordination mechanisms of knowledge sharing

According to Binz-Scharf (2003), the coordination mechanisms of knowledge sharing system could be classified into two different level: Individual and organizational level coordination. Both coordination forms can be done by formal or informal mechanisms. The formal type of organizational coordination mechanism defines as "hierarchy"; on the other hand the informal type of the mechanism defines as "social networks".

The hierarchical structures are used in order to coordinate knowledge processes in complex organizations comprised of multiple specialized units (Tsai, 2002). The analysis of hierarchical structure as a coordination mechanism of communication has played an important role in organizational research (Powell, 1990). Tsai (2002) also discussed the role of centralization in hierarchical structure, and deemed that it might restrain knowledge sharing in different units unless they were required to do that by the higher authority.

Intra-organizational networks "subsume relations between and among actors under a governance structure that handles conflict resolution and channel behavior," (Fountain, 1999). Knowledge networks are characterized by fluidity of actors and their ties. This fluidity can be defined as "all actors within an observable knowledge network have their own cognitive knowledge networks, which refer to their perceptions of the overall observable knowledge network" (Contractor et al., 2000).

The organization and the coordination of knowledge sharing is also influencing from two different factors that we can mention as institutional and organizational level factors.

2.2 Institutional and organizational factors of knowledge sharing

Knowledge sharing requires a focus on more than simply the transfer of the specific knowledge and needs to focus on structuring and implementing the arrangement in a way that bridges both existing and potential relationship issues, and examining the form and location of the knowledge to ensure its successfully achievement (Cummings, 2003). The knowledge sharing can be supported by acting on certain contextual and organizational variables that influence knowledge flows (Binz-Scharf, 2003).

The early literature demonstrates knowledge sharing dilemma grounded on the analysis of institutional (macro-level) variables (Cummings, 2003), and many insights into how broad environmental factors may affect knowledge sharing activities (Rousseau, 1985). Researchers have contributed to identify the patterns of national, institutional and organizational factors that best support knowledge transfer or knowledge diffusion in general among firms (Mansfield, 1975). The four main factors affecting knowledge transfers, which included the activities and modes, organizational profiles of the parties, broad environmental factors and technological absorptive capacity of host country, and emphasized the importance of national and organizational framework building (Contractor and Sagafi-nejad, 1981). The contextual variables affecting knowledge flows are the environment, organizational structural characteristics, and organizational cultural norms (Gupta and Govindarajan, 1991). The existence and richness of knowledge sharing channels could determine the attitude toward knowledge sharing behavior (Kwok and Gao, 2005).

The organizational-level research highlighted lower economic costs (Mansfield, Romeo and Wagner, 1979), time cost (Mansfield and Romeo, 1980), relationship ties (Hansen, 1999) business strategy (Grant, 1996), tacitness and embeddedness (Zander, 1991) recipient's absorptive capacity and causal ambiguity (Szulanski, 1996), facilitating activities (Lall, 2000), and categorized subsidiaries (Gupta and Govindarajan, 1991) as key variables.

3. Development of China's current knowledge sharing system

In order to illustrate the building of knowledge sharing system in China this article focused on the main practices at institutional and organizational level.

3.1 From the institutional perspective

Academics, policy-makers and practitioners are increasingly interested in the roles that public sectors playing in the effective management of knowledge sharing (Currie and Suhomlinova, 2006). In pursuit of systematic improvement, public sector institutions transform themselves into intermediaries supporting the sharing process. In this vein, the Chinese government becomes an active participant in the national knowledge system, providing assistance of the sharing of the knowledge flow between the government institutions, local players and firms. It includes first, the establishment of an enabling environment via appropriate laws, especially IP rights and regulations for knowledge sharing. Second, the development of an internal "horizontal" knowledge sharing system between different governmental functional institutions at strategic level. Third, the founding of vertical sharing coordination between different levels of players within the government institutions. To implement this strategic aims the Chinese government introduces various actions. For example, the government has promulgated and implemented the National Knowledge Management Standard GB/T23703. As well, in July 2009, the Chinese Ministry of Science and Technology (MOST) and other five ministries had jointly launched the National Technical Innovation Engineering General Plan. Moreover, in December 2009, they issued Building and Implementing Measures on Promoting the Development of Industry Technology Innovation Strategic Alliance (Trial). Additionally, in January 2010, MOST issued Notice on Choice of a Batch of Industrial Technology Innovation Strategy Alliance Pilot Work.

At the local level, the local institutions also take active roles in the process of building the knowledge sharing structure. In this sense, they set up special funds for stimulating industry-university-research cooperation, mainly using for innovation programs and R&D collaborations. For example, the subsidies and incentives for S&T intermediary service institutions (e.g. quality measuring, testing and certification services, and technology consulting and technology trade intermediary services) are established. Moreover, the subsidies for investment into industry-university-research cooperation programs are put in place. The subsidies for S&T business incubators development are introduced.

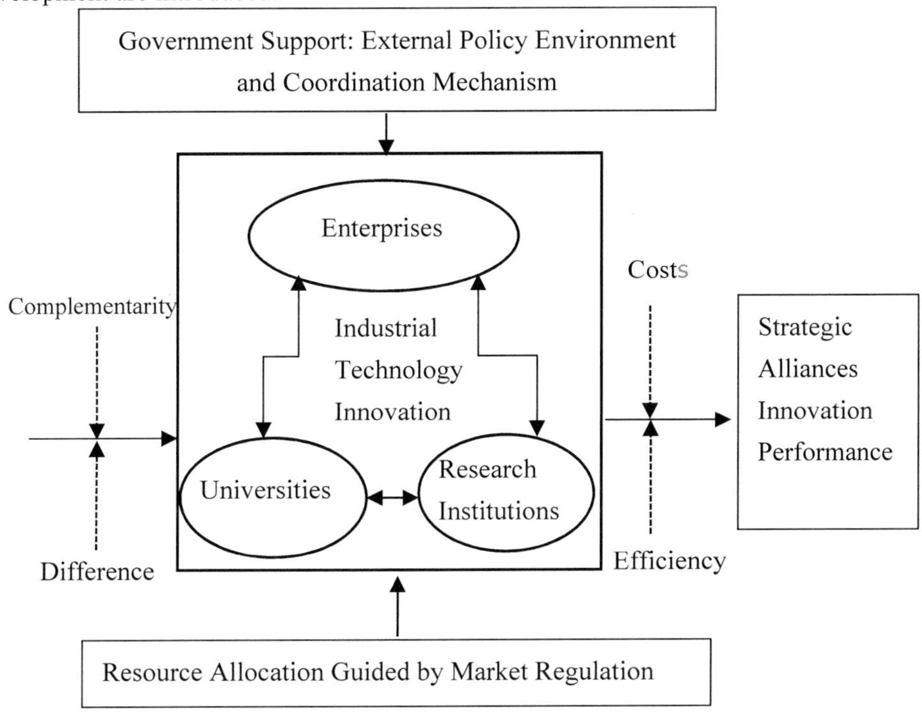

Fig. 1. Knowledge Sharing Mechanism: Framework.

Hence, the China's knowledge sharing system from the institutional perspective, mainly generalizing the government`s role in the system by optimizing the external policy environment and coordination mechanism (see Figure 1). The main players in this knowledge system are enterprise, university, research institutions which interactions are analyzed in an organizational perspective.

3.2 From the organizational perspective

3.2.1 General view of knowledge sharing modes

1) At the level of organizations the Chinese current knowledge sharing is implemented via a knowledge alliance, which takes tree main forms: project program, industry technology alliance and the scientific research alliance. Table 1 illustrates a general comparison between these tree possibilities:

The project cooperation program is a temporary organized knowledge sharing structure established by relevant parties to accomplish a specific scientific research aims in a given time span. The goal is to create a technological consortium between the firms setting up the mechanism of sharing in the research and production activities. The organization has strong structure, procedures, pertinence, timeliness, etc. China's current project teams are usually aiming at a certain key technology or to tackling key problems and the teams are always disbanded once the task is completed. This form of sharing lacks of consistency and continuity. Therefore, this organization type is suitable for a specific period of time within specific areas of knowledge sharing, but is not conducive to real-time updates and further optimization.

Tab. 1. Comparison of three organizational models

Organizational Model	Project cooperation program	Industry Technology Alliance	Scientific Research Based alliance
Members in Organization	Enterprises, Universities, Research institutions		
Feature	Advantage: Targeted, time-sensitive, Shortcoming: consistent, persistent	Complementary, advantages, benefit sharing, risk sharing,	Long-term stability
Shared Method	Market-oriented mainly	Government subsidies and market-oriented combination	Market-oriented mainly
Shared Degree	Lesser extent	Deeper level	Deepest

Tab. 2. Central Enterprise Electric Vehicle Industry Alliance Mechanism

Organizational Goals	Immediate objective: to promote harmonization of technical standards in the industry; Long-term objective: to master the core technology of electric vehicles, to create an internationally competitive Chinese electric car companies and brands.
Main Task	Through the integration of resources, technical cooperation, to achieve key industrial breakthroughs in common technologies and core technologies; through effective division of labor within the alliance and a reasonable convergence of resources, to achieve intellectual property sharing; through technology diffusion and transfer to accelerate the commercialization of the technological achievements, and enhance industrial competitiveness.
Organizational Funding	Dues and an initial capital of 1.3 billion Yuan is extracted from State-owned Capital Gains by the SASAC (State-owned Assets Supervision and Administration Commission).
Organization and Management	SASAC unified leadership, setting up the council, the secretariat, the Electric Drive Committees.
Knowledge Sharing Mechanism	Members share common technologies developed by the Alliance; personality technologies are developed by companies under market principles; diffusion of common technologies outside the Alliance take the for value way to transfer to other companies.

2) The industry technology alliance is established by enterprises, universities, research institutions or other organizations in order to promote inter-organizational knowledge sharing and industrial technology innovation capability. This form supports the joint development, complementary advantages, benefit and risk sharing. The knowledge sharing organization is based on common agreement. In general, when the technological alliance is based on government funding, the members in this alliance should share common technologies with each other or even with other organizations outside the alliance under certain conditions. Table 2 illustrates the case of the Central Enterprise Electric Vehicle Industry Alliance, which is government founded. But, if the

alliance is based on a cost-sharing mode between the members and partly subsidized by government, the members can require a paid transfer to other organizations outside the alliance. Yet in the industry technology alliances, the tacit and structural knowledge inside the alliance is difficult to share.

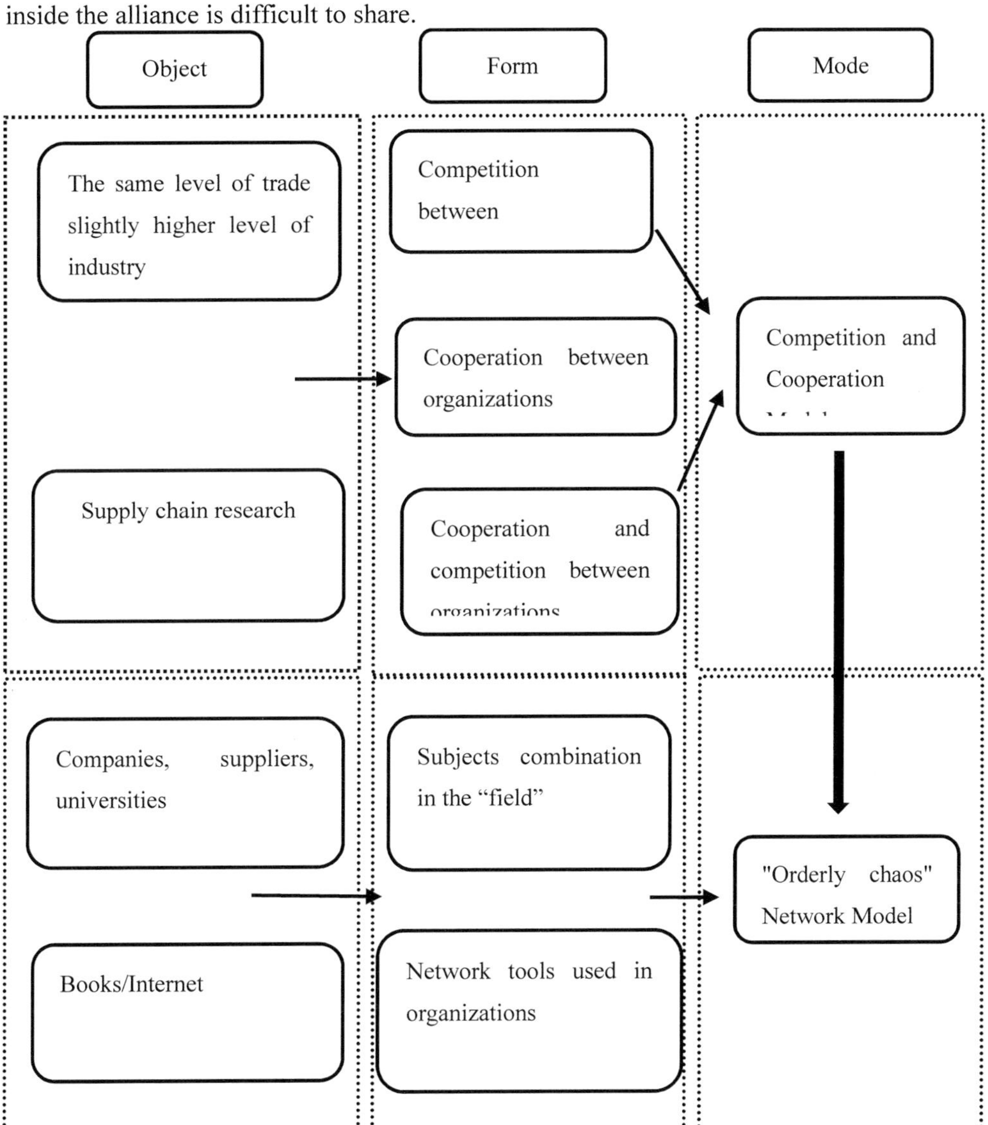

Fig. 2. Knowledge Sharing Models in High-tech Enterprises.

3) The scientific research based alliance is funded by the government or firmes with main task of knowledge sharing and research activities. Generally, this form needs the joint investment of enterprises, research institutions and universities funds to establish a collaborative R&D institutions, laboratories or engineering and technology center. The

scientific research alliance is a long-term and stable organizational structure, which makes it quite suitable for knowledge sharing.

In addition, China currently also has established industry clusters as the Pearl-River-Delta area of the home appliance industry with relatively free organizational structure of knowledge sharing.

In China, the main problem of these knowledge sharing forms is how to balance between cooperation and competition. Generally speaking, there exist two typical knowledge sharing models in High-tech enterprises, especially competition and cooperation model; and orderly chaos model (see in Figure 2). They are applicable for enterprises at different stages of their life cycle.

1) Competition and cooperation model is suitable for High-tech enterprises to generate new knowledge in their infancy and development stage. The infancy stage is characterized as the intensive use of tacit knowledge, when the High-tech enterprise generally could not achieve the amount of new explicit and tacit knowledge it needs to generate new knowledge by merely relying on the internal R&D efforts. Thus it is quite common to use the competition and cooperation model to fulfill its need in different knowledge to increase the accessibility and possibility of new knowledge and ensure that the emergence of new knowledge is sooner than other competitors in the same industry.

2) "Orderly chaos" network model is more suitable for the knowledge sharing in the mature stage. With the continuous expansion of the R&D activities, the amount of knowledge in the enterprises becomes a chaos. In the chaos, tacit knowledge is easily overlooked while explicit knowledge is difficult to summarize. This "Orderly chaos" network model is used to find an orderly procedure in chaotic environments and to optimize organizational knowledge sharing.

In the following part, the case of Huawei illustrates the evolution of the knowledge sharing system and its history.

3.2.2 Chinese agricultural knowledge sharing

Agricultural knowledge sharing refers to the process of the consistently realization of innovation by mutually sharing, learning and transforming the agricultural knowledge mastered by agricultural producer, agricultural researcher and agricultural companies through some certain media.

3.2.2.1 The disturbance to the three main bodies of agricultural knowledge sharing

(1) Agricultural producer

Agricultural producers are the units or the individuals who are directly engaged in agricultural production, they are also the direct users and practitioners of agricultural knowledge. However, once in the face of disasters, they will be at a loss for their own limitations in that it's difficult for them to have a well acknowledge of the agricultural planting knowledge system.

(2) Agricultural researcher

There always exists the problem of inapplicability between the research results and the actual planting situation through The theoretical research and experimental investigation on agriculture by this group of people.

(3) Agricultural corporation

They are an organized agricultural team focused on agricultural cultivation,it is difficult for them to obtain comprehensive agricultural planting knowledge from outside even though the exchange and sharing of the agricultural knowledge can be easily realized within the team.

3.2.2.2 The types of the sharing of agricultural knowledge

(1) The knowledge sharing in a small range

The knowledge sharing in a small range mainly refers to the mutual exchange of experience among individual farmers. This kind of knowledge sharing is usually carried on through face to face teaching between father and son, master and apprentice.

The main shortcoming of the knowledge sharing is its small range and its little influence, besides of that, there is often the risk of being lost. Even so, it is also the main way of knowledge sharing in nowadays Chinese farming.

(2) The knowledge sharing in mature organizations

The knowledge sharing in mature organizations mainly means the sharing of the planting knowledge within the agricultural organizations and companies. Sometimes, it also expands their knowledge reservation by employing experts specialized in agriculture externally.

(3) The knowledge sharing on the overall internet

It means to build an Internet knowledge sharing platform through a third party to realize the knowledge sharing among the three main bodies—agricultural producer, agricultural researcher and agricultural corporation and to realize the maximum efficiency of the agricultural knowledge sharing.

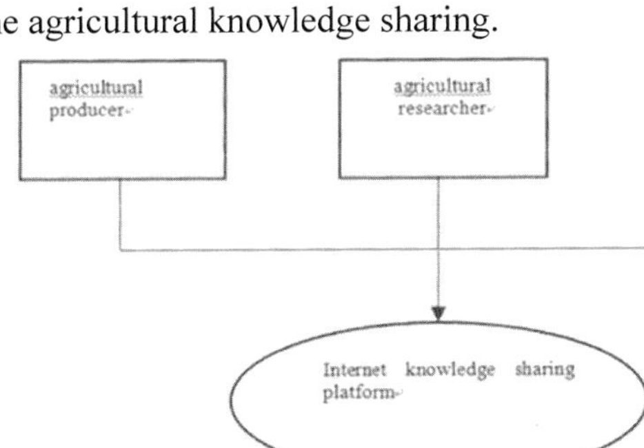

Fig. 3. The future Agricultural Knowledge Sharing System.

3.2.2.3 The future of agricultural knowledge sharing

The famous saying by Confucius that there must be one good enough to act as a teacher as three men walking together has fully reflected the importance of learning advanced knowledge and experience from others, especially in such an era of knowledge-based economy. The improvement of the system of agricultural knowledge sharing has a pivotal influence on the future development of agriculture of our country because it can better improve the backward status quo of China's agricultural development. As the concept of "Internet + agriculture" is put forward in our country, we believe that the agricultural knowledge sharing platform will become a reality in the near future and the development of agricultural knowledge sharing can be promoted from the technical level.

4. Conclusion

The latest academic and managerial works highlight the major role of knowledge sharing system in the management of innovation activities especially in the case of an open innovation model. Traditionally, the debates dedicated on the Chinese innovation system focuses on the barriers for open model (Savitskaya et al., 2010). Nevertheless, the implementation of the open model in China passes upon the development of various measures supporting the knowledge sharing system at institutional and organizational level. At institutional level the government aims to build a sustainable intermediary platform for knowledge sharing. In this sense, the possible ways to support knowledge sharing in China are:

- to establish the government portal website to the enterprise and public government information providing public services and policy advice;
- to build the government knowledge management center, to specialized knowledge mining, discovery, application, innovation and protection;
- to establish a government knowledge feedback platform, fully adapted to the outside opinion in order to test the efficiency of knowledge sharing.

Through the building of a knowledge platform at national level the government can expand the optimization of the information process and to emphasize on the knowledge mining, knowledge communication and knowledge transfer as well as to develop the information service and improve the knowledge management mechanism.

References

1. Abjanbekov, A., and Alvarez Padilla, A. E., (2004) From knowledge transfer to knowledge translation: Case study of a telecom consultancy, International Master's Programme in Strategy and Culture, 2004:6.
2. Alavi, Maryam, and Dorothy E. Leidner, (2001) *Review: Knowledge management and knowledge management systems: Conceptual foundations and research issues.* MIS quarterly (2001): 107-136.
3. Binz-Scharf, Maria Christina, (2003) Exploration and exploitation: Toward a theory of knowledge sharing in digital government projects. Diss. Universität St. Gallen.
4. Bogers, M., (2012) Knowledge sharing in open innovation: An overview of theoretical perspectives on collaborative innovation. Open innovation at Firms and Public Administrations: Technologies for Value Creation, 1-14.
5. Bogers, M., and West, J., (2010) Contrasting innovation creation and commercialization within open, user and cumulative innovation. User and

Cumulative Innovation (July 13, 2010).

6. Bosua, R., and Scheepers, R., (2007) "Towards a model to explain knowledge sharing in complex organizational environments", Knowledge management research and practice, 5(2), 93-109.
7. Brauner, E., and Becker, A., (2006) *Beyond knowledge sharing: the management of transactive knowledge systems*, Knowledge and Process management, 13(1), 62-71.
8. Buganza, T., and Verganti, R., (2009) *Open innovation process to inbound knowledge: Collaboration with universities in four leading firms*, European Journal of Innovation Management, 12(3), 306-325.
9. Bukowitz, Wendi R. and Williams, Ruth L., (1999) The Knowledge Management Fieldbook. FT Press.
10. Chen, D., and Li-Hua, R., (2011) *Modes of technological leapfrogging: Five case studies from China*, Journal of Engineering and Technology Management, 28(1), 93-108.
11. Chen, J., and Qu, W. G., (2003) "A new technological learning in China", Technovation, 23(11), 861-867.
12. Chesbrough, H. W., (2003) Open innovation: The new imperative for creating and profiting from technology. Harvard Business Press.
13. Contractor, F. J., and Sagafi-Nejad, T., (1981) *International technology transfer: Major issues and policy responses*, Journal of International Business Studies, 113-135.
14. Contractor, N., Carley, K., Levitt, R., Monge, P. R., Wasserman, S., Bar, F., and Kunz, J., (2000) Co-evolution of knowledge networks and 21st century organizational forms: Computational modeling and empirical testing. Urbana-Champaign, IL: University of Illinois.
15. Cummings, J., (2003) Knowledge sharing: A review of the literature. World Bank, Washington, DC.
16. Currie, G., and Suhomlinova, O., (2006) *The impact of institutional forces upon knowledge sharing in the UK NHS: the triumph of professional power and the inconsistency of policy*, Public Administration, 84(1), 1-30.
17. Davenport, E., and Hall, H., (2002) "Organizational knowledge and communities of practice", Annual review of information science and technology, 36(1), 171-227.
18. Deng, P., (2009) *Why do Chinese firms tend to acquire strategic assets in international expansion?*, Journal of World Business, 44(1), 74-84.
19. Dosi, G., Nelson, R., and Winter, S. (Eds.), (2000) The nature and dynamics of organizational capabilities. Oxford University Press.
20. Enkel, E., Gassmann, O., and Chesbrough, H., (2009) *Open R&D and open innovation: exploring the phenomenon*, R&D Management, 39(4), 311-316.
21. Fountain, J. E., (1999) A note on the critical incident technique and its utility as a tool of public management research. In Annual meeting of the Association of Public Policy and Management (pp. 4-6).
22. Gassmann, O., and Enkel, E., (2004) Towards a theory of open innovation: three core process archetypes. In R&D management conference (Vol. 6).

23. Grant, R. M., (1996) *Toward a knowledge-based theory of the firm*, Strategic management journal, 17(S2), 109-122.
24. Gupta, A. K., and Govindarajan, V., (1991) *Knowledge flows and the structure of control within multinational corporations*, Academy of management review, 16(4), 768-792.
25. Hansen, M. T., (1999) *The search-transfer problem: The role of weak ties in sharing knowledge across organization subunits*, Administrative science quarterly, 44(1), 82-111.
26. Hartley, J., and Rashman, L., (2007) How is knowledge transferred between organizations involved in change?, Managing change in the public services, 173-92.
27. Hung, S. W., (2009) *Development and innovation in the IT industries of India and China. Technology in Society*, 31(1), 29-41.
28. Huysman, M. H., and De Wit, D., (2002) Knowledge sharing in practice (Vol. 4). Springer Science and Business Media.
29. Huysman, M., and De Wit, D., (2004) *Practices of managing knowledge sharing: towards a second wave of knowledge management,* Knowledge and process management, 11(2), 81-92.
30. Islam, A. M., (2012) *Methods of open innovation knowledge sharing risk reduction: a case study*, International Journal of e-Education, e-Business, e-Management and e-Learning, 2(4), 294-297.
31. Kim, L., and Nelson, R. R. (Eds.). (2000), Technology, learning, and innovation: Experiences of newly industrializing economies, Cambridge University Press.
32. Kirschbaum, R., (2005) *Open innovation in practice*", Research-Technology Management, 48(4), 24-28.
33. Kline, S. J., and Rosenberg, N., (1986) An overview of innovation. The positive sum strategy: Harnessing technology for economic growth.
34. Kwok, S. H., and Gao, S., (2005) "Attitude towards knowledge sharing behavior", Journal of Computer Information Systems, 46(2), 45-51.
35. Lall, S., (2000) Technological change and industrialization in the Asian newly industrializing economies: achievements and challenges. Technology, learning, and innovation: Experiences of newly industrializing economies, 13-68.
36. Lichtenthaler, U., (2011) *Open innovation: Past research, current debates, and future directions*, The Academy of Management Perspectives, 25(1), 75-93.
37. Liebowitz, J., and Yan, C., (2004) Knowledge sharing proficiencies: the key to knowledge management. In Handbook on Knowledge Management 1 (pp. 409-424).
38. Liebowitz, S., (2002) Rethinking the networked economy: The true forces driving the digital marketplace. AMACOM Div. American Mgmt Assn, Dallas.
39. Liu, X., and Buck, T., (2007) *Innovation performance and channels for international technology spillovers: Evidence from Chinese high-tech industries*, Research Policy, 36(3), 355-366.
40. Llopis-Córcoles, Ó., (2011) Understanding Knowledge Sharing in Organizations: Multi-level Research Through a Social Cognitive Perspective, researchgate. net.
41. Mansfield E, Romeo A.,(1980) "Technology transfer to overseas subsidiaries by

US-based firms", The Quarterly Journal of Economics, 1980: 737-750.

42. Mansfield, E., (1975) *International technology transfer: forms, resource requirements, and policies*, The American Economic Review, 372-376.
43. Mansfield, E., Romeo, A., and Wagner, S., (1979) *Foreign trade and US research and development*, The Review of Economics and Statistics, 49-57.
44. McDermott, R., (2000) *Why information technology inspired but cannot deliver knowledge management*, Knowledge and communities, 41(4), 21-35.
45. Nonaka, I., and Takeuchi, H., (1995) The knowledge-creating company: How Japanese companies create the dynamics of innovation. Oxford university press.
46. Powell, W. ,(1990) *Neither market nor hierarchy: Network forms of organization,* Research in Organizational Behavior, 12, pp. 295–336.
47. Riege, A., (2005) *Three-dozen knowledge-sharing barriers managers must consider*, Journal of knowledge management, 9(3), 18-35.
48. Rothwell, R., (1994) *Towards the fifth-generation innovation process*, International marketing review, 11(1), 7-31.
49. Rousseau, D. M., (1985) *Issues of level in organizational research: Multi-level and cross-level perspectives*, Research in organizational behavior, 7(1), 1-37.
50. Savitskaya, I., Salmi, P., and Torkkeli, M., (2010) *Barriers to open innovation: Case China*, Journal of technology management and innovation, 5(4), 10-21.
51. Spencer, J. W., (2003) *Firms' knowledge-sharing strategies in the global innovation system: empirical evidence from the flat panel display industry*, Strategic Management Journal, 24(3), 217-233.
52. Szulanski, G., (1996) *Exploring internal stickiness: Impediments to the transfer of best practice within the firm*, Strategic management journal,17(S2), 27-43.
53. Tsai, W., (2002) *Social structure of "coopetition" within a multiunit organization: Coordination, competition, and intraorganizational knowledge sharing*, Organization science, 13(2), 179-190.
54. Verspagen, B., and Duysters, G., (2004) *The small worlds of strategic technology alliances*, Technovation, 24(7), 563-571.
55. Wang, S., and Noe, R. A., (2010) *Knowledge sharing: A review and directions for future research*, Human Resource Management Review, 20(2), 115-131.
56. West, J., Salter, A., Vanhaverbeke, W., and Chesbrough, H., (2014) *Open innovation: The next decade*, Research Policy, 43(5), 805-811.
57. Zander, U., (1991) Exploiting a Technological Edge: Voluntary and Involuntary Dissemination of Technology, Doctoral dissertation, Institute of International Business, Stockholm.

Photosynthetic response of greening seedling of four tree species to low temperature stress

T. T. Zhou[a], L. Xue[b,*] and Z. M. Wang[c]

College of Forestry and Landscape Architecture, South China Agricultural University, Guangzhou 510642, P. R. China;

[a]1152110287@qq.com; [b]forxue@scau.edu.cn; [c]wangzhuomin1@126.com

**Corresponding author*

Photosynthetic indexes of *Cassia bakeriana, Syzygium hainanense, Ilex rotunda* and *Phoebe sheareri* seedlings were determined under simulated low temperature stress using artificial climate chambers. The results showed that with increasing low temperature time, the net of photosynthetic rate (P_n), stomatal conductance (G_s) transpiration rate (T_r) and stomatal limitation (L_s) of seedlings of the four tree species decreased; intercellular CO_2 concentration (C_i) of *C. bakeriana* fluctuated and then increased, *S. hainanense* continuously increased, *I. rotunda* and *P. sheareri* decreased followed by an increase. Two days after low-temperature release, P_n, T_r, G_s and L_s of seedlings of the four species increased. C_i of *C. bakeriana, S. hainanense,* and *P. sheareri* increased, whereas *I. rotunda* decreased. The photosynthetic indexes of seedlings of the four tree species was evaluated using principal component analysis, indicating that cold-resistance ability decreased in the order of *C. bakeriana*>*P. sheareri*>*I. rotunda*>*S. hainanense*

Keywords: Seedlings; Low-resistance; Photosynthetic Index; Principal Component Analysis

1. Instructions

As one of the important environmental factors, temperature plays an important role during plant growth process. Low temperature has an influence on the photosynthetic characteristics of plants [1], such as net photosynthetic rate (P_n), stomatal conductance (G_s), transpiration rate (T_r) and intercellular CO_2 concentration (C_i) [2]. Therefore, the photosynthetic characteristics become an important index for evaluating plant cold-resistance and have been extensively studied [3-7]. *Cassia bakeriana* is famous ornamental tree species with large and beautiful flowers, blooming in the spring; *Syzygium* hainanense is a good greening tree species with developed roots and refractory. Ilex rotunda is an ideal ornamental tree with dense leaves and beautiful flowers. Phoebe sheareri is a good greening tree species with large leaves and dense shade. Knowledge of the photosynthetic characteristics of these tree species under low temperature stress is insufficient. This study is aimed at providing the reference for selection of cold-resistance greening species from the view of photosynthetic characteristics.

2. Material and method

2.1. *The testing site*

Study was conducted in the building of College of Forestry and Landscape Architecture, South China Agricultural University, Guangzhou City (113°18′E, 20°6′N), belonging to subtropical monsoon climate with characteristics of warm and rainy as well as sufficient sunlight. The annual average temperature, the most cold month and the most warm month temperature are 21. 8°C, 13. 3°C and 28. 1°C, respectively. Annual rainfall is 1714. 4 mm, occurring mainly from April to September, and annual average relative humidity is 79%.

2.2. *Experimental material*

One-year-old seedlings of *C. bakeriana, S. hainanense, I. rotunda, P. sheareri* came from Shenzhen Techand and Ecology & Environment CO. LTD. At the beginning of test, general characteristics of the seedlings of four tree species were shown in Table 1.

Tab. 1. General characteristics of the seedlings of the four species

Tree species	The average diameter/cm	The average seedling height/cm	The average crown width/cm
Cassia bakeriana	0.38±0.03	30.5±3.2	20.3±3
Phoebe sheareri	0.99±0.06	33.9±2.3	22.9±3.1
Ilex rotunda	0.5±0.01	34.5±2.5	14.5±2.6
Syzygium hainanense	0.56±0.2	54.9±1.3	25.9±0.2

2.3. *Test method*

Putting experimental seedlings with similar size into the RXZ intelligent artificial climate box. Lighting time was from 8:00 to 17:00 with light intensity of 120 mmol (photons) $\cdot m^{-2} \cdot s^{-1}$ and relative humidity was 80% ~ 85%. The seedlings were treated with 6±0.5°C low temperature (low temperature treatment 0 d as the control), temperature of intelligent artificial climate box was reduced with $6°C \cdot h^{-1}$. When the temperature dropped to 6°C, five leaves of each seedling (the third to eighth functional leaves from the seedling top) were marked, each marked leaf was selected to determinate Photosynthetic indexes after low temperature treatment of 0 d, 2 d, 4 d and 6 d. Seedlings treated with low temperature treatment of 6 d were moved from intelligent artificial climate box to the outdoor to restore growth for 2 d, photosynthetic indexes were determined with triplicate for each index.

2.4. *Determination method*

Li-6400 portable photosynthesis system analyzer produced by LI-COR Company from America, to used to determinate P_n, G_s, C_i and T_r of the seedlings at 9:00 ~ 11:30 and then their L_s was calculated according to L_s=1-C_i/C_a (C_a is atmospheric CO_2 concentration).

2.5. *Data analysis*

All statistical data were analyzed using Excel 2010 and the Statistical Analysis System (SAS 8.1).

3. Results

3.1. *Net photosynthetic rate*

With increasing stress time, the net photosynthetic rate (P_n) of seedlings of four species decreased sharply (Fig. 1). The P_n of the seedlings increased in different range but was significant lower than controls 2 d after low-temperature release (P<0.05).

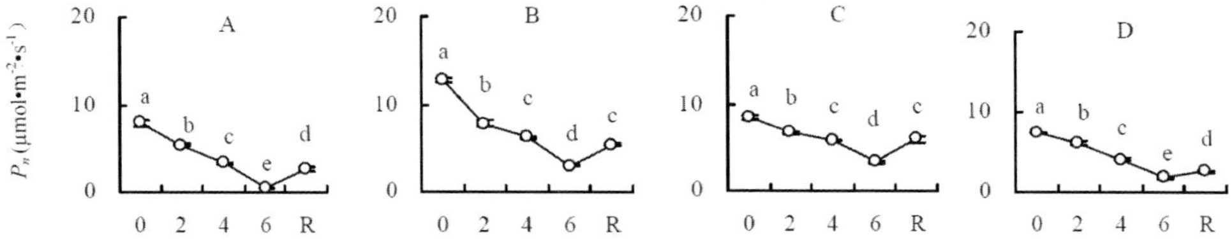

Fig. 1 Effect of low temperature stress on the net photosynthetic rate (Pn) in the seedling A. *Cassia bakeriana*; B. *Syzygium hainanense*; C. *Ilex rotunda*; D. *Phoebe sheareri*

3.2. *Stomatal conductance*

With increasing stress time, the stomatal conductance (G_s) of seedlings of four species decreased sharply (Fig. 2), among which decrease range of G_s of *P. sheareri* was the largest with a value of 95.6%, and *I. rotunda* was the second largest with a value of 74% for 6 d after low temperature treatment. The G_s of *C. bakeriana, I. rotunda, P. sheareri* had a significant increase 2 d after low-temperature release (P<0.05).

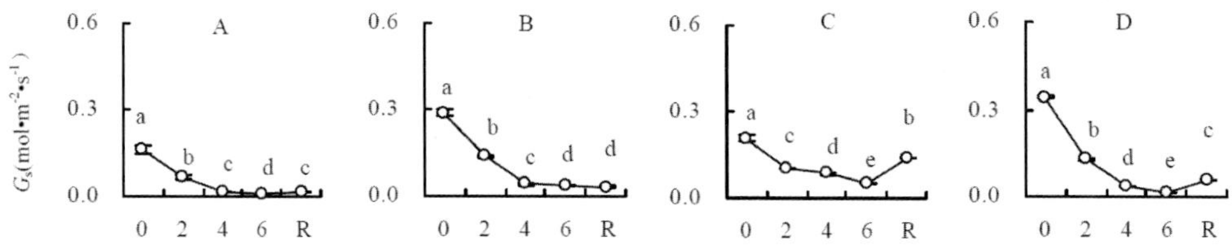

Fig. 2 Effect of low temperature stress on the stomatal conductance (Gs) in the seedling leaves A. *Cassia bakeriana*; B. *Syzygium hainanense*; C. *Ilex rotunda*; D. *Phoebe sheareri*

3.3. *Intercellular CO_2 concentration*

With increasing stress time, the intercellular CO_2 concentration (C_i) of *C. bakeriana* fluctuated and then increased, *S. hainanense* increased, *I. rotunda* and *P. sheareri* decreased followed by an increase (Fig. 3). The C_i of *C. bakeriana*, *S. hainanense* and *P.*

sheareri had a significant decrease ($P<0.05$), whereas *I. rotunda* had a significant increase 2 d after low-temperature release ($P<0.05$).

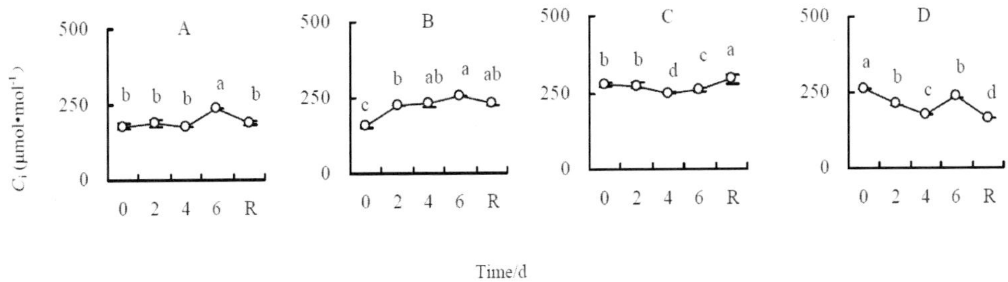

Fig. 3 Effect of low temperature stress on the Intercellular CO2 concentration (*Ci*) in the seedling leaves A. *Cassia bakeriana*; B. *Syzygium hainanense*; C. *Ilex rotunda*; D. *Phoebe sheareri*

3.4. *Transpiration rate*

With increasing stress time, the transpiration rate (T_r) of seedlings of four species decreased (Fig. 4). The T_r of seedlings of four species increased but was still significant lower than the controls 2 d after low-temperature release ($P<0.05$).

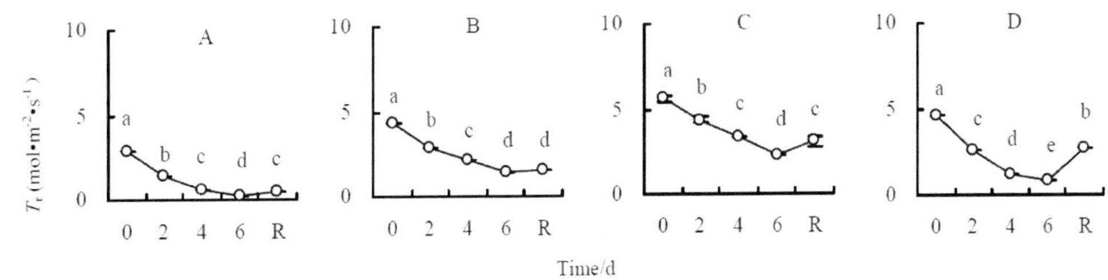

Fig. 4 Effect of low temperature stress on the transpiration rate (*Tr*) in the seedling leaves A. *Cassia bakeriana*; B. *Syzygium hainanense*; C. *Ilex rotunda*; D. *Phoebe sheareri*

3.5. *Stomatal limit*

With increasing stress time, the stomatal limit (L_s) of seedlings of four species decreased and was lower than the controls for 6 d after low temperature treatment (Fig. 5). The L_s of seedlings of four species had a significant increase 2 d after low-temperature release ($P<0.05$).

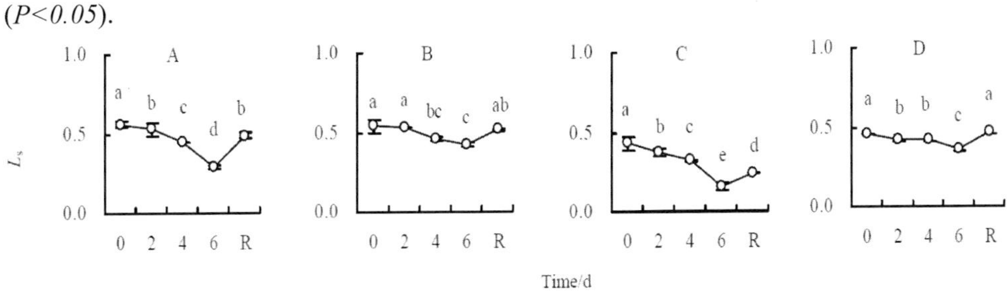

Fig. 5 Effect of low temperature stress on stomatal limit (Ls) in the seedling leaves A. *Cassia bakeriana*; B. *Syzygium hainanense*; C. *Ilex rotunda*; D. *Phoebe sheareri*

3.6. *Comparative evaluation on cold-resistance of seedlings*

In this study, principal component analysis was used to evaluate cold-resistance ability of seedlings of four species. The calculating score decreased in the order of *C. bakeriana* > *P. sheareri*> *I. rotunda* > *S. hainanense* (Table 2).

Tab. 2. Score of cold-resistance of seedlings of the four species

Species	Score	Ranking
C. bakeriana	1.56	1
P. sheareri	1.37	2
I. rotunda	1.09	3
S. hainanense	0.48	4

4. Discussion

In this study, the low temperature has obvious influence on photosynthesis of the seedlings. With increasing the low temperature stress time, the P_n, L_s, G_s and T_r of seedlings of four species decreased, the C_i of *S. hainanense* increased, whereas *C. bakeriana* and *I. rotunda* increased for 4 d low temperature stress, indicating that decline in P_n was caused by non-stomatal limitation under low temperature stress. The reasons for this decline may be disorders of metabolism of active oxygen, destruction of structure of biofilm and chloroplast. Moreover, stomata block CO_2 diffusion and water conduction between plant leaves and atmospheric, resulting in C_i accumulation, decreasing the active of mesophyll cells [8] and influencing photosynthetic rate [9]. Two days after low-temperature release, increased in P_n, T_r and G_s of seedlings of four species and decreased in C_i of *C. bakeriana, S. hainanense* and *P. sheareri* seedlings indicated that these seedlings had certain ability of cold-resistance.

The six photosynthetic indexes of seedlings of the four tree species was evaluated using principal component analysis, indicating cold-resistance ability decreased in the order of *C. bakeriana*>*P. sheareri*>*I. rotunda*>*S. hainanense*, which suggested that *C. bakeriana* and *P. sheareri* were as the first-choice tree species growing in the cold regions.

Acknowledgments

The study was partially supported by Foundation of Guangdong Forestry Department, China (No. 4400-F15050).

References

1. S. H. Guo, L. Xue, R. Zgang and Z. Y. Yang, Photosynthetic response of four species seedlings to low temperature stress. *J. South China Agri. Univ.* 33, 373 (2012).

2. Y. R. Shao and J. X. Xu, L. Xuye, C. Q. Wu and G. C. Lu, Effects of low temperature stress on physiological-biochemical indexes and photosynthetic characteristics of seedlings of four plant species. *Acta Ecol. Sin.* 33, 4237 (2013).
3. B. Martin, D. R. Ort and J. S. Boyer, Impairment of photosynthesis by chilling-temperatures in tomato (*Lycopersicon esculentum* cultivar Rutgers). *Plant Physiol.* 68, 329(1981).
4. Nir G, Ratner K, E. Gussakovsky and Y. Shahak, Photoinhibition of photosynthesis in mango leaves: effect of chilly nights. Acta Hortic. 45, 288(1997).
5. Y. H. Guo, Z. Q. Cai and K. F. Cao, Effects of nocturnal low temperature on photosynthesis of seedlings of two coffee species. Chin. J. Ecol. 24, 478 (2005).
6. D. J. Allen, K. Ratner, Y. E. Giller, E. E. Gussakovsky, Y. Shahak and D. R. Ort, An overnight chill induces a delayed inhibition of photosynthesis at midday in mango (*Mangifera indica* L.). *J. Exp. Bot.* 51, 1893 (2000).
7. J. Zhou, L. F. Yang, F. G. Hao and Y. You, Photosynthesis and chlorophyll-fluorescence of *Magnolia grandiflora* seedlings under low temperature stress. *Acta Bot. Boreal.-Occident. Sin.* 29, 136 (2009).
8. X. L. Sui, S. L. Mao, L. H. Wang, W. Li, B. X. Zhang and Z. X. Zhang, Effects of low temperature on photosynthesis of sweet pepper under low light. *J. Nucl. Agri. Sci.* 22, 880 (2008).
9. G. D. Farquhar and T. D. Sharkey, Stomatal conductance and photosynthesis. *Ann. Rev. Plant Physi.* 33, 317 (1982).

The influence of biological fertilizer on crop growth research

Xiaonan Chi[1], Qing Li[1], Yu Fu[2], Shiwei Wu[1]*

[1]*Experimental Center of Shenyang Normal University, Shenyang 110034, China*

[2]*AVIC Shenyang Aircraft Corporation, Shenyang 110034, China*

E-mail: wswcn1975@163.com, 626553148@qq.com

Liao Star 20 as test materials, using three different concentrations of photosynthetic bacterial manure (respectively diluted 500 times, 400 times, 400 times) into the same amount of solution in the 500 ml beaker, the results were compared with equal parts water. After training in jointing stage of plant height, root length. After growth to the appropriate degree of indoor KaoZhong, measuring plant fresh weight and dry weight per plant. The results show that the different concentration of photosynthetic bacterial manure can obviously increase the rice growth conditions, the effect of diluted 400 times, 500 times to optimal; Photosynthetic bacterial manure can increase after jointing stage and leaf chlorophyll content, increase the photosynthetic rate, thus improve the biological yield, spraying 500 times, 400 times, 300 times the photosynthetic bacterial manure, respectively, compared with weight by 29%, 39.9%, 13.6%.

Keywords: Photosynthetic bacterium; Field management; Number of tillers; Plant Height

1. Guidelines

Biological fertilizer from the initial agent of rhizobia to the bacteria fertilizer, and then to today's biological fertilizer, evolution has biological fertilizer is explained from the name of the process of gradual development. At present, more than 70 countries on international production, application and popularization of biological fertilizer, at present there are about 500 enterprises produce millions of tons of about biological fertilizer applied in production. This is compared with the same fertilizer production and consumption can't, but it has started to play a role in agricultural production, has obtained the certain economic benefit and social effect. With the deepening of the research and application of need, we should continuously expand the development of new varieties of biological fertilizer, its developing trend is as follows:

By a single species to compound strains. In the process of the transformation, it is not simple to pursue the improvement of nutrient supply, to the pursuit of multi-functional, such as disease resistance, insect, 5406 strains can enhance the disease resistance of the crops, such as reducing the use of chemical pesticides; Joint application of bacteria strains can make some or several performance from the level of the original to step up, make the composite or joint microbes play a mutual, collaborative, symbiosis, strengthening, isotopic effect, eliminate the occurrence of mutual antagonism. The second is to prolong the survival time of microorganisms in the soil, microbial survival, the longer the longer the term of a special effects of biological fertilizer. The length of the survival time of microorganisms in the soil, mainly depends on the level of available carbon source in the soil, can add a can be decomposed in the biological fertilizer containing carbon compounds in the soil of the strains, to constantly supply other microbial carbon source nutrition, prolong the survival time of microorganisms in the soil.

By the simple biological bacterium agent to compound fertilizer. From the past simple silicate bacterium agent, soil phosphorus activator, rhizobia agent to organisms such as bacteria, azotobacter agent agent and nutrient elements (N, P, K and other elements, trace elements) composite compound fertilizer, organic fertilizer, antibiotics such as transformation. This transformation is helpful to realize the combination of biological fertilizer and biological medicine, enhancement of multi-functional fertilizer effect.

By a single dosage form to diversified varieties.In order to adapt to different conditions, in addition to biological fertilizer carrier liquid dosage forms, lime powder, and granular dosage forms, freeze-dried formulations, dosage forms, such as mineral oil cover. As the development of biotechnology and biological fertilizer application in agriculture, brewing a good development space, from junior to senior, from low to high efficiency and to develop in the direction of industrialization.

At present in large amounts of chemical fertilizers and pesticides caused the fall of produce quality, in order to develop pollution-free, pollution-free benign ecological agriculture as soon as possible, we choose photosynthetic bacterial manure, is a kind of light as energy, carbon dioxide or organic compounds as carbon source nutrition and breeding of microorganism, rice a hydroponic experiment was carried out. Its researches products rich in physiological active substances required for plant growth. With strong symbiotic nitrogen fixation, value-added benefits, the effect such as disease-causing bacteria, it can enhance photosynthesis in plants and to promote crop growth, development, disease resistance, early maturity. We set rice photosynthetic bacterial manure application contrast test, into the light and researches provide scientific basis for large area [1].

2. Materials And Methods

2.1. *The experimental materials*

500 mL beaker (4); Water; The thermometer. Scale; Electronic balance. Photosynthetic bacteria microbial self fertile by the laboratory (select for adding plant hormone microbial test), microbial red, pH 8.5 9.0, have a fishy smell; Test for siting tieling county of liaoning province liao waters of farmland for experimental base, the soil physical and chemical properties and climatic conditions in [2] in Table 2.1, experiment for rice crop, as liao star 20 varieties. In May 2015 - September test time.

Tab. 2.1 the physical and chemical properties of the tested soil and climate conditions

Content Index	Output Indicators
The name of the species	Rice Soil
Organic Content	2.20%
Total Nitrogen(g/Kg)	1.13
Alkeline-N(mg/kg)	87.62
PzOs(mg/kg)	26.41
Kppm (mg/kg)	80.58
5 - September rainfall(mm)	675
Annual frost-free period(d)	146

2.2. *Methods of test*

2.2.1. *The experiment content and design*

With clear blister 12 hours. The rice seeds were divided into four copies of the seed planted in a beaker yarn online.

Processing A: no fertilizer, cultures of water, cover in the beaker with gauze as A medium.

Processing B: photosynthetic bacterial manure with water will be diluted to 300 times, with A equal amounts into the beaker as culture medium, repeat the steps to complete A.

Processing C: photosynthetic bacterial manure to 400 times diluted with water, with A equal amounts into the beaker as culture medium, repeat the steps to complete A.

Processing D: photosynthetic bacterial manure with water will be diluted to 500 times, with A equal amounts into the beaker as culture medium, repeat the steps to complete A.

After the above processing, the photosynthetic bacterial manure dilution as culture medium, at room temperature for a hydroponic, record the daily growth (roots, plant height), morning and evening temperature difference, repeated training three times, and randomized comparison.

2.2.2. *Rice hydroponic post-processing*

On April 2, at the end of culture, seed, sowing, April 10 to continue cultivating record update broth, April 14.

Every cup plant were randomly selected 10 strains, wet weight and dry weight of weighing, and recording the data.

2.2.3. *Field management*

In proportion to the photosynthetic bacteria bacterium fluid and irrigation water mixed with basal into the paddy fields, marked as B, C and D respectively.

Testbed in tieling county of liaoning province liao waters of farmland modern agriculture demonstration park, the village can be compared to the test. Each dealing with a pool, single single irrigation. Other management with regular.

3. Results And Discussion

Experiments respectively under different processing of rice germination rate, growth period, plant height, tiller number and measure the various production factors, discussed the influence of different processing on rice growth and yield.

3.1. *Effects on rice seedling germination rate*

Experimental results show that treatment B, C, D respectively than handling A germination rate increase 13.6%, 39.9%, 29%, shows that using photosynthesis researches on rice has effect on increasing yield. That, on the basis of do in water culture, together with 400 times dilution of the effect of photosynthetic bacterial manure production is the most significant.

3.2. *The influence of agronomic traits of rice*

Can be seen from Table 3.1, the more significant effects on plant height, germination rate.

Tab. 3.1 Yield analysis results

Content processing	Plant height (cm)	Rhizome (cm)	Wet weight (20 plants/g)	Per gram dry weight (20 cases)	Germination increase production rate
A	11	9	3.62	0.98	
B	12	10	3.64	1.23	13.6%
C	16	12	3.90	1.39	39.9%
D	14	11	3.85	1.25	29%

3.2.1. *Effects on rice growth period*

Every growth period of rice has its different morphology, growth and development, physiological characteristics. Seedling stage and tillering stage of increasing vegetation is rice, main root length, lay a foundation for the rice grain number number; Heading stage of rice stem long ear, is to determine the critical period of rice grain number; T hose mature rice grouting, determine the seed setting rate and grain weight. The growth period of rice under different processing conditions record the results are shown in Table 3.2.

Tab. 3.2 the influence of different processing on rice growth period

Treatment	A(contrast)	B	C	D
Seedtime	May 20	May 20	May 20	May 20
Turning green date	May 30	May 29	May 28	May 29
Beginning tillering period	June 5	June 5	June 5	June 5
Active tillering stage	June 30	June 30	June 30	June 30
Heading stage	August 10	August 9	August 8	August 10

Can be seen from Table 3.2: the processing rice in different growth period difference, the difference is only 1 to 2 days, shows that the addition of photosynthetic bacteria agent on rice growth period neither role ahead of time, no hysteresis, is with the void growth and development at the same time to mature.

3.2.2. *Effects on yield components and yield of rice*

Rice production factors mainly include the number of ears, total grain Numbers per spike, grain number and empty flat rate, etc., is the direct factor analysis of rice production, no correlation between various factors, but can reflect rice yield and yield in different degrees; Rice production is all kinds of influence factors on the ultimate effect of rice, also direct embodiment of economy, all the measures are for the purpose of increasing production.

Next results investigation plan in early October, are given for the influence of different processing on the rice yield factor table and the influence of different processing on rice yield. Going to further perfect the report.

3.2.3. *Affect growth looks*

Through the field observation records of live, as shown in Figure 3.1, the use of photosynthetic bacterial manure seedling tiller, leaf growth better, and can extend the leaf function period, prevent premature aging and lodging resistance. Basal due to add microbial agents, to promote the soil microbial breeding, activate the soil microorganisms and soil nutrient, promote the fast green rice points, increase the harvest of spikelets, and increase production.

4. Conclusion

Photosynthetic bacteria bacterium fluid and irrigation water mixed in the basal fertilization on rice, while bacteria agent had no obvious effect on rice growth period, but increases the tillering number, plant height of rice, among them 400 times dilution agent of photosynthetic bacteria to the influence of slightly stronger; After rice mature, into the photosynthetic bacteria compared to total grain number increasing rice production, but due to the empty flat rate is higher, to combining the elements of NPK fertilizer. Basal due to add microbial agents, to promote the soil microbial breeding, activate the soil microorganisms and soil nutrient, promote the fast green rice points, increased the harvest of spikelets and, in turn, increase output, and photosynthetic bacteria tap cheaply, photosynthetic bacteria in the application in rice.

Photosynthetic bacterial manure have obvious effect to increase production, effective tillers, grain, rice can be improved so as to achieve production effect.

Photosynthetic bacterial manure can't take the place of NPK element, with in the bottom of NPK one-time used together on the basis of photosynthetic bacterial manure fertilizer effect is better.

Acknowledgments

This work was supported by Shenyang Science and Technology Project (F14-231-1-35), and creation and innovation of college students training plan project of Shenyang Normal University(201510166254).

References

1. Sun Junlan Li Qiaojun. Photosynthetic bacterial manure application on wheat. At the beginning of the wheat research, 2002:35-36.
2. Tieling county TuFeiZhan. Liaoning tieling county farmland soil moisture bulletin, 2015, 2015: the second stage.

A rapid spectrophotometric method for the quantitative and quantitative detection of total coliforms

Lin Huang, Chaozheng Zhang, Xiufeng Shi, Mingjie Li, Shuang Liu, Yawen Cheng, Hongxi Zhu, Zhixiang Li, Yihan Liu*

Key Laboratory of Industrial Fermentation Microbiology (Tianjin University of Science & Technology), Ministry of Education, 300457
Tianjin, China
Tianjin Key Laboratory of Industrial Microbiology, 300457
Tianjin, China
College of Biotechnology, Tianjin University of Science and Technology, 300457
Tianjin, China

To develop a rapid, accurate and sensitive method for determining the number of total coliforms, a new approach was established based on the performance of triphenyltetrazolium chloride (TTC) medium combined with spectrophotometric detection. Bromocresol purple was used as the main indicator and the culture mixture was grown for 8 h at 36 °C. When the color of the culture reached stability, the intensity at 420 nm indicated the total coliform numbers in samples. The results of the new method were correlated to the actual content of bacteria (correlation coefficient, R^2 = 0.982). One-hundred and forty samples were tested by the new method and the traditional most probable number (MPN) method; the results of the two methods were compared. The new method was credible and accurate within the 95% confidence interval of the MPN method. Compared with the MPN method, the sensitivity of detection was increased. The TTC/spectrophotometric method saves testing time and produces results with a higher confidence level than the MPN method and may therefore be useful for the detection of total coliforms in many fields.

Keywords: Total coliforms; spectrophotometric method; most probable number method.

1. Introduction

Total coliforms (TC) are a group of bacteria, which are found in intestines of humans and warm-blooded animals, in soils and water [1]. The concentration of TC is defined as an indicator of potential public health risk in food and drinking water and implies that pathogenic bacteria and viruses may be present [2]. The event of food-pollution and food-poisoning by coliforms pollution has been increased in recent years.

The most probable number (MPN) fermentation tube procedure has been used as the traditional method for TC detection over past decades in the world [3, 4, 5]. This method is simple to use, but need at least an over night growth and additionally 24-48 h for confirmation, resulting in time-consuming. Therefore, a number of rapid methods for TC detection of drinking water, waste water and food have been developed in recent years. For instance, a membrane filtration (MF) technique on a basis of incubation the

* Corresponding author.
Mailing address: No.29, 13th. Avenue, Tianjin Economic and Technological Development Area, Tianjin, China 300457; Tel.:+86 022 60601958; Fax: +86 022 60602298; E-mail address: lyh@tust.edu.cn(Yihan Liu)

bacteria-filter on a selective medium was developed for TC detection [6], which process was more simpler and analytical time was shorter than MPN method (within 24 h). European Union (EU), US Environmental Protection Agency (EPA) and American Public Health Association (APHA) have been used MF technique for water quality analysis in past few years. But this method was not suitable for detection, when the number of background heterotrophic bacteria was increased. Chromogens and fluorogens substrates, e.g., o-nitrophenyl-3-D-galactopyranoside (ONPG), p-nitrophenyl-3-D-galactopyranoside (PNPG), and 4-methylumbelliferyl-p-D-galactopyranoside (MUGal), were used for detecting β-galactosidase and β-glucuronidase enzyme existing in TC and *E.coli* within 24 h [7], which technique was applied by Colilert® method and PMEU method [8]. These methods have more sensitivity and specificity than MPN method and MF method, but it is expensive and often with some false positives. A polymerase chain reaction (PCR) technique also has been used for special detecting TC and *E.coli* by amplification of *lac*Z gene and genes specially for *E.coli* (*uid* A, *dct* A, *dcu* B, *frd* A, *dcu* S or *dcu* R), respectively [9,10]. This method contained the steps of MPN test in a short time and with high sensitivity. But it was not appropriate for detection, when samples contained low levels of TC. Based on these existing methods, there were some commercial productions for TC detection, like ChromocultR coliform agar (CCA, Merck, Germany), FluorocultR LMX broth and Petrifilmk *E.coli* count plates (PEC, 3M Center, St. Paul, MN, USA) and so on.

According to the report's methods and comparing with the commercial productions for TC detection, they were relatively rapid and sensitive, whereas were not cost-effective for widely application in many fields comparing with the MPN method. In China, a number of food industries and local laboratories were prior to use MPN protocol due to simply operations and economic cost. It is an emerging requirement of a rapid, accurate, sensitive, easy to use and economic method for TC detection in large-scale of industrial processes and quality inspections.

To address these problems, the triphenyltetrazolium chloride (TTC) detecting medium and spectrophotometry was combined for establishment a new TC detected method. After cultivation for 8 h, the intensity of the color change of the culture supernatant indicated the number of TC in samples. TC numbers could be determined in many kinds of samples using this new method. The new method was fast (within 8 h), simply to use, sensitive, economic. And it is the subject of a patent application in China.

2. Materials and methods

2.1 Bacteria, Sample Sources and Collection

E. coli, *Staphylococcus aureus*, *Salmonella enteritidis*, *Lactobacillus bulgaricus*, *Lactobacillus acidophilus*, *Micrococcus luteus*, *Pseudomonas aeruginosa*, *Acinetobacter haemolyticus/genospecies4*, *Proteus mirabilis* and *Proteus vulgaris* were obtained from Tianjin University of Science and Technology's center of culture collection (TCCC). Pollution-free eggs, biscuits and non-fermented soy products were collected from the supermarket. Each sample was weighed to 25 g and transferred into a 500-ml flask containing 225 ml sterilized saline, then blended for several minutes. Then the mixture was used in 1:10 dilutions for testing. Mineral water, purified water and liquid samples were selected from the supermarket. These liquid samples were used as stock solutions for later detection. Tableware samples were collected according to the China National Standard GB 14934-94 rapid slip method [11] from a college dining room at different dining times. All samples were diluted with sterile saline to maintain the number of microorganisms in the dilutions in the range 1 to 1000 colony forming units (CFU) per ml. 1 to 100 CFU/ml concentrations of bacterial standard liquid were made by using the laboratory stock of *E. coli*.

2.2 Medium Development

Various ingredients, indicators and pH values were examined during the optimization of the liquid medium. The materials and parameters are listed in Table 1. After all ingredients were dissolved in 1 l distilled water, 9 ml of the mixture was added to each detection tube and was autoclaved for 15 min at 115°C, then 0.5% (w/v) sterilized TTC solution was added into the medium.

Tab. 1. Parameters of medium optimization

Ingredients	Mass-Volume Concentration/levels
Lactose	0.5 %, 1 %, 1.5 %, 2 %, 3 %, 4 %
TTC	0.5 %, 1 %, 2 %
Bromocresol purple	0.014 ‰, 0.035 ‰
Phenol red	0 %, 0.03 %
pH	6.5, 6.6, 6.7, 6.8, 6.85, 6.9, 7.0, 7.2, 7.3, 7.49

2.3 Absorption Optimization

One ml of *E.coli* standard liquid (20 CFU/ml) was added to a tube containing 9 ml test-medium. After cultivation at 36 °C for 8 h, the liquid was centrifuged at 7,000 × *g* for 5 min at room temperature, and then the supernatant was scanned in a UV/visible spectrophotometer to determine the best absorption value for measurement. The supernatant was also scanned at intervals to determine the optimum incubation time.

2.4 Establishment of Standard Curves

A loopful of *E. coli* was transferred to a 100 ml triangular flask containing 50 ml Luria–Bertani (LB) broth and incubated at 36°C for 24 h. Then, 1 ml of the overnight culture was diluted with sterile saline. The number of CFU in each dilution was counted by the violet red bile agar (VRBA) method [12]. One ml aliquots containing *E.coli* at different concentrations were transferred to testing tubes containing 9 ml of optimized detection medium supplemented with 0.5 % (w/v) TTC solution. After cultivating at 36 °C for 8 h, the mixture was centrifuged at 7,000 × *g* for 5 min (room temperature). The supernatant was scanned at the optimum absorption value in a UV/visible spectrophotometer. Thus, the intensity of the color change of the culture supernatant can indicate the number of TC in the samples.

2.5 *Specificity Test of the New Method*

Non-coliform strains, *S. aureus*, *S. enteritidis*, *L. bulgaricus*, *L. acidophilus, M. luteus*, *P. aeruginosa*, *A. haemolyticus/genospecies4*, *P. mirabilis* and *P. vulgaris* were used to characterize the specificity of the new method for TC detection. One milliliter of standard liquid harboring 20 CFU/ml of a mixture of *E. coli* and one kind of non-coliform strains, each present of non-coliform strains at varying percentages (0 %, 20 %, 40 %, 60 %, 80 % and 100 %) in respective samples, was tested by the spectrophotometric method to determine the TC number.

2.6 *Sample Detection*

Mineral water, purified water, pollution-free eggs, biscuits, non-fermented soy products and tableware samples were selected for detecting TC by the MPN method and the spectrophotometric method.

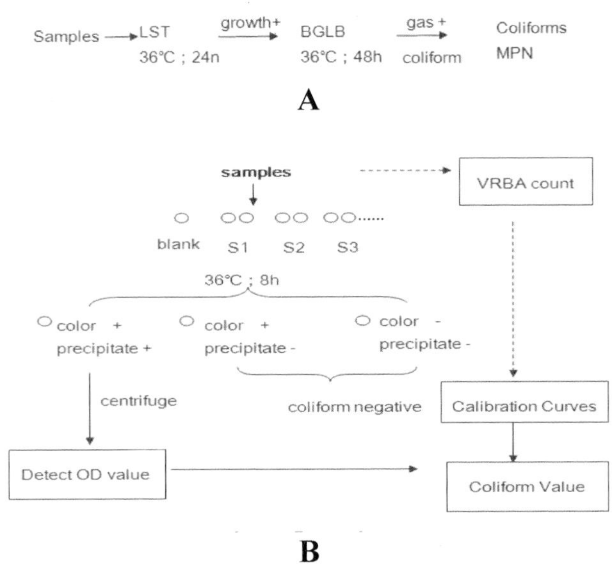

Fig. 1. A diagram illustrating the procedures followed in the study.
A The MPN method; **B** The spectrophotometric process

The MPN method [13] involves analysis of three tubes per dilution for coliforms (Fig. 1A). (i) Incipient experiment: 1 ml mineral water (for example) was transferred into a tube containing 9 ml lauryl sulfate tryptose (LST) and an inverted Duchenne tube. Each dilution was performed in triplicate. Three consecutive dilutions of samples (thus nine tubes in total) and a control tube, which contained the same volume of LST without the addition of any sample, were inoculated at 36°C for 24-48 h; a positive tube was defined by gas-production in the Duchenne tube. (ii) Recurrence experiment: a loopful of liquid from a tube that was judged positive in the incipient experiment was transferred into another tube containing 9 ml brilliant green lactose bile (BGLB) and an inverted Duchenne tube, and incubated at 36°C for 24-48 h; tubes that were gas-positive indicated the presence of coliforms. The coliform number in the sample was determined using the corresponding MPN table. Other samples followed the same process and all experiments were performed in triplicate.

Spectrophotometric method: coliforms produce acid by using lactose, which can make bromocresol purple turn yellow against a light purple background. At the same time, TTC in the medium will accept hydrogen from gas production and will be converted to triphenyltetrazolium formazide (TTF). This forms a red precipitate, indicating a positive culture. In the assay, 1 ml mineral water (for example) was transferred to a testing tube containing 9 ml optimum detection medium supplemented with 0.5% (w/v) TTC solution. The testing tube was cultivated at 36°C for 8 h. Then the mixture was centrifuged at 7,000 × *g* for 5 min (room temperature). The supernatant was scanned in a UV/visible spectrophotometer. As Fig. 1B shows, the intensity of the color change of the culture supernatant indicates the number of TC in samples. Other samples followed the same process. All experiments were performed in triplicate.

3. Results and Discussion

3.1 Development of a Spectrophotometric Method for TC Detection

In order to obtain the best testing medium for TC detection, five key components were optimized, namely the concentrations of lactose, mannitol, bromocresol purple, TTC and the pH of the testing medium. We found that the stable phase of the *E.coli* strain was reached most quickly when the composition of the medium per 100 ml of distilled water was as follows: 0.5 g lactose, 2 g tryptone, 0.5 g D-mannitol, 0.75 g NaCl, 0.02 g sodium dodecyl sulfate, 0.01 g sodium deoxycholate, 0.014 g bromocresol purple, and 0.5 g TT, with a pH of 6.9.

Analysis suggested that the composition of the testing medium, including lactose, tryptone and NaCl (growth ingredients) along with 0.5 % (w/v) D-mannitol supplementation could improve the osmotic pressure of cells and promote the metabolism of lactose. In a previous study, it was found that D-mannitol could prevent fading of the yellow color of the indicator bromocresol purple [14]. Furthermore, the presence of mannitol in the growth medium was helpful for confirming *E. coli*. Other research has shown that mannitol in the medium does not affect the false-negative or false-positive rates [15]. Therefore, we concluded that including mannitol in the medium was appropriate for TC testing in this study. The addition of sodium deoxycholate to the medium can increase cell membrane permeability and thereby inhibit the growth of Gram-positive bacteria [14].

TC strains which utilize lactose produce acid, which causes the color of a bromocresol purple solution gradually to change to yellow. The intensity of the color change of the solution could be used to indicate the numbers of TC present. This is the basis of our detection method. Optimum medium was prepared at the optimum pH (6.9) for growth of coliforms. This is also close to the color change point of bromocresol purple when it is used to verify acid production [16]. Prior to this study, other research on TC coliform detection has focused on the use of TTC agar medium or spectrophotometric detection [14,17]. However, these two different methods have hitherto never been combined. In order to determine the appropriate absorption wavelength for the study, 10 ml of culture liquid containing *E.coli* strain (20 CFU/ml) was incubated at 36°C until the color of the liquid was stable and red precipitate gathered at the bottom of testing tubes. By this means, 420 nm was determined to be the appropriate wavelength with which to measure experimental absorption values (Fig. 2A). The solution was tested at 420 nm every 30 min whilst being cultured. The results are illustrated in Fig. 2B. After cultivation for 8 h, the absorption value at 420 nm was essentially stable; therefore, 8 h was chosen as the appropriate incubation time for our method. Note that absorption did not reach a suitable stable plateau at the other absorption maximum in Fig. 2A (~638 nm).

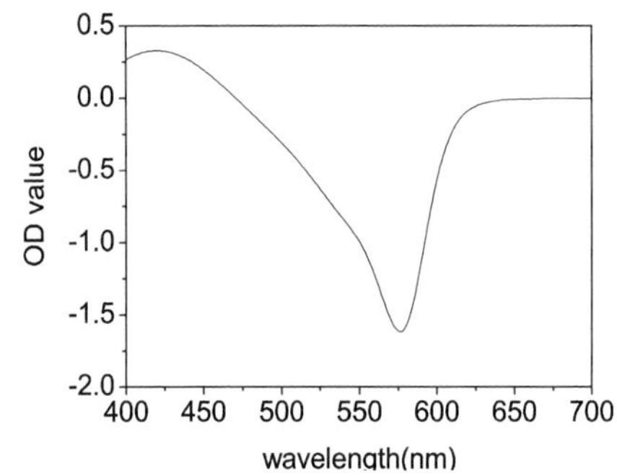

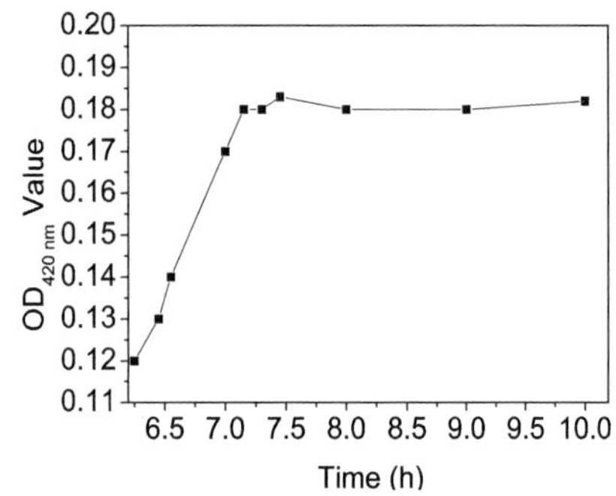

Fig. 2. Optimized absorption and cultivation time. **A.** OD values scanned from 400 nm to 700 nm; **B.** OD_{420nm} values of the culture according to the duration of culture

3.2 Establishment of a Standard curve for the Spectrophotometric Method

Standard liquid containing different concentrations of *E.coli* strain were used to investigate the relationship between the number of coliforms and the absorbance of the culture solutions (Fig. 1B).

Under optimum conditions, the OD_{420nm} of different concentrations of *E.coli* standard liquid were measured and used to produce a standard curve to be used to determine the concentrations of *E.coli* in samples in which its concentration is not known. Our results showed good agreement between the predicted and the experimental counts of *E.coli* strain ($R^2 = 0.982$; Fig. 3), as in equation (1).

$$Y = 14.7X - 1.294 \quad (1)$$

where the parameter X represents the absorption values of samples and Y represents the log of TC numbers. When the TC number increases, the acid generated in the culture liquid will also increase, as will the absorption value. Consequently, the standard curve provides a satisfactory model for predicting TC. To determine the relative deviation of the new method, 1 ml of standard liquid containing 20 CFU of *E.coli* was added to a tube containing 9 ml of testing medium and incubated at 36°C for 8 h. After centrifugation, the $OD_{420\ nm}$ values of the supernatant were examined. The assay was repeated six times. The results are shown in Table 2. The relative standard deviation (RSD) of the new method was approximately 3.0%.

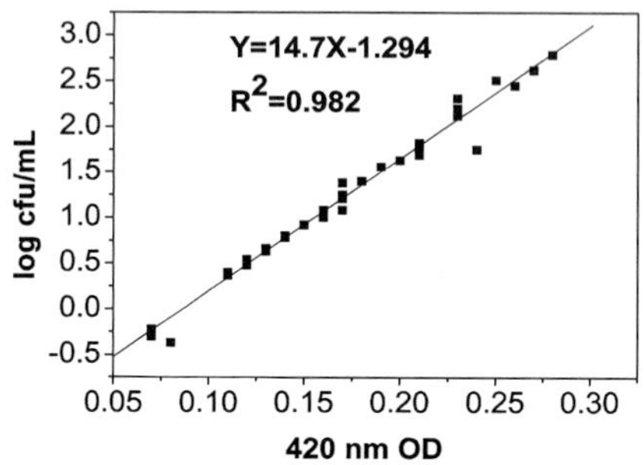

Fig. 3. Correlation between the number of coliforms and OD_{420nm} values. The calibration curve was established by using the OD_{420nm} values as the abscissa and the log of the number of coliforms numbers as the ordinate.

Tab. 2. Results of precision experiments (n = 6)

Assay number	1	2	3	4	5	6
OD_{420nm}	0.172	0.174	0.172	0.173	0.173	0.172
Concentration (CFU/ml)	17.1	18.4	17.1	17.7	17.7	17.1
RSD (%)	3.0%					

3.3 Analysis Specificity of the Spectrophotometric Method

Coliforms are generally described as Gram-negative, rod-shaped enterobacteriaceae that ferment lactose, producing acid and gas [18]. *S. aureus*, *S. enteritidis*, *L. bulgaricus*, *L. acidophilus, M. luteus*, *P. aeruginosa*, *A. haemolyticus/genospecies4*, *P. mirabilis* and *P. vulgaris* were chosen as non-coliform strains with which to analyze the specificity of the new method. *S. enteritidis, P. aeruginosa*, *A. haemolyticus/genospecies4*, *P. mirabilis* and *P. vulgaris*, are Gram-negative bacteriums, which do not use lactose. *S. aureus*, *L. bulgaricus*, *L. acidophilus* and *M. luteus* are Gram-positive bacteria, which can ferment lactose and produce acid. Samples of liquid containing *E. coli*, *S. ureus*, *S. enteritidis*, *L. bulgaricus*, *L. acidophilus, M. luteus*, *P. aeruginosa*, *A. haemolyticus/genospecies4*, *P. mirabilis* and *P. vulgaris* strains at different concentrations, with a total final bacterial concentration of 20 CFU/ml were established. The results of TC detection were not affected by different concentrations of *S. ureus*, *S. enteritidis*, *L. bulgaricus*, *L. acidophilus, M. luteus*, *P. aeruginosa*, *A. haemolyticus/genospecies4*, *P. mirabilis* and *P. vulgaris* strains, respectively (data not shown). The results indicated that bacteria which do not use lactose, have no effect on the test results and that Gram-positive bacteria which use lactose are inhibited by deoxycholate in the testing medium [19]. Therefore, we conclude that the new method is specific for TC detection.

3.4 *Sample Testing by MPN and the Spectrophotometric Method*

Pollution-free eggs, non-fermented soy products, biscuits, tableware samples, mineral water and purified water (140 samples in total) were used to determine TC numbers by the MPN and the spectrophotometric methods. The detection results for each by the spectrophotometric method were within the 95% confidence intervals obtained using the MPN method (detail data not shown).

TC numbers in the samples tested in our studies were relatively small. Hence, the results from the MPN and spectrophotometric methods could not be compared with each other by statistical methods. However, the results for these samples can be compared by national standards and industrial standards [11, 20-24] (Table 3) to investigate differences between the two methods (Fig. 4). As the investigative findings show for all the different sample types, the unqualified rates, which are defined as the ratio of samples not complying with national or industrial standards to total samples, demonstrated that the sensitivity of the new spectrophotometric method exceeded that of the MPN method.

Tab. 3. The details of samples for TC detection

Sample Name	NO. of Sample	National Standard/Industrial Standard	Detection Limit
Pollution-free Eggs	20	NY5039-2005 [20]	$\leq$ 100 MPN/100 ml
Non Fermented Soy Products	15	GB2711-2003 [22]	$\leq$ 150 MPN/100 g
Biscuits	15	GB7100-2003 [23]	$\leq$ 30 MPN/100 g
Tableware Samples	30	GB14934-94 [11]	$\leq$ 3 MPN/100 cm^2
Mineral Water	30	GB 8538-2008 [24]	$\leq$ 3MPN/1 ml
Purified Water	30	GB17324-2003 [21]	$\leq$3MPN/100 ml

Some investigators found that the MPN method demonstrates a Poisson distribution. However, the extremes of the distribution were not established in many of these studies [25]. Therefore, MPN tables could not be used to determine some concentrations, particularly at the extremes of the distribution range [26]. A major disadvantage of the MPN method, besides being time consuming and cumbersome, was that it could not be used to measure samples that did not reach the threshold of sensitivity. For instance, in our research, the MPN method result for one pollution-free egg sample was TC <3 per ml, while hygiene monitoring standards require the TC number to be <1 per ml. In this situation, the MPN method was inadequate for testing needs. A crucial advantage of the spectrophotometric method, besides short testing time, was accurate measurement. Thus, the results would reduce the fraction of defective (non-testable) samples. Detection of TC numbers could report actual results and provided a warning when samples were close to detection limits. This would be beneficial for preventing problems of food security. The new method had already been successfully applied in tableware detection and water quality monitoring.

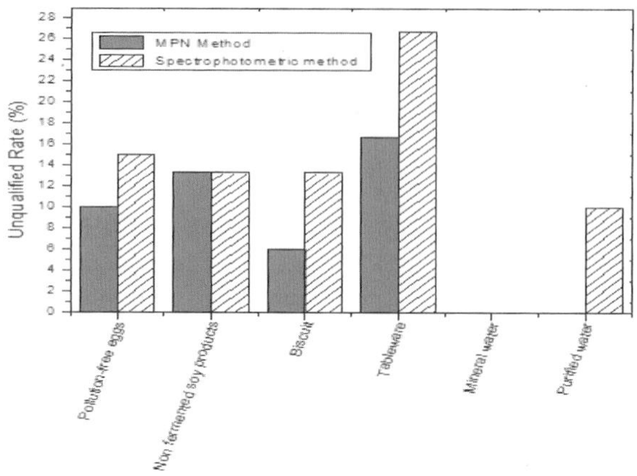

Fig. 4. Unqualified rates of six kinds of detected samples using the MPN method and the spectrophotometric method

4. Conclusion

A rapid new method for TC detection based on spectrophotometry was developed. Predicted results of the new method showed good agreement with experimental results ($R^2 = 0.982$). Comparing test results from the new method with the MPN method for six types of sample (total 140 samples), the spectrophotometric method was demonstrated to be suitable for quantitative and qualitative determination of TC, and it will be useful in many fields of TC monitoring.

Acknowledgements

This work was supported by the National High-Tech Research and Development Plan ("863" Plan) (2012AA022108), the National Natural Science Fund of China(31571805), and the Tianjin Research Program of Application Foundation and Advanced Technology (14JCYBJC23800).

References

1. J. Victoria, B. S. Colclasure, J. Thomas, M. S. Soderquist, B. S. Thomas Lynch, B. S. Nina Schubert, et al., Coliform bacteria, fabrics, and the environment, *Am. J. Infection Control,* 43, 154-158 (2015).
2. Gabriel. *Microbiology of Drinking Water Production and Distribution,* Wiley Blackwell, 2014, http://www.wiley.com/go/permission. Accessed August 2014.
3. R. J. Blodgett and W. E. Garthright, Several MPN models for serial dilutions with suppressed growth at low dilutions, *Food Microbiol.* 15, 91–99 (1998).
4. A. Hoszowski, A. D. E. Fraser, B.W. Brooks and E. M. Riche, Rapid detection and enumeration of *Salmonella* in chicken carcass rinses using filtration, enrichment and colony blot immunoassay, *Int. J. Food Microbiol.* 28, 341–350 (1994).

5. G. Moore, C.Griffith, A comparison of traditional and recently developed methods for monitoring surface hygiene within the food industry: an industry trial, *Int. J. Environ. Health. Res*. 12, 317-329 (2002).
6. M. A. Grant, A new membrane filtration medium for simultaneous detection and enumeration of *Escherichia coli* and total coliforms, *Appl. Environ. Microbiol*. 63, 3526-3530 (1997).
7. S. Bascomb, Enzyme tests in bacterial identification, *Methods Microbiol.* 19, 105-160 (1987).
8. E. Hakalehto, A. Heitto, L. Heitto, Fast coliform detection in portable microbe enrichment unit (PMEU) with Colilert® medium and bubbling, *Pathophysiology*, 20 257–262 (2013).
9. D. Li, B. Liu, M. Chen, D. Guo, X. Guo, F. Liu and L.Wang, A multiplex PCR method to detect 14 *Escherichia coli* serogroups associated with urinary tract infections, *J. Microbiol. Methods*, 82, 71-77 (2010).
10. Z. M. Reza, A.Mohammad, K. Salomeh, A. G. Reza, S. Hossein, S. Maryam, et al., Rapid detection of coliforms in drinking water of Arak city using multiplex PCR method in comparison with the standard method of culture (Most Probably Number), *Asian Pacific J. tropical biomedicine*, 4, 404-409 (2014).
11. GB 14934-94. Hygienic standard for disinfection of dinner and drinking set [S]. China.
12. H. Ogihara, M. Ogawa, B. J. Skura and S. Nakai, Evaluation of chromogenic enzyme substrate mediums, chromocult coliform agar (CCA) and XM-G, by detection of freeze-, heat-, high-pressure-injured coliforms, and coliforms in food samples. *Food Sci. Technol. Research*. 10, 168-173 (2004).
13. A.S. Lepeuple, S. Gilouppe, E. Pierlot and M. R. de Roubin, Rapid and automated detection of fluoresceent total bacteria in water samples, *Int. J. Food Microbiol.*92, 327-332 (2004).
14. D. J. Reasoner, J. C. Blannon, E. E. Geldreich, Rapid seven-hour fecal coliform test. *Appl. Environ. Microbiol*. 38, 229-236(1979).
15. T. K. Graczyk, D. Sunderland, G. N. Awantang, Y. Mashinski, F. E. Lucy, Z. Graczyk, L. Chomicz and P. N. Breysse, Relationships among bather density, levels of human waterborne pathogens, and fecal coliform counts in marine recreational beach water. *Parasitol. Res*. 106, 1103-1108 (2010).
16. A. Mavridou, E. Smeti, G. Mandilara, P. Boufa, M. Vagiona-Arvanitidou, A. Vantarakis, G. Vassilandonopoulou, O. Pappa, V. Roussia, A. Tzouanopoulos, M. Livadara, I. Aisopou, V. Maraka, E. Nikolaou, G. Mandilara, Equivalency testing of TTC Tergitol 7 agar (ISO 9308-1:2000) with five culture media for the detection of *E. coli* in water samples in Greece, *Water Sci.Technol.* 61, 67-76 (2010).
17. Y. Ding, Y. G. Yin, Rapid detection of total coliforms counts in foods by spectrophotometry. *Sci. Technol. of Food Industry*. 33, 83-87 (2012) (in Chinese).
18. F. Molina, E. López-Acedo, R. Tabla, I. Roa, A. Gómez and J.E. Rebollo, Improved detection of *Escherichia coli* and coliform bacteria by multiplex PCR, *BMC biotechnology*, 15, 48(2015).

19. J. D. Phillips, M.W. Griffiths, Estimation of Gram-negative bacteria in milk: a comparison of inhibitor systems for preventing Gram-positive bacterial growth, *J. Appl Bacteriol.* 60, 491-500 (1986).
20. NY5039-2005. Agricultural industry criteria for Pollution-free eggs [S]. China.
21. GB17324-2003. Hygienic standard of bottled purified water for drinking [S]. China.
22. GB2711-2003. Hygienic standard for non-fermented bean products and gluten [S]. China.
23. GB7100-2003. Hygienic standard for biscuits [S]. China.
24. GB 8538-2008. Methods for examination of drinking natural mineral water [S]. China.
25. B. Shen, Y. Q. Wang, J. H. Xu, Data evaluation about the TEMPO instrument detection coliform in food using MPN theory, *Journal of Inspection and Quarantine*.19, 39-40 (2009) (in Chinese).
26. J. L. Oblinger and J.A. Koburger, Understanding and teaching the most probable number technique. *Journal of Milk and Food Technology (JMFT)* 38, 540-545 (1975).

The investigation on the risk of indoor environment in Wuhan city, China

Y.Z Yuan, B.J Huang*, S. Chen, Z.Q Min and M. Jiang
School of Chemical and Environmental Engineering, Jianghan University
Wuhan, 430056, China
**huangbijie1982@163.com*

The study, based on questionnaires, with residents of five typical urban areas----Dongxihu district, Xinzhou district, Caidian district, Jiangxia district and Qingshan district----- in Wuhan as the objects, is aimed to investigate five methods of reducing indoor air pollution, including natural ventilation, plant purification, active carbon adsorption, air cleaners use and curtains installation. The study has recorded the residents' adoption of these five methods, made an analysis of factors which affect residents' recognition of the methods, and evaluate the residents' health risks caused by indoor air pollution. The result shows that the acceptance of the methods varies greatly among the residents. Most of them accept Method A, followed by Method C, Method D, Method B and Method E successively. The acceptance of the method also varies from age to age. Group 2 (the age between 20 to 29) accepts least, followed by group 3 (the age between 30 to 39), group 1 (the age between 12 to 19) and group 4 (the age between 40 to 65). The residents do not think highly of these five methods, especially those people who are at the age of 20 to 29. There is a relatively low rate on the understanding about these five methods. Survey shows that the residents who know least belong to group 1, followed by group 3, group 2 and group 4 successively.

Keywords: Indoor Air Pollution; The Methods of Reducing Pollution; Environmental Risks.

1. Introduction

"Indoor air pollution" has become today's direct killer of the broad masses of the people health, followed the "soot pollution" and "photochemical pollution" [1]. In recent years, as the country's GDP raises continuously and people's living standards improve significantly, the use of various complex building and decoration materials, as well as pesticides, detergents, organic detergent and other organic life and poor air exchange capacity from indoor to outdoor and the relative density of residential and public places can cause indoor air pollution. According to the statistics, the acute poisoning accidents caused by air pollution are as many as four hundred every year. The number of poisoning is as high as fifteen thousand people. Therefore, the research on indoor air pollutant to the crowd health is getting more and more attention.

The current indoor air pollution that causes health risks, mainly includes the composition of indoor air pollutants, the concentrations of indoor air pollutant, the distribution and the toxicity of pollutants. Although many researchers have been working on exploring the ways and technology to reduce indoor air pollutants But the public are lack of acceptance of the methods, In view of this, we study in the form of a questionnaires, count the degree of acceptance of the public of the methods, analyze and discuss the factors that affect the degree of acceptance, and evaluate the environmental risks of indoor air pollutants to residents, which can provide basic data and theoretical basis for controlling technology of indoor air pollution, evaluation of equipment effect, indoor air improvement and people's health protection.

2. Current research situation of indoor air pollution

The study on indoor air pollution began in 1960s in Europe, but our country began to pay wide attention to this in 1990s [2]. The main contents of the research included the detection of indoor pollutants, the distribution characteristics of indoor air pollutants, component of indoor air pollution, the relationship between indoor air pollution and health and the governance of indoor pollution. The testing included detection method and detection technology. As for the detection method of indoor, air testing mainly used the traditional chemical detection method, such as spectrophotometry and chromatography, etc. These methods were of high accuracy, but the operation was complex, the cost of equipment was high and the supplies of material was much great [3-6].

In terms of detection technology, "an indoor environmental monitoring system" was developed by Feng Aiming et al [7]. Miao Liuxi and Gong Xuemei developed “the indoor environment real-time monitoring and sensing transmission system” [8]. Zhang Yinping et al from Tsinghua University, developed a passive sampler for detecting the concentration of indoor air pollutants [9] and so on. But its cost was beyond the affordability of the people, resulting in people’s reluctance to take the initiative to trouble the testing department. There are many studies on the distribution regulation of indoor pollutants, such as Lee, et al.’s study on the environmental conditions of different indoor places in Hongkong [10], Ohura et al’s study on the distribution of indoor environmental pollutants in Shizuoka city and Hangzhou city [11]. Fitzger-ald et al’s study on the distribution and pollution level of PCBs which was from old buildings in the upstream of New York Hudson River [12]. In the aspect of indoor air pollutant composition, Zhang Xuewei [13] found that the indoor air pollutants mainly included formaldehyde, benzene series, radon total volatile organic compounds (TVOC) and ammonia. In terms of the studies on indoor environment pollutant and toxicology, Polizzi studied the relationship between the volatilization of Trichoderma atroviride and sick building syndrome [14], Vincent et al studied the relationship between air conditioning and sick building syndrome [15] and Bull-inger et al. studied the effect of sick building syndrome on men and women[16]. In the terms of indoor air pollution and health, studies have found that the pollutants mainly included formaldehyde, benzene series, volatile organic compounds, ammonia and radon. Because of the different concentrations of them and the exposure time, these pollutants had a different degree of damage to people, including pathological building syndrome, building related diseases and Chemical substance allergy [17-18]. In the treatment of indoor air pollution, many researches mainly focused on the traditional methods including ventilation, adsorption, ozone purification and negative ion purification. Most of them had the defects of the secondary pollution and low purification efficiency; new control technology included photocatalyst, activated carbon fibre and plant purification method and so on. It has the characteristics of strong pertinence, high purification efficiency and no secondary pollution [19-23]. Based on the present study, indoor air pollutants and the health threat to human health have been determined and a series of measures to reduce and prevent indoor air pollution has been proposed, However, whether the public accept these measures and what the degree of acceptance is and what are the factors that impact the degree of acceptance, the current study is far less. Thus, we investigate the adoption of the methods which can reduce indoor contaminants in five typical cities of Wuhan city in the form of a questionnaire survey, and evaluate the health risks of indoor air pollution in Wuhan.

3. The questionnaire survey of the adoption of the methods in reducing indoor air pollution from the residents in the typical urban area in Wuhan city

3.1 Survey field and Methods of data statistics

In 2005 Jia'an Fang launched an indoor pollution monitoring for those newly decorated houses. A set of objective monitoring indicators included formaldehyde, benzene, toluene, xylene, radon, the total number of colonies. The results showed that the degree of pollution was different in summer and winter. The degree of pollution in summer was 4.91 times than in winter so that we chose to research in July and August. According to the distribution of regional per capita in Wuhan, we selected the urban residents of the five typical areas (Dongxihu, Xinzhou, Caidian, Jiangxia and Qingshan district) in Wuhan as the survey population. The respondents were divided into five stages, including group 1 (the age from twelve to nineteen), group 2 (the age from twenty to twenty-nine), group 3 (the age from thirty to thirty-nine), group 4 (the age from forty to sixty-five). A total of 1080 questionnaires were distributed and 881 copies were recovered. Many studies have indicated that the measures of reducing indoor air pollution are mainly in natural ventilation, absorption, air purification, etc.

So we chose the ways of natural ventilation as method A, plant purification as method B, buying materials like active carbon adsorption as method C, using air cleaners as method D and the installation of curtain as method E, which are our content of investigation. Each method has four types of subjects, and some of which has multiple choices. Finally, we used Excel to carry on the data processing, and combined the form and the graph to count the acceptance of each method, then we analysis the influence factors in acceptance, in the end we evaluated the health risk condition of indoor air pollution in Wuhan city.

3.2 Survey results

3.2.1 Residents' acceptance of the five methods has larger difference

On the choice of the method, we set for multiple choices. The investigation results showed that method A was selected by most people (Table 1), which accounted for 86.5% of the total number. Then the next one were method C and method D, and the last one were method B and method E. This reflected that the residents' acceptance of the five methods was relatively different, which might be related to the promotion of the each method. Method A has been the most original and simple one, which everyone can do it easily. While method C and method D can also be done in a family which has a better economic condition, on the other hand the sales of them is gradually improving in the market. However method B and method E are comparatively uncommon in the market, also the price is relatively expensive. In addition, people in different age have different acceptance to the five methods. Investigation results showed that the acceptance in group 2 (the age from twenty to twenty-nine) and group 3 (the age from thirty to thirty-nine) were much higher than in group 1 and group 4 (the age from forty to sixty-five), which is because of different living environment. Most people aged from 20 years-old to 30 years-old (in group 2 and group 3) have been married or have children or are about to get married, so they pay more attention to indoor air quality. While people in group 1 (the age from twelve to nineteen) are students, whose consciousness on environmental quality is not strong enough. People in group 4 are not sensitive to those fresh things so that they prefer to method A. Therefore, for people in group 1 and group 4, they must strengthen the awareness of preventing indoor air pollution; for those 5 different methods, we should increase the publicity on method B, D, C, E.

Tab. 1. Public acceptance to the five methods in Wuhan

Method	Group 1	Group 2	Group 3	Group 4	Total number
A	150	216	273	123	762
B	10	34	11	22	77
C	22	60	26	19	127
D	19	54	26	19	118
E	9	41	11	10	71

3.2.2 *Urban residents' understanding degree on the five methods is low*

The percentage of the people who do not understand the 5 methods means that people only know that the method can purify air, but do not know why it works. They lack the knowledge of methods. The investigation results show that the percentage of the people who do not understand the 5 methods is high. People in different age have different understanding of it (Table 2). The percentage from high to low is group 4, group 1, group 3, group 2.

Tab. 2. The percentage of the people who do not understand the 5 methods

Method	Group 1	Group 2	Group 3	Group 4
B	17.79%	15.14%	6.90%	36.67%
C	23.75%	21.74%	19.57%	9.09%
D	33.75%	30.95%	38.46%	45.83%
E	35.56%	30.5%2	32.74%	56.78%

As for method B, D, E, people in group 4 have the lowest understanding than the other groups, which may be related to their acceptance to fresh things. The development of method B, D, E is a little bit late than method C and they have a certain scientific and technological content, which led to the limitations of being chosen; the percentage of not understanding of group 1 is almost the same among each method, which might be related to their limited knowledge of indoor air pollution and their weak protection awareness on indoor environment. On the other hand, almost all the people in group 1 are students, they are not so convenient to receive information than group 3. The percentage of not understanding of group 2 and group 3 is almost the same among each method. The reason may be that they live in the information age so that they can obtain the relevant knowledge by Internet, mobile phones and various information platform, which means the degree of understanding to methods is relatively higher. On the other hand, most people in group 2 have their own family or are about to get married, so they pay much attention to their living environment including the problem of indoor air pollution.

3.2.3 Urban residents' recognition degree of 5 methods are not high

People in different age do not give too much approval of the effective in these 5 methods (Fig. 1). People in group 2 are the highest one in the percentage of not recognition. Because they are the one who have a child or want to get married, in the future they will have a much higher demand for indoor air quality and it is likely for them to use a clean product. That is to say, the government should increase the intensity of supervision, satisfy the people's needs, and then expand the market of purification products, which will contribute to the development of indoor environmental protection. On the other hand, the low recognition on the effect of method might be related to the products itself. Now the boom of environmental protection has been increasing continuously, some criminals catch the opportunity to develop environment-friendly products. However many products are directly sold in the market without testing on effect, which disrupt the market order so that the quality of products could not be guaranteed. It also might be that as a matter of fact the indoor environment is very clean, but the quality of outdoor environment is not good enough. After a poignant contrast, the quality of the indoor environment could not be recognized by the public under the influence of the outdoor environment. This conclusion is consistent with the conclusion of the physical model used by Xiao Mingxing. They think that there is a close relationship between the concentration of indoor and outdoor pollutants. The outdoor air has a great influence on the quantity and distribution of indoor pollutants. The change trend of the two is basically consistent under normal air tightness. In other words, if the concentration of outdoor pollutants is high, the indoor pollutant concentration will increase at the same time. On the other hand, it is related to the people themselves. The survey show that the understanding degree of urban residents on these five methods is not high, especially to plant purification. More than 50% people in each group say that they do not know which specific plants can absorb indoor pollutants. If people do not understand how the method works may also lead to the results, because people cannot find the right way to clean the air, which will lead to the low effective utilization rate of the product, so we should strengthen the propaganda of each method.

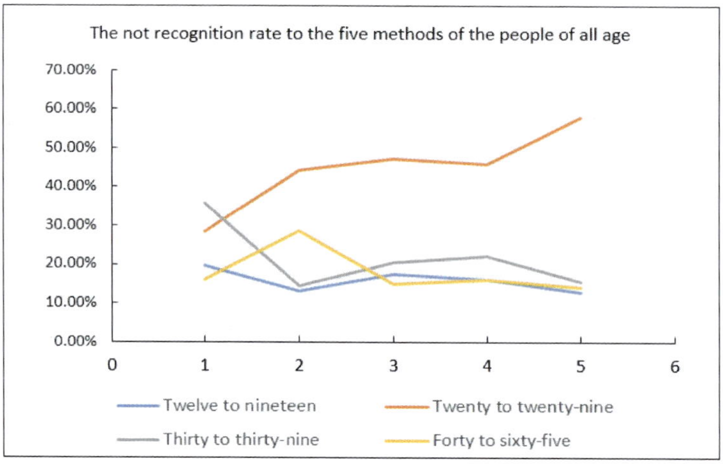

Fig. 1. The percentage of not recognition on five methods of the people in different age Number 1, 2, 3, 4, 5 in x-axis means methods A, B, C, D, E)

4. Conclusions

1. Urban Residents from typical districts in Wuhan have large difference in the acceptance of the five methods. The order of acceptance from high to low is method A, method C, method D, method B, method E. On the other hand the acceptance is also varies in different ages, the survey results show that people in group 2 and group 3 are more likely to accept the methods than group 1 and group 4.
2. The degrees of approval of the residents to the effectiveness produced by five methods is not high, among them people in group 2 have the highest disapproval. As for method B, C, D, E they are much higher than other groups. Also they are the one who will have a much higher demand for indoor air quality and indoor environmental quality in the future. Therefore, the indoor air purification products can be sold in the market before strict testing, which is in order to ensure the quality of the products.
3. The degree of not understanding to the five methods is relatively high among people in different age. The order from high to low is group 4, group 1, group 3, group 2. But physical resistance of the people in group 4 and group are relatively weak and they are susceptible to infection, so propaganda should be focus more on these two groups of people.
4. To some extent, the acceptance to the five methods could reflect the residents' awareness of reducing indoor air pollution. According to the degree of acceptance, it can be drawn that people in different age all have the awareness of reducing indoor air pollution. But the intensity of consciousness is not high, of which people in group 3 is much higher. And then followed by people in group 2. However the awareness of environmental protection is weak in group 1 and group 4. Therefore we conclude that indoor air pollution is a great environmental risk to residents in Wuhan city.

Acknowledgements

This work was supported by Scientific Research Foundation for excellent student, Jianghan University and Scientific Foundation for Talented Scholars (1009-06590001), Jianghan University. The students who come from the college of chemistry and environmental engineering in Jianghan University, included Jing Zhan, Meng Liu, Bingxin Xia, Yuhao Ge, Ke Luo, Yan Ru Yang, Xing Chen, Bichun Ma, Bing Bing Zhao participate in the investigation. I'm grateful to these students who provide assistance in the research process.

References

1. State Environmental Protection Administration of science and technology standards division, China Environmental Science Society, indoor environment and health [M]. Beijing: China Environmental Science Press. 2002.
2. Zhang Yuanhang, Tang Xiaoyan, Shao Min. Beijing: Higher Education Press, 2006.
3. SUZUKI Y, NAKANO N, SUZUKI K. Portable sick house syndrome gas monitoring system based on novel colorimetric reagents for the highly selective and sensitive detection of formaldehyde [J]. Environ Sci Technol, 2003, 37(24): 5695-5700.
4. Guo Zhanjing, Jia Jinhai, Zhang Xiaolin. Survey of air quality in the newly decorated rooms of 69 [J]. Modern preventive medicine, 2007, 34 (15): 2883-2884.
5. Zhao Hong, Liu Wenjun, white light. Determination of ammonia in indoor air pollutants analysis method for [J]. China environmental monitoring, 2003, 19 (3): 32-34.
6. Tian Yuanchun, Sun Yabing, Feng Jingwei. Research on the [J], HCHO and SO2 pollution in Nanjing City, city, 2007, 27 (2): 190-194.
7. AI Ming Feng, Hou Dexin, Shu Liang Ye, et al. A kind of indoor environment monitoring system: China, CN201716041U [P], Hangzhou City, Zhejiang Province Xiasha Higher Education Park source street. 2011-01-19:12.
8. Miao. Liuxi, Gong Xuemei. Indoor environment real time monitoring and sensing transmission system: China, CN102213604A [P]. Shanghai Jing'an District Xiangyang Bei Lu 6nong, room 103, No. 43, 2011-10-12.
9. Xu Qiujian, Zhang Yinping, Li Xinxiao, et al. indoor air pollutant concentration detection of passive adsorption sampling device: China, CN101852691A [P]. Haidian District, Beijing, 1, 2010-10-06.
10. LEE S-C, GUO H, LI W-M, et al. Inter-compar-ison of air pollutant concentrations in different indoor environ-ments in Hong Kong [J]. Atmospheric Environment, 2002, 36(12):1929-1940.
11. OHURA T, AMAGAI T, SHEN X, et al. Comparative study on indoor air quality in Japan and China: Characteristics of residential indoor and outdoor VOCs[J]. Atmospheric Environ-ment, 2009, 43(40) : 6352-6359.
12. FITZGERALD E F, SHRESTHA S, PALMER P M, et al. Polychlorinated biphenyls (PCBs) in indoor air and in ser-um among older residents of upper Hudson River communities[J]. Chemosphere, 2011, 85(2) :225-231.

13. Zhang Xuewei. The present situation of indoor air pollution and the prevention and control of [J]. Science and technology vision, 2012 (6): 153-158.
14. POLIZZI V, ADAMS A, PICCO A M. Influence of environmental conditions on production of volatiles by trichoderma atroviride in relation with the sick building syndrome [J]. Building and Environment, 2011, 46(4): 945-954.
15. BULLINGERM, MORFELD M, VON MACKENSENS. The Sick-Building-Syndrome-Do women suffer more? Das Sick-Building-Syndrom-Leiden Frauen mehr [J]. Zentralblatt für Hygiene und Umweltmedizin, 1999, 202(2-4): 235-241.
16. Wei Chi, Zhuang Zhixiong, Yuan Jianhui. Low dose formaldehyde induced MRC-5 damage tolerance differential expression gene was studied by [J]. Health research, 2005, 34 (3): 265-268.
17. Yang Zhenyu, Ji Li. Indoor air quality evaluation and related research [J]. Environmental science and technology, 2003, 26 (6): 61-63.
18. Chen Xiaochun, Zhou Dong, Qiao Yi. Study on sick building syndrome ([J].) environmental health engineering, 2003 (2): 101-103.
19. Sheng Gong, Xiao Rong Huang, Luo. Indoor air purification. Environmental pollution control technology and equipment, 2004, 5 (4): 55 - 57.
20. Cai Jian, Hu, Zhang Yan. Study on the adsorption properties of modified activated carbon fibers to formaldehyde adsorption properties of [J]. Environmental science and technology, 2004, 27 (3): 16-17.
21. Wang Jiajia, Shi Bing, Liu Xiaodong et al. Study on the indoor air purification ability of 3 species of woody plants [J]. Northern horticulture, 2007 (11): 142-143.
22. Ding Zhen, Daniel Chan, Lin Ping et al. Study on the photocatalytic degradation of formaldehyde in gas phase by nano [J]. TiO2 environmental science research, 2006, 19 (4): 75-79.
23. ICHIURA H, KITAOKA T, TANAKA H. Removal of indoor pollutants under UV irradiation by a composite TiO2zeolite sheet prepared using a papermaking technique [J]. Chemosphere, 2003(50): 673 -678.

The deficiencies and directions of rural settlements consolidation in China

Dongmei Li, Dongyan Wang
Earth Sciences College, Jilin University,
Changchun, Jilin, 130061, China
E-mail: dmli14@mails.jlu.edu.cn, wang_dy@jlu.edu.cn

Rural settlements consolidation is the best choice to carry out the new-type urbanization strategy, but current rural settlements consolidation is insufficient, which may restrict the rural settlements consolidation in the process of new urbanization. In order to explore the current situation and development trend of rural settlements consolidation research, this paper firstly gave a general review of rural settlements classification research from different major angles, then pointed out the major deficiencies and pressures currently existing in rural settlements consolidation based on the analysis of the related literatures. Based on this, we gave some appropriate suggestions for rural settlements consolidation in the process of new-type urbanization, so as to provide references for relative researches. The paper concludes that the deficiencies and pressures currently existing in rural settlements consolidation are typical and common. In the new period, rural settlements consolidation should be people-oriented and integrate smoothly into rural land comprehensive improvement project. In order to realize the high quality rural settlements consolidation and steadily push forward the new-type urbanization, It should develop a new mode of community to promote the population agglomeration; It should give priority to peasants' rights and interests and gradually realize intensive land; It should timely introduce rural land financial mechanism to augment support of rural residential renovation funds; It should focused on rural public infrastructure to improve the living environment and people's well-being, at the same time, it should pay attention to the landscape construction and ecological protection, too.

Keywords: Land Use; Rural Settlements Consolidation; New-type Urbanization; Literature Review.

1. Introduction

Urbanization was the inevitable course to modernization, thus, the CPC central committee and the state council of China attached great importance to the development of urbanization. On the eighteenth national congress of the communist party of China, the new urbanization strategy was put forward, which emphasized the significance of the integration of urban and rural development, the improvement of people's livelihood and the construction of ecological civilization. Since 2000, China's urbanization rate increased quickly, but quality was not good. The problems brought by rapid urbanization was increasingly apparent[1-2], such as: simultaneous increase in urban and rural construction land scales with occupying cultivated land, inefficient land use in cities and towns[3] and faster land urbanization than population urbanization[4], which were all universal phenomena in many Chinese cities. Fang's research (2013) showed that Chinese urbanization was facing increasingly serious land supply bottlenecks in the future[5]. On the basis of urban construction and food security, villagers should be transformed to citizens, and the ecological environment should be improved at the same time. Given the current situations of land use, the rural residential land renovation would be the best choice.

The new urbanization strategy was human-centered, so it paid attention to social justice and protection of the farmers' rights and interests. Under the circumstance, rural settlements consolidation should improve the rural public service and ecological construction, but a series of problems existing would be the resistance. What's more, the natural factors, geographical factors, spatial distribution characteristics and rural household types jointly influenced the characteristics of rural settlements, regulatory directions or measures varied according to different characteristics of rural settlements to get the best effect. And the rural settlements consolidation should adjust measures to local conditions, too. On the basis of above analyses and references to relevant literatures, this study firstly analyzed influencing factors according to the characteristics of villages , then gave a generalization of the common deficiencies existing in rural settlements consolidation, and probed into the improvement directions at last, so as to provide references for related researches.

2. Influencing factors

According to different classification standards, influencing factors of rural settlements could be classified into different types. This paper summarized the types of the influencing factors from four below angles: the natural environment, location, economic conditions and spatial distribution characteristics of rural settlements.

2.1 Natural environment factors

The natural environment factors contained topography, ecological environment and the resources endowment, *etc*. Researches on rural settlements in arid areas were common in China, such as the northwest China arid area and arid oasis region. In terms of other ecological factors, Chinese researches included the rural settlements in karst landform areas, ecologically fragile areas and geological disasters areas. The rural settlements consolidation in ecological fragile districts should pay special attention to the impact on the environment. According to resource endowments, Chinese scholars' researches on the rural residential land consolidation are major in mining areas, the three gorges reservoir areas and farming-pastoral zones. Rural settlements consolidation in resource-based regions should consider the reasonable use and effective protection of resources. According to landform, Chinese landscapes were roughly divided into mountain, plain, plateau, basin and hilly area. In corresponding, the rural settlements could be divided into 5 types, rural settlements in plateaus, rural settlements in plains, rural settlements in mountains (mountain area, middle hills region), rural settlements in hilly areas and rural settlements in basin. Some settlements contained more than a kind of regional landform, such as the rural settlements in low hilly area of the loess plateau[6]. Different ecological environments resulted in different characteristics of the regional rural settlements attributed to landforms. Land reclamation in plains were less restrictive, but slope would be constraints to hilly and mountain areas.

2.2 *Geographical factors*

According to the relative position to city, rural settlements could be roughly divided into villages inside city, suburban rural settlements and exurban rural settlements. Villages inside city were the products of urbanization processes, including "villages in city" located in the internal downtowns and located in the urban fringe areas. Because of different sizes of city, suburb villages could be divided into rural settlements on the outskirts of big cities and near small or medium-sized cities. From the view of sociology, villages inside city referred to the paradox which villages had already located inside city from the regional perspective, but they still belonged to the traditional farming villages in terms of the social attribute[7]. As we all know, cities generally had strong effects on suburban rural settlements closed to cities. In addition to extra cost of time, suburban rural settlements could enjoy urban public services and infrastructures, while the exurban rural settlements should solve by self.

2.3 *Economic conditions*

Existing research (Zhang, 2012) showed that the differences in accumulation of natural assets and other assets, future livelihood could be used to divide rural households[8], and the proportion of rural household types could be used as the classification basis of rural settlements. With references to related researches (Wang, 2011), as shown in Table 1, the rural households could be divided into agricultural professional farmers, agricultural diversified farmers, multiple occupations farmers, non-agricultural diversified farmers and non-agricultural professional farmers[9]. Correspondingly, according to that whether the proportion of household type is more than half or not, the rural settlements can be divided into dominating agricultural professional settlements, auxiliary agricultural professional settlements, dominating agricultural diversified settlements, auxiliary agricultural diversified settlements, dominating multiple occupations settlements, auxiliary multiple occupations settlements, dominating non-agricultural diversified settlements, auxiliary non-agricultural diversified settlements, dominating non-agricultural professional settlements and auxiliary non-agricultural professional settlements.

Tab. 1 The classification standard for household

household type	classification standard
Agricultural professional farmers	Natural assets value is greater than 0.40, rest of the asset value is much higher than average
Agriculture diversified farmers	Natural assets value is between 0.25 to 0.40, the rest of the asset value is low
Multiple occupations farmers	Natural assets value between 0.10 to 0.25, the rest all assets in the average
Non-agricultural diversified farmers	Natural assets value is less than 0.10, the rest of the assets above average
Non-agricultural professional farmers	Natural assets value is 0 or social output value is above 0.80

Notes: from reference 9.

Depending on the degrees of regional asset accumulation, some scholars also divided rural settlements into the rural settlements in developed areas and in poor areas.

2.4 Spatial distribution characteristics

According to the spatial layout characteristics of rural residential areas, based on variable coefficient of the Voronoi Diagram (Cv) and its suggested standard values[10] (Table 2), Zhang, et al (2005) and Li, et al (2011) divided spatial distribution modes of rural settlements into even distribution, randomized distribution and aggregated distribution[11] or dispersed distribution, zonal distribution and clustered distribution[12], and scholars also adopted the combination of landscape index analysis and spatial analysis methods to classify the rural settlements.

Tab. 2 The recommended classification standard concerning spatial distribution

spatial distribution types	recommended threshold	suggestive value
evenly distribution	less than 33%	29%
randomized distribution	between 33% to 64%	57%
aggregated distribution	greater than 64%	92%

Notes: from reference 10.

Rural residential land, of course, could not belong to only one type. Comprehensive rural settlements often influenced by many factors. And natural environment, geographical factors, economic conditions or their own characteristics is only part of the conditions of rural settlements. As foreign rural land renovation is often originated from changing the unreasonable layout, thus, rural settlements consolidations in China need to base on fully understand of the status quo[13].

3. The deficiencies and pressures of rural settlements consolidation

Rural settlements consolidations were the activities of reconstruction, merge and reuse of villages by using engineering technology and adjustment of land property rights, made the rural construction land gradually concentrated and intensive, in order to realize the scientific and rationalized land use and improve farmers' productivity and living conditions as well as the rural ecological environment[14]. So rural settlements consolidation should adjust number and layout from the macroscopic angle and alter scale and internal structure from the microscopic angle, respectively[15]. However, the rural settlements consolidation still had deficiencies and pressures to be urgently solved, particularly since the “new national urbanization planning (2014~2020)” was issued.

3.1 The main aim was getting potentialities, and consciousness of ecological protection was weak

With the increase of population, the demand of construction land in China increased, resulting that the contradiction between supply and demand of construction land intensified. However, Liu’s study[16] (2010) showed that the national per capita of rural settlements was 228 m^2, far in excess of the national average standard of 150 m^2. Unable to settle down in cities, as migrant labors, farmers were reluctant to give up dwellings in countries even if which were idle[17]. As income increased and the willingness to improve living conditions rose, some farmers abandoned the original houses and built new houses on better areas, so the scales of idle and waste homesteads increased, and problems emerged, such as hollow villages[18]. However, subjecting to the demand of construction land, rural settlements consolidation in recent years focused on realizing potentialities, which neglected the ecological capacity and rural land use sustainability, as a result, the natural ecological barrier in the countryside might be destroyed.

3.2 The rural settlements was scattered and close to the cultivated land, but land consolidation was lack of adjust measures to local conditions

Rural settlements were formed over a long period, their spatial patterns were the wisdom of the ancients, but they were also affected by the natural, social and economic factors and other limited conditions[19]. Subjecting to limited production conditions, farmers were willing to arrange villages close to arable lands. Thus, majority of China's rural settlements were scattered and spatial consistent with arable lands[20]. However, differences of agricultural climate areas, levels of economy, even the rural population, or the density of the agricultural labor force could impact on the farming radius. Farmers provided poor management to cultivate lands located on the outer radius, resulting the low yield of farmlands, which was not an intensive way of land use[21]. Sixth national census bulletin showed that the rural population in 31 provinces, autonomous regions and municipalities in China mainland (including the serviceman) reduced by more than 133 million from 2000 to 2010. The statistics data of annual statistical bulletin of national bureau also showed that the population of permanent residents in rural areas decreased almost linearly from 2011 to 2014. For rural land consolidations in China, we paid more attention to the rural land intensive utilization or not, but the consideration about the decrease of rural population was deficient. As a result, the scattered rural settlement layout concentrated, and the original cultivated radius also changed, if there was no comprehensive consideration, it would cause the waste of cultivated land.

3.3 The sociological feature of the rural settlements was apparent, but the public participation in land consolidations was low

The historical rural settlements were geographical space and social network among acquaintance formed by blood, in-laws and clans[22], so rural settlements showed unique sociological characteristic which promoted the unity of villagers. Nowadays, the land compensation for farmers was not reasonable, which combined with different individual expectation and the violence of execution. If the interest of one family was under threat, the whole family even the whole villagers would resist together, resulting social contradictions and conflicts increased. Rural land consolidations in China, however, were mostly government-led. In the actual execution, top-down way of administrative decision led to that the right of ordinary people is easy to ignore. Due to the lack of legal system and platform, however farmers, the real beneficiary, had rarely substantive participation in rural settlements consolidations, which was not appropriate[23]. In addition, fund demand of land management was great, and its investment cycle was long, but it mainly relied on national financial fund, if the public participation was not enough, it would cause great pressure on the government[24]. These were not conducive to the rural settlements consolidations.

3.4 Rural residential structure had the characteristics of regional differentiation, and rural infrastructure was imperfect

Existing study (Jiang, 2007) showed that the level of productivity, the distance between villages and cities and other geographical factors together led to regional differentiation of rural land, and the land use structure and concentration ratio of rural settlements were also different[25]. In China, the proportion of residential land was large in the homestead, but land for basic service facilities was insufficiency, it caused imperfect infrastructure in a way. With the increase of rural income, differences between urban and rural areas had no obvious changed, its deep reason was the imbalance of urban and rural public services. The unequal possession of resources and cumulative assets in the market jointly widened the gap between city and country[26]. Urban and rural public services were not balanced performance in education, health care, employment services, and other aspects, land-lost farmers' social insurance was the most obvious example. Urban and rural residents with the same public demand could not enjoy the same public service and opportunity[27]. Inadequate public services limited the farmers' consumption demand, then inadequate consumption demand limited investment demand of traffic, communication, power supply, gas supply, water supply, education, medical and other aspects, which restricted the improvement of the rural public infrastructure, then rural development fell into a vicious circle [28].

4. The development directions of rural settlements consolidations

There is no denying that land reclamation had its particular dilemma: it should increase the scales of cultivated land as many as possible to guarantee the local land for construction, whereas it also affected or even damaged landscape[29]. In order to better implement the new strategy of urbanization, rural settlements consolidations need to face up to the existing deficiencies and chose measures, based on the characteristics of rural settlements. This study discussed the development directions of rural settlements consolidations, based on the national new urbanization planning (2014~2020).

4.1 Improving the rural environment and developing a new mode of community to promote the population agglomeration

For the integrated development of city and countryside, increasing construction land index should no longer be the main target of rural settlements consolidations. Even though, the rural residential areas with big renovation potentialities, especially the hollow villages, could be considered as the main objects of rural settlements consolidations[30]. Under the background of the new type urbanization, promoting rural population agglomeration and improving rural living environment should be as control targets and contents. Villages inside city could be changed into neighborhood committees to realize the urbanization. In environs (suburbs and outer suburbs), new rural communities is a suitable pattern for rural development practice in China. Without breaking the existing administrative system, decentralized villages could be centralized by relatively concentrated investment in infrastructure[31]. Hollow village renovation avoided waste of rural land, such as disorder construction and repeated investment. And, in terms of improving the rural environment, we need to build green infrastructure to ensure the accessibility of the villagers to green ecological environment, at the same time, we should combine the construction of new rural community with modern ecological concept of residential design. It would not only promote the centralization of rural population, but also improve the villagers' life quality and infrastructure sharing.

4.2 Respecting the farmers' rights and interests, and adopting elastic exit mechanism to gradually promote the land consolidation

Current problems of rural land consolidation in China were mainly caused by land compensation. Either the "violent demolition" or "nail household", were the results of the land compensation below the farmers' expectations, which was due to the neglect of farmers' rights and interests. As the sole life-support, the land was vital to peasants. If the actual economic compensation for land-lost cannot guarantee farmers' life level is not falling, it is bound to cause a lot of social problems. Except in accordance with the national standard, the compensation for rural settlements should also give consideration to personal actual demand, especially for settlements with higher economic expectations or near the urban. The attention to farmers' rights and interests, not only required the reasonable economic compensation and compensation procedure, but also required suitable way of compensation combined with actual situation. If farmers were willing to become urban residents and exit the homesteads, returning the rural land and obtaining monetary compensation should be allowed. If farmers need to continue to live in, they could be resettled with the "exchange house" and get compensation for the differences.

4.3 Introducing rural land finance combined with rural land exchange to expand financial source of rural settlements consolidation

In current mode of the rural settlements consolidation, all levels of government, village collectives and farmers formed a series of commissioned-agent relationships[32]. How to guaranteed the public participation, and alleviated the pressure of the money? We could try to combine rural settlements consolidation with rural real estate exchange and timely introduce rural land finance. Through the establishment of sample demonstration zone, the enthusiasm of farmers to participate in rural settlements consolidation was aroused. Farmers could manage the homestead usufruct by themselves within the legal scope, at the same time, the depth and fairness of the public participation in the process of land regulation could be guaranteed. Farmers could use land income for rural settlements consolidation projects, which could alleviate the pressure of the national capital. So, the current rural settlements consolidation should timely introduce rural land financial business, such as the land use right mortgage model, to enlarge the funds source of rural settlements consolidation.

4.4 Strengthening the rural basic public services to benefit farmers combined with the protection of rural landscape

Lv's study (2008) showed that[27] public services in China was mainly decided by the government supply ability, however, because of the lack of expected economic returns, the provided preference of all levels of government public service were inadequate. Thus, strengthening rural basic public services should be from the aspects of industry guidance. Combined with the actual situation, tourism industry could be developed. On one hand, it was conducive to protection and construction of rural landscape; on the other hand, it would promote regional development, then to the public service. The improvements of traffic condition, health care and education resources were especially need. From the perspective of the rural industrial development and infrastructure construction, the equal social security system of rural area should firstly be established[33], then the equalization of basic public services between city and country realized with it, to service rural areas and benefit farmers.

5. Conclusions

In general, rural settlements consolidation was a complicated and comprehensive engineering. The rural settlements consolidation had a certain background of historical evolution, for example, Tong's study (2011) found that villages inside city conformed to the general evolution rule of urban spatial morphology in a certain extent, so the villages inside city were suggested as city spatial forms to guide their rational development and should be combined with the city industry development and residents' demand for housing[34]. Rural settlements consolidation also had a certain realistic background, such as the new urbanization strategy which was not a simple increase of urban population and the expansion of urban land. Under the background of new type urbanization, rural settlements consolidation should, balance various types of urban and rural land, save and centralize land, perfect the basic public service facilities, improve rural production and the ecological landscape, realize the population urbanization and improve people's well-being. To inner urban areas, we should strengthen internal reform of villages, improve existing facilities and restructure the internal living environment. On the edge of the cities, we should base on the interaction between urban and rural areas to promote more intensive utilization of land resources. In the traditional agricultural regions, we should focus on the industry development path. In economically developed areas, rural residential land management should guide the reasonable use of land. But for poor areas, the focus of the land management should be promoting the development of regional economy. In conclusion, in the new period, the rural settlements consolidation should timely improve the existing deficiencies and meet the requirements of the new urbanization to achieve quality rural settlements consolidation.

Acknowledgments

We thank the 2012 Jilin province science and technology development projects.

References

1. Zhou Y, Sun X Z. Rethinking and Countermeasures on China's Urbanization Road. *China Population, Resources and Environment*, 2012, 22(4):56-59.
2. Lu D D, Yao S M, Li G P, *et al.* Comprehensive Analysis of the Urbanization Process Based on China's Conditions. *Economic Geography*, 2007, 27(6): 883-887.
3. Ni J. Land intensive utilization in the process of urbanization in our country and its countermeasures. *Economic Review*, 2006, 11: 40-42.
4. Chen F G, Zhang H G, Wu QT, *et al.* A Study on Coordinate Development between Population Urbanization and Land Urbanization in China. *Human Geography*, 2010, 25(5): 53-58.
5. Fang C L, Ma H T. New City District Development and Intensive Land Use in the Context of New-type Urbanization. *China Land Sciences*, 2013, (7): 4-9.
6. Jiao B B, Shi P J, Liu C F, *et al.* The Distribution of Rural Settlements in Relation to Land Form Factors in Low Hilly Land on the Loess. *Plateau Resources Science*, 2013, 08:1719-1727.
7. Wang Y P, Wang Y L, Wu J S. Urbanization and informal development in China: Urban villages in Shenzhen. *International Journal of Urban and Regional Research*, 2009, 33(4):957-973.
8. Zhang Y Y, Wang C, Wang L P, *et al.* Cultural Landscape Features of Different Types of Rural Settlement in a Shallow Hilly Region: A Case Study in Xingba Village. *China Land Sciences*, 2012, 11:45-53+90.
9. Wang C, Wang L P, Li X Q, *et al.* The Source of the Forward-security of Farmers' Livelihood and Settlement Integration: Based on the Survey of 477 Farmers in Bailin Village, West Suburbs of Chongqing. *Acta Geographica Sinica*, 2011, (8):1141-1152.
10. Duyckaerts C, Godefroy G. Voronoi tessellation to study the numerical density and the spatial distribution of neurons. *Journal of chemical neuroanatomy*, 2000, 20(1): 83-92.
11. Li Y Q, Qi W, Wang D, *et al.* Research on Spatial Distribution Characteristics of Rural Settlements in Mountainous Areas at County Level Based on GIS:A Case Study in Qixia City. *Geography and Geo-Information Science*, 2011, 03:73-77.
12. Chen Z J, Li M C, Liu Y X. A Gis Based Research on Spatial Distribution of Rural Settlements in Tonglu County. *Resources and Environment in the Yangtze Basin*, 2008, 17(2): 180-184.

13. Chen X H, Zhang X L, Liang D. The Practice of Rural Development and Build in Foreign Countries' Urbanization and its Enlightenments to China. *World Regional Studies*, 2005, 03:13-18+49.
14. Chen B M. Overview of land resources [M]. *Beijing: China environmental science press*, 1999: 294-297.
15. Liu Y, Wu C F, Yang Z R. Review and the Outlook of Residential Area Consolidation in Rural China. *China Land Science*, 2008, 22(3): 68-73.
16. Liu Y S, Liu Y. Progress and prospect on the study of rural hollowing in China. *Geographical Research*, 2010, 29(1) : 35-42.
17. Luo M Z, Lu Y X, Lu Z X. Rural Migrant Workers into the City, the Land Circulation and Ecology of Its Migration —— Based on the Questionnaire Survey and Analysis of Guangdong Province. *Rural Economy*, 2012, (2): 109-113.
18. Wang H L. The formation reasons and countermeasures of the rural "hollow village". *Rural Economy*, 2005, 9: 21-22.
19. Jin Q M. The History and Current T rends of Research on Rural Settlement Geography in China. *Acta Geographica Sinica*, 1988, 04:311-317.
20. Liu M H, Dai Z Z, Qiu D C, *et al.* Influencing Factors Analysis and Rational Distribution on Rural Settlements in Mountains Region. *Economic Geography*, 2011, 03:476-482.
21. Jiao Y M, Hu W Y, Su S H, *et al.* Spatial Pattern and Farming Radius of Hani's Settlements in Ailao Mountain Using GIS. *Resources Science*, 2006, 03:66-72.
22. Zeng S S, Zhou G H. Discrimination on Concepts Related Rural Settlements. *Yunnan Geographic Environment Research*, 2011, 03:26-31.
23. Zhao J N, Hong T L. Analysis of public participation present situation of our country land consolidation. *Acta Agriculturae Jiangxi,* 2010, 04:204-206.
24. Liu H M, Wang P, Wang K Q. Research on capital operation of rural land consolidation. *Public Finance Research*, 2011, (12): 20-22.
25. Jiang G H, Zhang F R, Zhou D Y, *et al.* Analyzing the Land Use Structure Characteristics of Rural Residential Area in Beijing City. *Resources Science*, 2007, 02:109-116.
26. Guo X F. Demonstration and Analysis based on Multi-Factors of Urban-Rural Income Gap in China. *China Population, Resources and Environment*, 2005, 15(4):17-21.
27. Lv W, Wang W T. Study on Equalization of Basic Public Services in China, Based on the Analysis of Public Demand and Government Ability. *Public Finance Research*, 2008, (5):10-18.
28. Zeng G L. The urban and rural public infrastructure spending disparity caused consumption gap. *Consumer Economics*, 2011, (4): 36-40.
29. Wang J. Land Consolidation Calls for Landscape Ecological Construction. *China Land Science*, 2011, 25(6):15-19.
30. Liu Y S, Yang R, Li Y. Potential of land consolidation of hollowed villages under different urbanization scenarios in China. *Journal of Geographical Sciences*, 2013, 23(3): 503-512.

31. Xu K S, Liu Y S. Theories and a Critical Review on Central Village Construction under Urban and Rural Harmonious Development. *Areal Research and Development*, 2011, 30(5): 7-11.
32. Liu Y. Inspiring Mechanism of Property Rights of Rural Residential Land Consolidation and Operating Measures. *Reform*, 2008, (006): 92-96.
33. Ma Y H, Zhang L J, Xu W H. Scientific Understanding of New-type urbanization and Advance Development of Realizing Integration Urban and Rural Area. *Urban Development Studies*, 2013, 20(7): 98-103.
34. Tong D, Feng C C, Deng J J. Spatial Evolution and Cause Analysis of Urban Villages: A Case Study of Shenzhen Special Economic Zone. *Geographical Research*, 2011, 03:437-446.

Molecular characteristics of L-galactose-1-phosphate phosphatase in cherry, a key enzyme involved in biosynthesis of AsA

Dong Liang, Ling Lin, Tingting Zhu, Hui Xia*
Institute of Pomology and Olericulture, Sichuan Agricultural University, Chengdu, Sichuan 611130, China
**Author for correspondence*
E-mail: susanxia_2001@163.com

The objective of this study was to investigate the characteristics of cherry L-galactose-1-phosphate phosphatase (GPP) and the relationship between its expression pattern and ascorbic acid (AsA) in cherry. cDNA of GPP, which was cloned from fruit of cherry, has an 813-bp open reading frame and encodes a protein of 270 amino acid residues. Both of its cDNA and amino acid sequence showed high similarity with GPP genes reported in other plants. AsA concentration was the highest in young fruit post-anthesis and then decreased steadily toward maturation. mRNA expression patterns of GPP showed a pattern of change similar to that of AsA accumulation in cherry fruit development. Also, AsA content and mRNA expression levels of GPP were detected in different tissues of cherry. The results showed that both AsA content and mRNA expression levels of GPP were higher in leaves than that in fruit, in young fruit was higher than that in mature fruit, and while in peel was higher than that in flesh.

Keywords: L-galactose-1-phosphate phosphatase; AsA; mRNA expression; Cherry.

1. Introduction

AsA (ascorbate) is the most abundant water-soluble antioxidant in higher plants and has important roles in promoting resistance to numerous environmental stresses such as ozone [1], water stress [2], excess light [3], and pathogen infection [4]. AsA also functions as a cofactor and participates in the regulation of several fundamental cellular processes [e.g., photoprotection, the cell cycle, and cell expansion] and important plant hormones biosynthesis [5-7].

De novo biosynthesis is believed to be a main reason for its accumulation in plant cells [8, 9]. To date, L-galactose pathway has been proposed to be the main route for AsA accumulation in various species, and several structural genes from this pathway have been proposed to be key regulators of fruit AsA concentrations in various species [10-12]. In this pathway, L-galactose-1-phosphate phosphatase (GPP) was considered playing a central role and can catalyzed the dephosphrization reaction of L-galactose-1-phosphate to produce L-galactose. Researchers have reported that GPP is a bifunctional enzyme that affects myo-inositol and ascorbate biosynthesis [13]. Studies have also indicated that GPP transcript abundance has good correlation with the rate of AsA accumulation in kiwifruit [14] and apple [15]. Moreover, it is the only one enzyme whose activity appears to be up-regulated in two strains of *Chlorella pyrenoidosa* AsA hyper-accumulating mutants [16]. Expression of GPP increased in leaves of *Arabidopsis thaliana* under continuous light, but decreased when plants are moved to dark conditions [17]. Transcript levels of GPP in tomato fruits also tend to decline when those plants are placed under shade [18]. Therefore, these reports have prompted us to investigate the novel regulatory mechanism for GPP expression in different fruit.

Cherry is one of the most popular and economically important fruit in the worldwide. Cherry fruits are widely recognized for rich sources of antioxidants and other nutritional and health-related metabolites, such as carotenoids, phenolic compounds, flavonoids, and AsA. Despite the characteristics of GPP in different plants had been investigated, the molecular properties of GPP in cherry were almost unknown.

To gain better investigate the molecular characteristics and regulatory mechanisms for GPP expression in cherry, as well as to understanding of the relationship between AsA amounts and transcriptional levels of GPP, we performed a systematic evaluation of AsA content, gene cloning and mRNA expression during fruit development. Our objective was to obtain a functional characterization of GPP so that this might possibly benefit future research on the regulation of gene expression.

2. Materials and Methods

2.1 Plant materials

Eight-year-old cherry (Prunus avium cv. Binku) were grown at a 3×4 m spacing in an orchard at the Hanyuan County, Sichuan province, China (29.51N, 102.62E). Fruits were harvested at 10±1d intervals following anthesis. The day on which the petals had just dropped off was designated as 0 DAA (days after anthesis). Fruits were immediately frozen in liquid nitrogen and stored at -80°C. On each collection date, at least fifth fruits, as one replication, were harvested from five trees, and five replications were done, respectively.

In addition, peel and flesh of fruits were harvested respectively on 10DAA (young fruit) and 50DAA (mature fruit). Leaves in the top and middle of branch were collected as young leaf and mature leaf on 20DAA. All these tissues were used to investigate mRNA expression of PacGPP and AsA content.

2.2 Measurement of ascorbic acid contents

Fruit (2g each) were homogenized in 8mL of ice-cold 6% (v/v) HClO4 and centrifuged at 12,000g for 20min at 2°C. Hepes buffer (0.1M; pH7.0) was added to the extracts at a 1:5 ratio (buffer:extract, v:v); K2CO3 (5M) was then added in until the pH reached 5.6. Extracts were centrifuged again at 12,000 g for 2min to allow the removal of precipitated K2ClO4, and the supernatants were used to assay total ascorbate (T-AsA) and AsA as described by Ma and Cheng [19]. That assay is based on the oxidation of AsA by ascorbate oxidase in an acidic solution. AsA was calculated as the difference in absorption at 265 nm before and after the addition of ascorbate oxidase. Its concentration was quantified by comparing the difference in absorption relative to a standard curve.

2.3 Gene cloning and mRNA expression

Total RNA was extracted from samples by the modified CTAB method [20], and DNase was used to clean out DNA before reverse-transcription reaction.

GPP cDNA fragments were amplified by reverse transcription polymerase chain reaction (RT-PCR) using gene-specific primers (Table1). The primers were designed from EST sequences in the Genome Database for Rosaceae (http://www.bioinfo.wsu.edu/gdr) (Table 1).

mRNA expression was evaluated by quantitative reverse transcription polymerase chain reactions (qRT-PCR) with gene special primers(Table1). The amplified PCR products were quantified by a CFX96 Touch™ Real-Time PCR Detection Systems (*Bio-Rad*), with a SYBR Premix Ex Taq kit (Takara). qRT-PCR experiments were done with four technical replications. All data were analyzed by the ddCT method, with CFX Manager™ Software. Actin transcripts were utilized to standardize the different gene cDNA samples throughout the text (Table 1).

Tab. 1 Primers used in this study

Primer name	Accession number	Primer and probe sequence (5_–3_)	Purpose
GPPs		CCGCATTTCCTTCTCTCATC	Cloning for GPP
GPPa		AGCTTCGAGCAAGATTACCG	
GPPs-real		GTACGTGGAGGAGGTGCATT	mRNA expression for GPP
GPPa-real		AGCCACTCATGCGAAGAGAT	
Actins	EC969944	CTTGCATCCCTCAGCACCTT	mRNA expression for reference gene
Actina		TCCTGTGGACAATGGATGGA	

2.4 Statistical analysis

Each sample was replicated at least four times, with each replication measured twice. Results were represented as means six standard deviation (SD).

3. Results

3.1 cDNA isolation of GPP in cherry fruit

Using degenerate primers, a 1034-base pair (bp) product was amplified using mature cherry fruit cDNA. This full-length cDNA, namely PacGGP, encoded a protein of 270 amino acids. When translated into protein, PacGPP was deemed to belong to inositol monphosphatase protein family (IPR000760). PacGPP demonstrated a high degree of homology with other GPP (Fig. 1). The amino acid identity were 78.3% similar to *Arabidopsis thaliana* GPP, 80.4% similar to *Nicotiana tabacum* GPP, 90.0% similar to *Malus* x *domestica* GPP, 90.1% similar to *Rosa roxburghii* GPP, and 97.8% similar to *Prunus persica* GPP respectively.

The phylogenetic analysis indicated that PacGPP is closely related to GPP of peach, apple and *Rosa roxburghii* (Fig. 2). PacGPP was clustered with *Prunus persica* GPP that are involved in the regulation of AsA biosynthesis in peach fruit [10]. These results suggested that PacGPP may play an important role in regulating AsA biosynthesis in cherry.

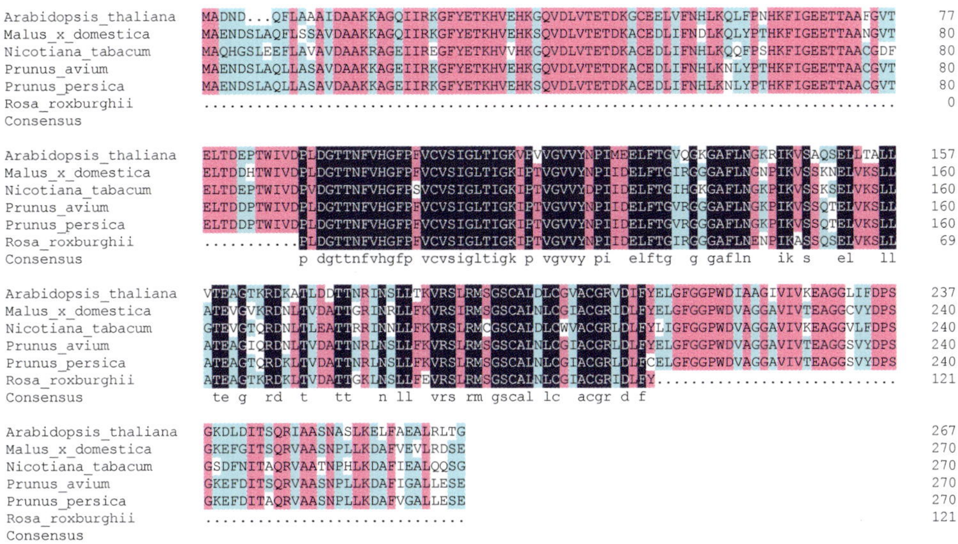

The accession numbers of these proteins were as follows in the GenBank database: *Arabidopsis thaliana*, NM_001125086.1; *Malus* x *domestica*, AY787586.1; *Nicotiana tabacum*, EU700060.1; *Prunus persica*, AB457584.1; and *Rosa roxburghii*, KC782567.1.

Fig. 1 Protein sequence alignment of PacGPP and the other known GPP in other species

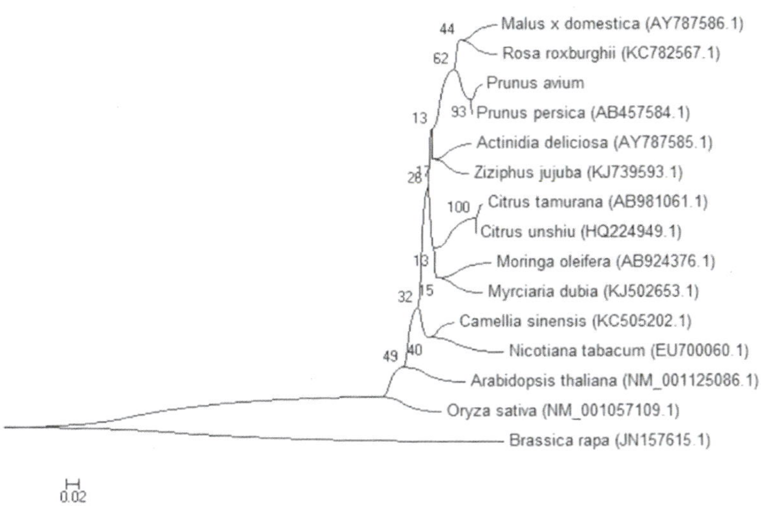

The tree was constructed using MEGA 5, neighboring-joining phylogeny testing, and 1,000 bootstrap replicates.

Fig. 2 Phylogenetic relationships between PacGPP and GPPs in other species

3.2 *Changes in T-AsA and AsA contents during fruit development and different tissues*

T-AsA and AsA contents were measured during fruit development and in different tissues. As shown in Fig. 3, The T-AsA and AsA concentration were highest at 0DAA (based on fresh weights), decreased slowly until 30DAA, and then remained constant until harvest.

In different tissues of cherry, leaves had higher remarkably AsA and T-AsA content than that of other tissues. The content of AsA and T-AsA in fruit was about 10% of that in leaves. The mature leaves had more AsA and T-AsA than that in young leaves, but in contrast, the content of AsA and T-AsA in young fruit was higher than that of mature fruit. Peel had higher AsA and T-AsA content than that of flesh, either in mature fruit or in young fruit (Fig. 4). All these data indicated that the AsA and T-AsA content was related with different tissues and growth status.

3.3 *Changes in mRNA expression of PacGPP during fruit development and different tissues*

Expression patterns of PacGPP during fruit development were examined using the real-time qRT-PCR. Relative expression levels of PacGPP showed progressively increased before 20 DAA and peaked in 20-DAA fruits then decreased greatly through fruit maturation (Fig. 5). From 30DAA to 50DAA, expression decreased by about 85%. Such expression pattern was similarly with the change of AsA content during fruit development.

The transcripts of PacGPP was ubiquitously detected in all six cherry tissues with the highest signal in the leaves and the lowest in the mature fruit, which was in agreement with the AsA content (Fig. 6). There were no obviously difference between peel and flesh of young fruits, peel and flesh of mature fruits.

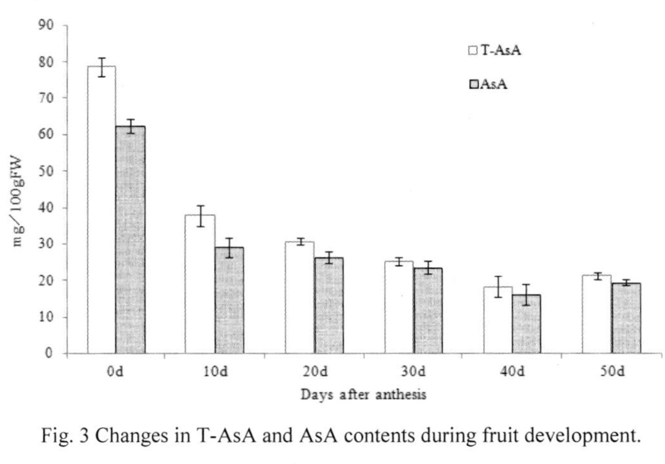

Fig. 3 Changes in T-AsA and AsA contents during fruit development.

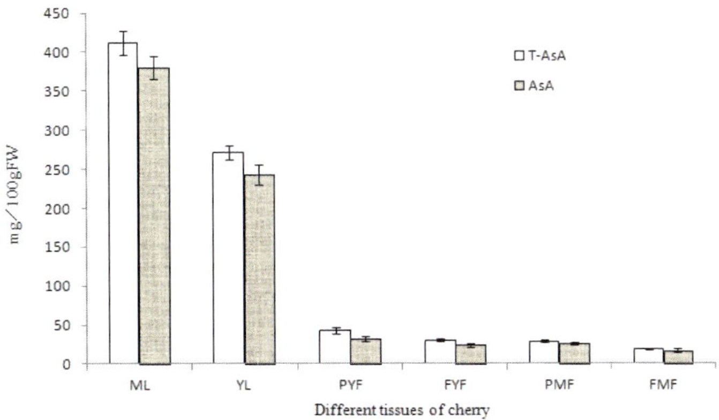

ML: mature leaf; YL: young leaf; PYF: peel of young fruit; FYF: flesh of young fruit; PMF: peel of mature fruit; FMF: flesh of mature fruit

Fig. 4 Changes in T-AsA and AsA contents in different tissues of cherry.

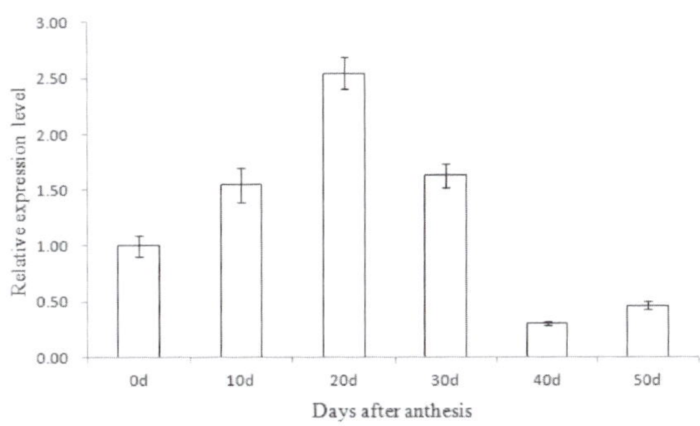

Fig. 5 Changes in relative mRNA expression for *PacGPP* during fruit development

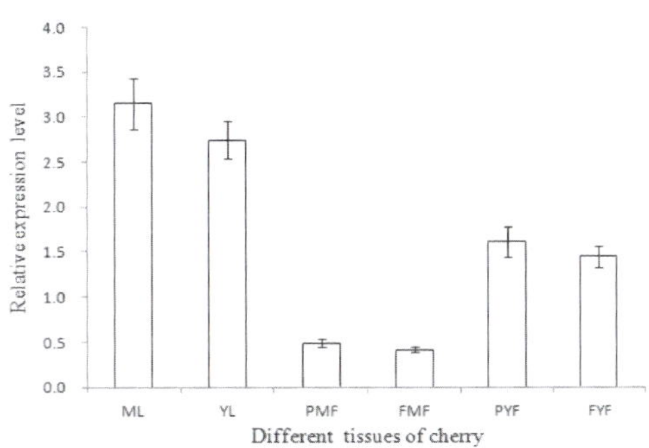

ML: mature leaf; YL: young leaf; PYF: peel of young fruit; FYF: flesh of young fruit; PMF: peel of mature fruit; FMF: flesh of mature fruit

Fig. 6 Changes in relative mRNA expression for *PacGPP* in different tissues of cherry fruit.

4. Discussion

Phenotypic analysis of mutant or transgenic plants provided strong supporting evidence for the operation of the L-galactose pathway as a main AsA biosynthetic route in plants [5-7]. In recent years, GPP, as a key enzyme in this pathway, have been identified and characterized in different kinds of fruits [10-12]. In the present study, the 1034-bp full-length PacGPP gene was isolated from the cherry fruit. Sequence analysis of the encoded PacGPP protein confirmed its high homology with GPP in other species (Fig. 1), e.g., *Arabidopsis*[5], apple[15], peach[10] and kiwifruit[21]. The encoded protein exhibited putative functional properties of inositol monphosphatase protein family by analysis of online bioinformatics website (Fig. 1). In previous studies, GPP of *Arabidopsis* [5] and kiwifruit [21] expressed in *E. coli* had very similar *myo*-inositol-1-Pase properties and highly specific ability to hydrolyze L-gal-1-P. These results suggest that PacMYB obtained in this study maybe a full-length and functional protein.

AsA concentration in plant cells is highly regulated by developmental processes. Our young fruit post-anthesis had the highest amounts of T-AsA and AsA, but those values then decreased steadily up to maturity (Fig. 3). This suggests that, during fruit development, the rate of AsA synthesis, based on fresh weight, is less than the rate of oxidation loss. Higher AsA concentration also has been reported in the young fruit of peach [10], bilberry [12], apple [15], acerola [22], kiwifruit [23], and blankcurrant [24].

During cherry fruit formation, we observed much greater transcription of GPP in young fruit (Fig. 5), similar with change pattern of T-AsA and AsA content (Fig. 3). This indicated that young fruit have a stronger ability to produce AsA. However, here we noted that increased mRNA expression of GPP in middle growth stage of fruit did not lead to a change in AsA concentration (Fig. 5), such results also found in peach fruit development [10]. Two pathways of ascorbic acid biosynthesis involve a phosphatase step: the L-gal pathway [11] and the *myo*-inositol pathway [25]. *Myo*-inositol metabolism is also of significance in the widespread phosphatidylinositol signaling pathway. Previous researches showed GPP had an affinity with L-gal-1-P and *myo*-inositol-1-P [5, 21]. Hence, we speculate parts of GPP were used to produce *myo*-inositol; this was the reason why there was higher mRNA expression level in middle growth stage of fruit. Further study is necessary for identifying the functioning of GPP in controlling AsA synthesis, especially research that involves ascorbic acid and *myo*-inositol biosynthesis.

Compared with leaves, the content of AsA and T-AsA in fruit was about 10% of that in leaves, and fruits had significantly lower GPP mRNA expression level than leaves (Fig. 6). Davey [26] and Li [17] also found AsA content was remarkable higher in leaves than that in fruits. These results indicated that leaves had higher capability of AsA biosynthesis than that in fruit of cherry. A lot of researches showed that AsA, GSH and Other antioxidant enzymes were mainly located in peel [27]. In this study, we also demonstrated that peel had higher content of T-AsA and AsA than that in flesh, and consistent with GPP mRNA expression level in different tissues.

References

1. M. Sanmartin, P.D. Drogoudi, T. Lyons, I. Pateraki, J. Barnes, A.K. Kanellis, Over-expression of ascorbate oxidase in the apoplast of transgenic tobacco results in altered ascorbate and glutathione redox states and increased sensitivity to ozone.

Planta 216:918-928. (2003)

2. Z. Wang, Y. Xiao, W. Chen, K. Tang, L. Zhan. Increased vitamin c content accompanied by an enhanced recycling pathway confers oxidative stress tolerance in Arabidopsis. J. Integr. PlantBiol. 52:400-409. (2010)
3. Y. Yabuta, T. Mieda, M. Rapolu, A. Nakamura, T. Motoki, T. Maruta, K. Yoshimura, T. Ishikawa, S. Shigeoka. Light regulation of ascorbate biosynthesis is dependent on the photosynthetic electron transport chain but independent of sugars in Arabidopsis. J. Expt. Bot. 58:2661-2671. (2007)
4. V. Pavet, E. Olmos, G. Kiddle, S. Mowla, S. Kumar, J. Antoniw, M.E. Alvarez, C.H. Foyer. Ascorbic acid deficiency activates cell death and disease resistance responses in Arabidopsis. Plant Physiol. 139:1291–1303. (2005)
5. M.W. Davey, M. Van Montagu, D. Inze, M. Sanmartin, A. Kanellis, N. Smirnoff, I. J. J. Benzie, J. J. Strain, D. Favell, J. Fletcher, Plant L-ascorbic acid: chemistry, function, metabolism, bioavailability and effects of processing. J Sci. Food Agric. 80: 825-860. (2000)
6. N. Smirnoff, P.L. Conklin, F.A. Loewus, Biosynthesis of ascorbic acid in plants: a renaissance. Annu Rev Plant Physiol. Plant. Mol. Biol. 52: 437-467. (2001)
7. G. Noctor, Metabolic signaling in defense and stress: the central roles of soluble redox couples. Plant Cell Environ. 29: 409-425. (2006)
8. P.L. Conklin, S. Gatzek, G.L. Wheeler, J. Dowdle, M.J. Raymond, S. Rolinski, M. Isupov, J.A. Littlechild, N. Smirnoff, *Arabidopsis thaliana* VTC4 encodes L-galactose-1-P phosphatase, a plant ascorbic acid biosynthetic enzyme. J Biol. Chem. 281: 15662-15670. (2006)
9. W.A. Laing, M.A. Wright, J. Cooney, S.M. Bulley, The missing step of the L-galactose pathway of ascorbate biosynthesis in plants, an L-galactose guanyltransferase, increases leaf ascorbate content. Proc. Natl. Acad. Sci. USA 104: 9534-9539. (2007)
10. T. Imai, Y. Ban, S. Terakami, T. Yamamoto, T. Moriguchi, L-Ascorbate biosynthesis in peach: cloning of six L-galactose pathway-related genes and their expression during peach fruit development. Physiol. Plant. 136:139-49. (2009)
11. C.L. Linster, T.A. Gomez, K.C. Christensen, L.N. Adler, B.D. Young, C. Brenner, Arabidopsis VTC2 encodes a GDP-l-galactose phosphorylase, the last unknown enzyme in the Smirnoff-Wheeler pathway to ascorbic acid in plants. J Biol. Chem. 282:18879-85. (2007)
12. G. Cocetta, K. Karppinen, M. Suokas, A. Hohtola, H. Hagman, A. Spinardib, I. Mignanib, L. Jaakolaa, Ascorbic acid metabolism during bilberry (*Vaccinium myrtillus* L.) fruit development. J Plant Physiol 169:1059-1065. (2012)
13. J. Torabinejad, J.L. Donahue, B.N. Gunesekera, M.J. Allen-Daniels, G.E. Gillaspy. VTC4 is a bifunctional enzyme that affects myoinositol and ascorbate biosynthesis in plants. *Plant Physiol.*, 150:951-961. (2009)
14. M.J. Li, F.W. Ma, D. Liang, J. Li, Y.L. Wang, Ascorbate Biosynthesis during Early Fruit Development Is the Main Reason for Its Accumulation in Kiwi. PLoS ONE 5(12): e14281. (2010)

15. M.J. Li, X.S. Chen, P.P. Wang, F.W. Ma, Ascorbic acid accumulation and expression of genes involved in its biosynthesis and recycling in developing apple fruit. J. AMER. SOC. HORT. SCI. 136(4):231-238. (2011)
16. A. Di Matteo, R.D. Hancock, H.A. Ross, L. Frusciante, R. Viola. Characterisation of *Chlorella pyrenoidosa* L-ascorbic acid accumulating mutants: Identification of an enhanced biosynthetic enzyme activity and cloning of the putative gene from *Arabidopsis thaliana. Comparative Biochem. Physiol.* 134:S155. (2003)
17. M.J. Li, F.W. Ma, P.F. Shang, M. Zhang, C.M. Hou, D. Liang, Influence of light on ascorbate formation and metabolism in apple fruits, Planta 230:39-51. (2009)
18. E. Ioannidi, M.S. Kalamaki, C. Engineer, I. Pateraki, D. Alexandrou, I. Mellidou, J. Giovannonni, A.K. Kanellis, Expression profiling of ascorbic acid-related genes during tomato fruit development and ripening and in response to stress conditions. J Exp. Bot. 60: 663-678. (2009)
19. F.W. Ma, L.L. Cheng, The sun-exposed peel of apple fruit has higher xanthophyll cycle-dependent thermal dissipation and antioxidants of the ascorbate-glutathione pathway than the shade peel. Plant Sci. 165: 819-827. (2003)
20. K. Gasic, A. Hernandez, S.S. Korban, RNA extraction from different apple tissues rich in polyphenols and polysaccharides for cDNA library construction. Plant Mo.l Biol. Rep. 22: 437a-37g. (2004)
21. W.A. Laing, S. Bulley, M. Wright, J. Cooney, D. Jensen, D. Barraclough, E. MacRae, A highly specific L-galactose-1-phosphate phosphatase on the path to ascorbate biosynthesis. Proc. Natl. Acad. Sci. USA 101: 16976-16981. (2007)
22. A.A. Badejo, Y. Fujikawa, M. Esaka, Gene expression of ascorbic acid biosynthesis related enzymes of the Smirnoff-Wheeler pathway in acerola (Malpighia glabra). J. Plant Physiol. 166:652-660. (2009)
23. S.M. Bulley, M. Rassam, D. Hoser, W. Otto, N. Schu¨nemann, M. Wright, E. MacRae, A. Gleave, and W. Laing, Gene expression studies in kiwifruit and gene over-expression in Arabidopsis indicates that GDP-L-galactose guanyltransferase is a major control point of vitamin C biosynthesis. J. Expt. Bot. 60:765-778. (2009)
24. R.D. Hancock, P.G. Walker, S.D.A. Pont, N. Marquis, S. Vivera, S.L. Gordon, R.M. Brennan, and R. Viola. L-Ascorbic acid accumulation in fruit of *Ribes nigrum* occurs by in situ biosynthesis via the L-galactose pathway. Funct. Plant Biol. 34:1080-1091. (2007)
25. A. Lorence, B.I. Chevone, P. Mendes, C.L. Nessler, Myo-inositol oxygenase offers a possible entry point into plant ascorbate biosynthesis. Plant Physiol. 134:1200-1205. (2004)
26. M.W. Davey, C. Franck, J. Keulemans, Distribution, development and stress response of antioxidant metabolism in Malus. Plant Cell and Environment, 27:1309-1320. (2004)
27. B. Lata, A. Trampczynska, M. Oles, Antioxidant content in fruit peel, flesh and seeds of selected apple cultivars during cold storage. Folia Horticulturae, 17: 47-60. (2005).

The evaluation of ecosystem and agricultural development in the Kyrgyzstan Republic

Hong Shen, Xiao Han, Lizhao Zhang and Wuzheng Su

Institute of Agricultural Economics and Scientific Technical Information, Xinjiang Academy of Agricultural Sciences,
Urumqi, Xinjiang, China, 830091
E-mail: 13899899151@126.com

The response of ecosystem to agricultural development can have good understanding of land use, the regional ecological environment change and maintaining the ecological balance, which has the vital significance to improve the utilization rate of land and agricultural development to promote the sustainable development of regional social economy. Advice and suggestions were given involving the climate change tendency in the country and problems in agricultural development. Proposals for carrying out cooperation in agriculture between the two sides were put forward.

Keywords: Kyrgyzstan Republic; Agricultural Development; Ecosystem; Land Use; Production.

1. Introduction

In the past decades, in order to meet increasing needs for food, fresh water and energy demand, people make the ecosystem experience the unprecedented changes for improving people's lives, at the same time, which have weakened the important services of the natural ecological system, such as cleaning air, preventing the disaster, filtering water source and so on. Although people have been aware of the importance of ecosystem to human society from various aspects, there are much dimness for the relationship with ecosystem and agricultural development. Furthermore, the exact estimation for ecosystem service is of utmost importance to environmental policy makers to recognize and improve ecological service functions, which makes ecosystems reflect biodiversity [1]. The evaluation of ecosystem and agricultural development in the Kyrgyzstan Republic can provide scientific basis for land use structure optimization, sustainable development of the regional social economy and ecological security of China as neighboring country. The response of ecosystem to agricultural development can have good understanding of land use, the regional ecological environment change and maintaining the ecological balance, which has the vital significance to improve the utilization rate of land and agricultural development to promote the sustainable development of regional social economy [2-3].

The ecosystem service function, which was firstly put forward in the year of 1974, has been paid attention by most of scholars. Based on previous work, Stephen has compared the meaning of the economic and ecological value and summarized the current problems and error existing in the research of ecosystem service value for promoting the advance of research [4]. Aris and Stijin [5] make the adjustment of land use structure to achieve the optimal level by means of the application on ecosystem service value. Some scholars have studied on the ecosystem service value change to the response of China's regional land use change and analyzed the ecological service value of various land use types in different regions of the world [6-8].

In the current paper, the research on ecosystem services functions and agricultural development in response to changes in land use, climate change and resource utilization is to understand changes in the ecological environment in the region. In order to maintain ecological balance and increase the rate of use of land, it is of utmost significance to promote sustainable economic development and social region.

2. Basic information of Kyrgyzstan Republic

2.1. *Geographical Location*

Kyrgyzstan is a landlocked country, which is located in the hinterland of Eurasia and the northeast of Central Asia. This country has bigger land areas about 199,900 km^2. The straight line distance is 925 km from east to west, but the longest one 453 km between north and south. It is bordered with Kazakhstan in the north, connected to the southwest of Tajikistan, also west of the border with Uzbekistan and along the Tianshan mountains. In the east and southeast, it is the border of China, which is the boundary with 1,096 km long with China.

Kyrgyzstan has many mountains about 93% of the land area, which is regarded as Mountain Country. Most of cultivated land is located in the region of the height of 1,200~1,600 km. However, almost all of them require artificial irrigation and have heavily influence of climatic conditions. Kyrgyzstan is located in the interior of the Eurasian and in the heart of central Asia, away from the ocean in that area at the edge of the desert location and complex mountain landscape. For dry continental climate zone, fragile climate, rainfall, drying air, clouds, strong solar radiation, extremely hot summer, dry, winter is too cold and have large temperature difference between day and night. Solar radiation is one of the important factors that affect the temperature of Kyrgyzstan. The difference in winter and summer, sunlight is longer, generally more than 10~12 h in June and December only 5~6 h.

2.2. *Mineral and Energy Resources*

The central Asian state of Kyrgyzstan is a young country, with long histories of Soviet rule and mining. From the year 2005, the share of output of mining sector of total production in Kyrgyz industry has been increasing and the country's mining industry has been strongly developed. It is considered the most important industry in Kyrgyzstan, with a good potential for growth. Kyrgyzstan is relatively rich mineral resources. In terms of the total amount, although it is less neighboring Kazakhstan and Uzbekistan, Kyrgyzstan still has a certain advantage. In country region, more than 2,000 to all kinds of properties were found, which make up of most of chemical element periodic table. The mineral resources extracted are only part of the industrial development. What's more, many resources reserves and distribution needs further exploration and research to determine the development prospect. According to the statistics, the advantage of the proved reserves mineral mainly includes gold, tungsten, tin, mercury, antimony, iron, etc. The reserves of major mineral resources and production of energy resources in Kyrgyzstan are shown in Table 1 and Table 2.

Tab. 1. Reserves of mineral resources (Units: ten thousand tons)

Minerals	Reserves
Iron	1479540
Manganese	595.64
Vanadium	2189.36
Aluminum	34900
Copper	206.68
Tin	1378.48
Antimony	35.58
Mercury	4.87
Arsenic	8.22
Gold	315.49
Silver	0.62

Tab. 2. Production of energy resources from the year 2000 to 2011

Project \ Year		2000	2005	2010	2011
Oil	(ten thousand tons)	7.71	7.72	8.28	8.99
Gas	(million m³)	32.2	25.1	22.8	26.6
Coal	(ten thousand tons)	42.49	33.53	57.5	83.07
Electricity	(million kW•h)	14931.3	14891.2	12062.7	15158

2.3. *Water Resources*

Kyrgyzstan is rich in water resources. The total internal renewable water resources is about 51.2 billion m^3, surface water produced internally is 48.8 billion m^3, groundwater produced internally is 11 billion m^3, and overlap between surface water and groundwater is 8.6 billion m^3. The usable water resources in Kyrgyzstan include rivers, lakes and groundwater. There are more than 25,000 rivers, with the total length of 500,000 km. Even though most of the rivers are less than 10 km, there are still 73 rivers over 50 km. Kyrgyzstan has numerous lakes, which is third in reserves in the CIS countries, only next to Russia and Tajikistan. In region there is much more beautiful scenery and it has the high value of tourism. In Kyrgyzstan, there are 1,923 lakes, with the area of 6,836 km^2. About 84% of the lakes are located in alpine areas with the elevation of 3,000~4,000 meters. Table 3 shows the comparison of total amount of water resources among Kyrgyzstan and its neighboring countries.

In 2010, the total amount of water withdrawal was 7.56 billion m^3, in which 95.7% withdrew from surface water and 4.3% withdrew from groundwater. The total amount of water consumption was 4.48 billion m^3, only accounted for 59.3% of water withdrawal. Most of the water was used for the purpose of agriculture, which was 93% of the total water consumption. Municipal water consumption accounted for 5%. Industrial water consumption only accounted for 2%. Compared to 2000, the total water consumption in 2010 had reduced 0.52 billion m³. Except for agricultural water consumption, both of the industrial and municipal water consumption had increased in 2010. Water consumption of agriculture, industry and municipality from the year 2000 to 2010 are shown in Table 4.

Tab. 3. Comparison of water resources among Kyrgyzstan and neighboring countries

countries	total internal renewable water (billion m³)	water resources per capita (m³)	water resources per hectare (m³)
Kyrgyzstan	512	9347	2561
Tadzhikistan	160	2326	1118
Uzbekistan	504	1790	1127
Turkmenistan	247	4899	506
Kazakhstan	1096	6792	402
Turkey	2136	2936	2724
Israel	18	236	818
Iran	1375	1863	788
China	28000	2080	2917

Tab. 4. Water consumption from the year 2000 to 2010

Project \ Year		2000	2005	2010
Total		49.76	44.85	44.78
Agriculture	Amount(billion m³)	47.48	41.35	41.63
	As % of total	95	95	93
Industry	Amount(billion m³)	0.48	0.59	0.91
	As % of total	1	1	2
Municipality	Amount(billion m³)	1.82	1.49	2.06
	As % of total	4	3	5

3. The influence of agricultural development on ecosystem

3.1. *The Plantation and Animal Husbandry*

After independence in 1991, once the Kyrgyzstan agricultural production fell sharply, but in 1996 under the impetus of Agricultural reform wave privatization of agricultural production gradually recovered and had relatively fast development. So the value of agricultural production increased year overall. The gross agricultural productions of planting, livestock, agricultural services, hunting-forestry and agricultural share of GDP from the year 2000 to 2011 are shown in the Table 5.

Tab. 5. Gross agricultural production from the year 2000 to 2011(Units: Million som)

Project \ Year	2000	2005	2010	2011
Planting	225.68	344.96	596.20	753.00
Livestock	178.19	278.29	528.75	710.82
Agricultural services	5.56	9.59	25.28	26.22
Hunting and forestry	0.56	0.96	0.45	2.18
Agricultural share of GDP (%)	34.2	30.5	18.5	18.0

Kyrgyzstan is mainly dominated by agricultural production of crops and livestock. And the cultivation of industrial development level of planting is slightly higher than that of the beast husbandry. From the year of 2000 to 2011, the planting and animal husbandry respectively account for about 54% and 44% of the production of basic agricultural output ratio, which remained stable. However, the agricultural services industry, forestry and hunting industry account for a low proportion, which generally remain at around 2% (see Table 6).

Tab. 6. Agricultural production structure distribution (Units: %)

Project \ Year	2000	2005	2010	2011	Average
Total	100	100	100	100	100
Planting	55.05	54.43	51.81	50.46	54.26
Livestock	43.46	43.91	45.95	47.63	43.90
Agricultural services	1.36	1.51	2.2	1.76	1.70
Forestry and hunting	0.14	0.15	0.04	0.15	0.14

The planting area of economic crops from the year of 1991 to 2011 is shown in the Table 7. The Kyrgyzstan economic crop planting area in 2011 was 104,000 hectares, which is 9.1% of the total area. Under cultivation, economic crop production was 32.66 million tons in 2011 in the Table 8. The Table 8 shows the production of economic crop from 2001 to 2011, especially cotton, sugar beet, tobacco and oil.

Tab. 7. The planting area of economic crops from the year of 1991 to 2011 (Units: one thousand hectares)

Crop type \ Year	1991	1995	2000	2005	2010	2011
Total	51.26	113.69	166.3	147.7	94.8	104.4
Cotton	25.91	33.25	46.3	45.5	26.7	37.4
Tobacco	19.90	8.54	5.5	5.6	4.1	4.1
Beet (sugar)	0.82	13.48	27.1	14.5	8.4	8.1
Oil seeds	4.63	58.42	86.9	81.5	55.3	54.6

Tab. 8. The economic crop production (Units: million tons)

Crop type \ Year	2001	2005	2010	2011
Total	40.88	50.78	28.38	32.66
Cotton	9.82	11.81	7.4	10.13
Tobacco	2.4	1.34	0.99	0.99
Beet (sugar)	28.66	28.88	13.92	15.88
Oil seeds	-	8.75	6.07	5.66

The Animal husbandry in Kyrgyzstan have a long history and has been providing people with the necessary food, such as meat, milk, eggs and food industries. Light industry and the pharmaceutical industry provide raw materials. At the same time, cattle, horses, mules and donkeys are treated as an important work in remote areas in the mountains of the Kyrgyzstan, which still plays a big role in the transportation industry. Kyrgyzstan has 9176100 hectares of meadows and pastures across the country. And most of the natural pastures belong to a good ecological environment. With good natural conditions for the development of animal husbandry, farming in Kyrgyzstan consists mainly of sheep industry, and contribute to, the cattle farming industry and the industrial structure and a small number of beekeeping and fishing. The total number of major livestock and its production in Kyrgyzstan from the year 2004 to 2011 are shown in Table 9 and Table 10.

Tab. 9. The number of major livestock from the year of 2004 to 2011 (Units: ten thousand heads)

Project \ Year	2004	2005	2010	2011
Cattle	103.49	107.48	129.88	133.86
Pigs	8.27	7.78	5.98	5.92
Goats and sheep	377.36	387.6	503.77	528.81
Horses	34.72	34.52	37.84	38.9
Poultry	451.09	427.9	474.99	481.53

Tab. 10. The production of major livestock from the year of 2004 to 2011

Project \ Year	2004	2005	2010	2011
Meat (ten thousand tons)	18.77	18.17	18.78	19.04
Milk (ten thousand tons)	118.47	119.76	135.99	135.81
Wool (ten thousand tons)	1.10	1.06	1.09	1.11
Honey (ten thousand tons)	0.13	0.13	0.16	0.12
Eggs (ten thousand)	29870	31750	37310	39280

3.2. *General Evaluation of Agriculture*

The early independence, the Kyrgyzstan agricultural production is minor damage to industrial production in the Kyrgyzstan. But agricultural output and industrial production continued to decline, and the decline is small smaller. Since 1996, the beginning of the agricultural production is not stable. Compared with output of the previous year, the agricultural output has increased by 15.3% in 1996. At the same time, Kyrgyzstan promotes agricultural development through a series of reforms. After falling and volatile agricultural growth, agricultural production remains the stable development. But agriculture in GDP declined from 29.9% in 2004 to 18% in 2011 (see Table 11).

Tab. 11. The key indicators of agricultural production in Kyrgyzstan Republic.

Indicators	2004	2005	2006	2007	2008	2009	2010	2011
Share of agriculture in GDP	29.9	30.5	28.7	26.9	23.5	22.1	18.5	18.0
Labor productivity (thousand soms/ person)	114.1	110.5	82.4	114.2	117.9	150.7	154.4	194.9
Agricultural output per hectare of agricultural land (thousand soms)	5.4	5.9	6.7	8.4	10.4	10.4	10.8	14.0
Arable cropping means	0.84	0.84	0.88	0.88	0.91	0.91	0.90	0.91

4. Summary

The summary of research on ecosystem and agricultural development in the Kyrgyzstan Republic is to find out data information ecosystem experienced changes, to improve people's live and natural ecological system and to develop technology, in order to demonstration the advantages of agricultural in Kyrgyzstan Republic. But there are some problems existing in agricultural development so that some measures should be taken to solve it for improving the ecosystem value. Some problems are briefly summarized as following: (1) cultivated lands are underutilized; (2) The current efficiency of water usage is relatively low due to their water-saving irrigation facilities; (3) The input of agriculture such as fertilizer and organic fertilizer are limited, which significantly impact the soil fertility and agricultural production.

Acknowledgement

The work was supported by Environmental protection and resource management in Kyrgyzstan comprehensive study and research (2010DFA92720-11).

References

1. National Statistical Committee of the Kyrgyz Republic. www.stat.kg.
2. Mogilevsky R, Atamanov A. Remittances and their impact on macro-economic situation of and financial sector development in the Kyrgyz Republic. In G. Kochendorfer-Lucius, & B. Plescovic (Eds.), Spatial disparities and development policy. The World Bank. (2009).
3. Akramov K, Omuraliev N. Institutional change, rural services,and agricultural performance in Kyrgyzstan. International food policy research institute discussion paper, 00904. Washington, DC: IFPRI. (2009).
4. Stephen C F, Robert C, Matthew A W. Economic and ecological concepts for valuing eeosystem services. *Ecol Econ*, (41): 375-392. (2002).
5. Aris G, Stijin R. Incorpotration the value of ecological networks into cost-bebefut abakysus to imorive spatial;;y exlicit land-use planning. *Ecol Econ*, (73): 66-74. (2012).
6. Li H L, Wang S L, Ji G L, et al. Changes in land use and ecosystem service values in Jinan, China. *Energy Procedia*, 2011, (5):1109-1115. (2011).
7. Li J C, Wang W L, Hu G Y, et al. Changs in ecosystem service values in Zoige Plateau, China. *Agricultural, Ecosystems and environment*, (139): 766-770. (2010).
8. Li T H, Li W K, Qian Z H. Variations in ecosystem service value in response to land use changes in Shenzhen. *Ecol Econ*, (69):1427-1435. (2010).

Density effect on physiological characteristics of broadleaved seedlings of *Elaeocarpus sylvestris*

J. Li[1], Z. Y. Lie[2], L. Xue[3*], W. L. Huang[4]

College of Forestry and Landscape Architecture, South China Agricultural University, Guangzhou 510642, P. R. China

[1]houis_56028923@qq.com, [2]876672171@qq.com, [3]forxue@scau.edu.cn, [4]278870610@qq.com

Taking 1 year-old *Elaeocarpus sylvestris* seedlings as test materials, density effect on physiological characteristics of the seedlings was studied using pot experiment. The seedlings were planted with four different densities: 1, 2, 4 and 8 seedlings of each bag (density I, density II, density III and density IV). Physiological indexes of the different density seedlings were determined. The results showed that with increasing density, changes in contents of chlorophyll, soluble sugar and MDA in the leaves of the seedlings were not significant, soluble protein content and SOD activity increased followed by a decrease, whereas proline content decreased.

Key words: seedling; physiological characteristics; density effect

1. Introduction

Density is one of the important selection pressures for plants in the nature [1], which is closely related to the level of environmental resources such as light, water and nutrients [2], affecting the growth, morphology and survival of plants. Density may maintain reasonable resources utilization for populations and maintain the stability of the population by regulation of the number and growth of population individuals [3]. Population density is the most important content of plantation management, which is directly related to afforestation quality, being the key factor affecting the structure and growth of stands, woodland utilization rate and stand productivity [4-5]. Foreign studies on density control began in 1960s [6]. Domestic scholars began density study from crops [7-8]. At present, most researches on plant competition focus on plant growth under different density. For example, Xue et al. [9] studied on the competition density effects of Chinese fir, Masson pine, poplar and Japanese red pine plantations by simulating the growth characteristics of tree species using the density effect models. There are relatively few studies on the effect of seedling density, such as Cicek et al. [10] studied the seedbed density effect on *Fraxinus angustifolia* seedlings, An et al. [3] reported the biomass accumulation of *Robinia pseucdoacacia* seedlings with different densities, Woeste et al. [11] analyzed the growth law of Woeste (*Juglans nigra*) seedlings and Fang et al. [12] reported the density effect on the survival and growth of seedlings of *Myricaria laxiflora* , Yan et al. [13] described the density effect on the survival and growth of *Quercus*

The study was partially supported by Foundation of Guangdong Forestry Department, China (No. 4400-F15050).

*Corresponding author, E-mail: forxue@scau.edu.cn

liaotungensis seedlings. The study of potted plant density are mainly conducted on the crops, such as Wu et al. [14] studied the competition effect of weedy rice on the physiological characteristics of the root system and yield of the cultivated rice using pot experiment method.

Seedling density has an important influence on the quality and yield of the seedlings and affects plantation productivity. Seedling density affects the physiological and biochemical characteristics of plant leaves. Current studies about plant physiological and biochemical characteristics under different density mainly focused on crops and medicinal herbs, such as *Zea mays* [15], *Triticum aestivum* [16], immature fruit of *Poncirus trifoliata* [17]. The photosynthetic pigment content in plant leaves is an important index to reflect the photosynthetic capacity of plants, while chlorophyll is an important pigment molecule in the photosynthetic pigment. Different seedling density affected chlorophyll content of leaves by changing light. So far, study about density effect on chlorophyll in plant leaves was mainly aimed at the economic crops with a few studies on chlorophyll of seedlings. Some studies showed that chlorophyll content decreased with increasing density [18]. Under stress conditions, the cells actively yield osmotic adjustment, such as soluble sugar, soluble protein and proline, to increase the concentration of soluble substances and to reduce the water potential, so that cells continuously absorb water from extracellular, thus plants can grow normally. Cell membrane plays a role in regulating cell system, which is the most sensitive part of cells responding to the environment. When plants are stressed by light intensity, temperature, water and other factors, the cell membrane permeability will change. With increasing density, the damage of plant cell membrane become more and more serious, and malondialdehyde (MDA) content of leaves increased [17]. Superoxide dismutase (SOD) activity is an antioxidant enzyme that plays an important role in the process of scavenging reactive oxygen speciest. Hu [19] found that with increasing density, the content of soluble protein decreased, the content of MDA increased, and the activities of SOD and peroxidase (POD) increased firstly and then decreased.

Elaeocarpus sylvestris is a fast-growing evergreen broad-leaved timber and landscape tree species with characteristics of strong adaptability, easy to breed, and good water conservation ability. The studies about E. sylvestris include morphological characteristics, physiological characteristics[20], tissue culture[21], genetic [22], whereas the seedling density effects on physiological characteristics of this species has not been reported. In this study, physiological characteristics of *E. sylvestris* seedlings under different densities were studied, which can enrich population knowledge of this species.

2. Materials and methods

2.1 Study Site

The experimental field site was located in Yuejin Nursery of South China Agricultural University in Guangzhou City, South China (113°21′E, 23°09′N). It belongs to the subtropical monsoon climate. The annual average temperature and the average relative humidity are 21.9°C and 77%, respectively. The average annual rainfall is 1899.8 mm, focusing on April to October.

2.2 *Study materials*

Taking 1 year-old *E. sylvestris* seedlings as experimental materials, and they were planted according to 1, 2, 4, 8 seedlings per bags, respectively, namely 10, 20, 40 and 80 trees m2. The experimental period is from March to December 2013. Mean ground diameter, seedling height and crown of the seedlings were 0.5±0.1, 45.01±5.2 and 15.9±2.4 cm, respectively.

2.3 *Study methods*

In September 2013, third to eighth functional leaves of each seedling were selected to measure physiological indexes. The contents of chlorophyll, soluble sugar, soluble protein were determined by spectrophotometry; anthrone colorimetric method and Coomassie brilliant blue method, respectively [23]. Proline content, SOD activity and MDA were determined by acidic ninhydrin method, nitroblue tetrazolium (NBT) photoreduction method and thiobarbituric acid method determination, respectively [24]. All indexes were measured in triplicate.

All statistical analyses were used Excel 2010 and the Statistical Analysis System (SAS 9.3).

3. Results

Chlorophyll content of seedling leaves in 1 and 8 seedlings bag^{-1} was slightly higher than other two density, but the difference in different density seedlings was not significant (Figure 1). Soluble sugar content in 4 seedlings bag^{-1} was slightly lower than other density seedlings, but the difference among the four density seedlings was not significant (Figure 2). The content of soluble protein in seedlings decreased in the order of 4 seedlings bag^{-1} >8 seedlings bag^{-1} >2 seedlings bag^{-1} >1 seedling bag^{-1}, among them soluble protein in 4 and 8 seedlings bag^{-1} was significantly higher than other density seedlings (P<0.05) (Figure 3). Proline content in *E. sylvestris* seedlings decreased in the order of 1 seedling bag^{-1} >2 seedlings bag^{-1} >4 seedlings bag^{-1} >8 seedlings bag^{-1}, that in 1 seedling bag^{-1} was significantly higher than 4 and 8 seedlings bag^{-1} (P<0.05) (Figure 4). SOD activity of seedlings decreased in the order of 4 seedlings bag^{-1} >8 seedlings bag^{-1} >2 seedlings bag^{-1} >1 seedling bag^{-1}, that in 4 and 8 seedlings bag^{-1} was significantly greater than other density seedlings, whereas that in 2 seedlings bag^{-1} was significantly higher than 1 seedling bag^{-1} (P<0.05) (Figure 5). The content of MDA in all densities seedlings was similar, and there was no significant difference among them (Figure 6).

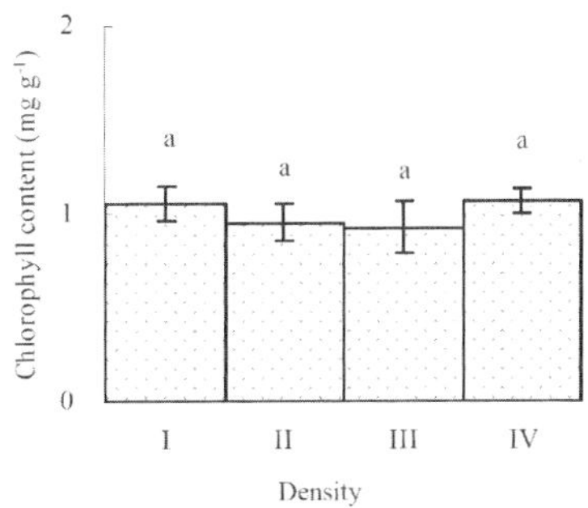

Fig. 1 Leaf chlorophyll content of the seedlings with different densities
I, 1 seedling bag-1; II, 2 seedlings bag-1; III, 4 seedlings bag-1; IV, 8 seedlings bag-1
The same letter indicates that the difference is not significant (*P*<0.05).

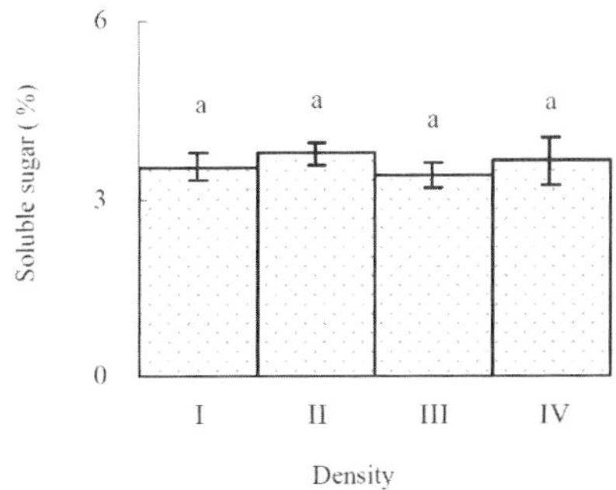

Fig. 2 Leaf soluble sugar content of the seedlings with different densities
I, 1 seedling bag-1; II, 2 seedlings bag-1; III, 4 seedlings bag-1; IV,8 seedlings bag-1
The same letter indicates that the difference is not significant(P<0.05).

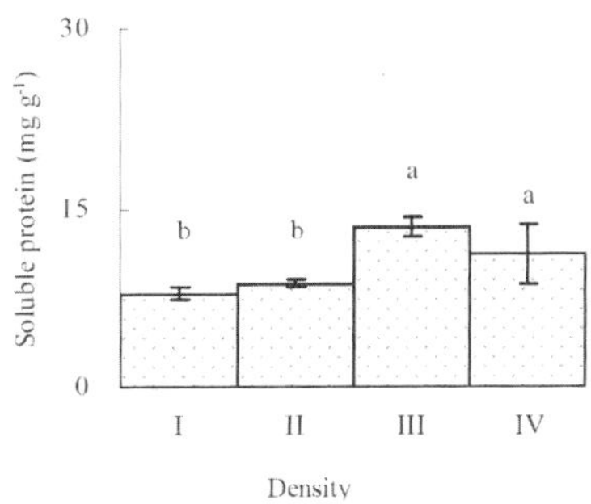

Fig. 3 Leaf soluble protein content of seedlings with different Densities
I, 1 seedling bag-1; II, 2 seedlings bag-1; III, 4 seedlings bag-1; IV, 8 seedlings bag-1 a and b are the results of multiple comparisons.
The same letter indicates that the difference is not significant (P<0.05).

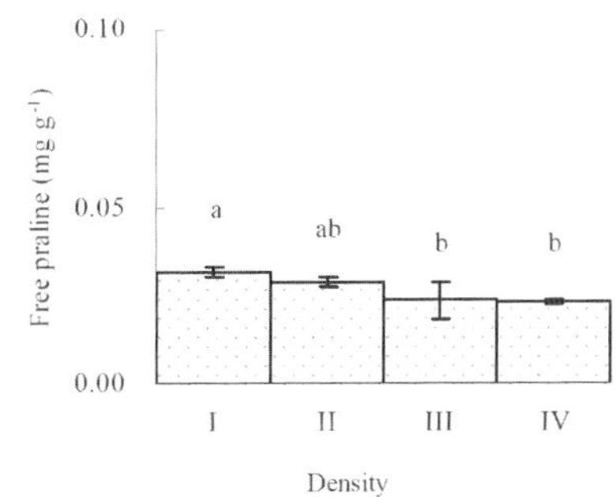

Fig. 4 Leaf proline content of the seedlings with different Densities
I, 1 seedling bag-1; II, 2 seedlings bag-1; III, 4 seedlings bag-1; IV, 8 seedlings bag-1 a and b are the results of multiple comparisons.
The same letter indicates that the difference is not significant (P<0.05).

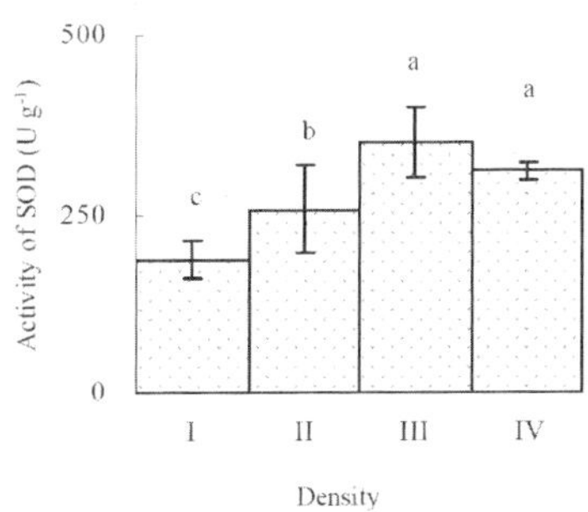

Fig. 5 Leaf SOD activity of the seedlings with different densities
I, 1 seedling bag-1; II, 2 seedlings bag-1; III, 4 seedlings bag-1; IV, 8 seedlings bag-1 a, b and c are the results of multiple comparisons.
The same letter indicates that the difference is not significant (*P*<0.05).

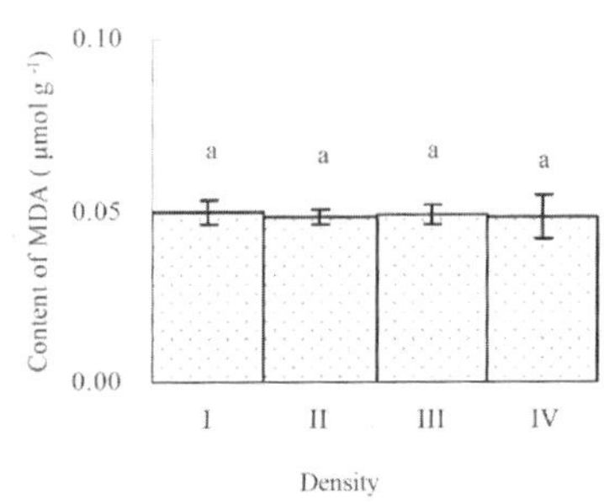

Fig. 6 Leaf MDA content of the seedlings with different densities
I, 1 seedling bag-1; II, 2 seedlings bag-1; III, 4 seedlings bag-1; IV, 8 seedlings bag-1
The same letter indicates that the difference is not significant (*P*<0.05).

4. Discussion

Chlorophyll content is an important indicator of leaf physiology, being directly or indirectly related to photosynthetic rate. Qi et al. [25] studied density effect on the photosynthesis of maize and found that the contents of chlorophyll a, b, a+b and carotenoid pigment decreased in maize with increasing density. Duan et al. [17] reported that chlorophyll content decreased with increasing density. In this study, density effect on chlorophyll content is not obvious, the reason may be that the seedling size is small, so that density has little effect on chlorophyll content.

Soluble sugar is an ideal osmotic adjustment substance in plant cells, which can alleviate the damage of membrane system under stress. Changes of soluble sugar content in the leaves of the seedlings were not obvious, because the seedling size is still small, so that the effect on the both osmotic adjustment substances is not yet reflected.

Soluble protein is one of the important osmotic adjustment substance under environment stress. Soluble protein may increase the water holding capacity of cells to maintain the low osmotic potential and resistance to damage. Hu [18] found that the content of soluble protein decreased in spring corn with increasing density, decrease range was great in less tolerant varieties. In this study, the soluble protein content increased first and then decreased with increasing density, which indicated that in a certain density range, the soluble protein increased to resist light and nutrient stresses. However, higher density seedlings may result in the degradation of protein and directly influence protein synthesis [18].

The proline mainly exists in the cytoplasm, being the important osmotic adjustment substance in cells. Proline has two functions in the anti-inverse: (1) maintaining the osmotic balance between the protoplast and the environment; (2) maintaining the integrity of membrane structure. Proline content decreased with increasing seedling density, indicating that high density may be disadvantage for osmotic adjustment of the seedlings.

SOD is a key enzyme in the protective enzyme system, which can effectively remove the free radicals in plants, thus delaying leaf senescence [18]. For the varieties with strong resistance, decrease range was small when the activity of SOD decreased, and increase range is great when the SOD activity increased [17]. With increasing density, SOD activity of *E. sylvestris* seedlings first increased and then decreased, showing that proper high density can improve SOD activity in order to resist environmental stress caused by high density, whereas when density is too high, seedlings resistance to stress decreases, as a result the activity of SOD decreases.

As a product of lipid peroxidation of plant membrane, MDA can binding protein and cause cross-linking for protein molecules-protein molecules, which damages the plasma membrane, cause serious damage to the enzyme and the membrane, and ultimately destroy the membrane structure [26-27]. Environmental stress can increase the content of MDA, and increase range in MDA content is small in tolerant plant types and great in non-tolerant plant types [18]. There was no significant difference in MDA content among different density *E. sylvestris* seedlings, which indicated that the content of MDA was not affected by the density effect of *E. sylvestris.*

References

1. W. Japhet, D. W. Zhou, H. X. Zhang and T. Yu. Evidence of phenotypic plasticity in the response of *Fagopyrum esculentum* to population density and sowing date. *J. Plant Biol.* **52**, 303 (2009).
2. H. Maherali and E. H. Delucia. Influence of climate-driven shifts in biomass allocation on water transport and storage in ponderosa pine. *Oecologia* **129**, 481 (2001).
3. H. An and Z. P. Shangguan. Effects of density on biomass and allometric pattern of *Robinia pseudoacacia* seedling. *Sci. Sil. Sin.* **44**, 151 (2008).
4. G. F. Qin, Z. C. Zhou, G. Q. Jin, W. C. Rong and T. L. Wu. Determining optimum planting density of Masson Pine fast-growing and high-yielding plantations for various cultivation objectives. *For. Res.* **12**, 620 (1999).
5. M. K. Yu, H. Qiu, L. S. Yang, L. Y. Xu and Y. L. Chen. Study on afforestation density o f *Pinus taeda. J. Anhui Agri. Univ.* **26**, 398 (1999).
6. S. B. Wang and S. L. Peng. Studies on the measuring techniques of interspecific association of lower-subtropical evergreen-broadleaved forests. I. The exploration and the revision on the measuring formulas of interspecific association. *Acta phyloecol. Geobot. Sin.* **9**, 274 (1985).
7. J. M. Li, F. Xie, C. J. Chen, S. Y. Zhang, R. H. Xiao and D. Z. Zhao. Interspecific association of dominant species in *Betula luminifera* natural forest communities. *Chin. J. Appl. Ecol.* **12**, 168 (2001).
8. L. X. Zhang, F. Zhang and T. L. Shangguan. Studies on relationship between species in plant communities of Luya Mountains. *Acta Bot. Borea.-Occident. Sin.* **21**, 1085 (2001).
9. L. Xue and A. Hagihara. Growth analysis on the C-D effect in self-thinning Masson pine (*Pinus massoniana*) stands. *For. Ecol. Manage.* **165**, 249 (2002).
10. E. Cicek, N. Cicek and N. Bilir. Effects of seedbed density on one-year-old *Fraxinus angustifolia* seedling characteristics and out planting performance. *New For.* **33**, 81 (2007).
11. K. E. Woeste, D. F. Jacobs and J. R. McKenna. Half-sib seed source and nursery sowing density affect black walnut (*Juglans nigra*) growth after 5 years. *New For.* **41**, 235 (2011).
12. F. Q. Chen, G. M. Xiong and Z. Q. Xie. Effects of density on seedling survival and growth of an endangered species *Myricaria laxiflora. Biodiv. Sci.* **13**, 332 (2005).
13. X. F. Yan, L. B. Zhou, K. W. Zhang and Y. F. Zhou. Cotyledon loss and its effects on survival and growth of *Quercus wutaishanica* seedlings under different densities. *Chin. J. Plant Ecol.* **36**, 831 (2012).
14. Y. Y. Wu, D. R. Ma, J. Y. Li, W. Ma and W. F. Chen. Effect of weedy rice on root morphological and physiological characteristics and yield of cultivated rice. *Acta*

Agri. Boreali-Sin. **5**, 150 (2010).

15. S. B. Chen. Studied on high oil corn physiology characteristics and yield traits in different densities. Master's thesis, China Agricultural University (2005).
16. F. Cha. Effects of planting densities on physiological characteristics and grain yield of two winter wheat cultivars with different spike type. Master's thesis, Henan Agricultural University (2007).
17. J. H. Li. Studies on the physiological characteristics of *Fruetus Poniciri Trifoliatae Immaturus* with different density cultivation. Master's thesis, Fujian Agriculture and Forestry University, (2008).
18. W. W. Duan, H. L. Li, K. Xiao and Y. M. Li. Effects of density on photosynthetic physiological characteristics and yield of maize. *J. Maize Sci.* **2**, 98 (2007).
19. M. Hu. Effect of plant density on photosynthetic senescence physiology and yield of spring corn. Master's thesis, Northeast Agricultural University (2009).
20. L. Xue, R. Zhang, R. C. Xi, S. H. Guo, Z. Y. Yang, B. Liu and R. P. Wei . Seasonal change of leaf morphological traits of six broadleaf seedlings in South China. *Acta Ecol. Sin.* **32**, 123 (2012).
21. D. X. Shi, Y. J. Gu, M. L. Wang, X. M. Deng and J. K. Zhang. Plant regeneration from in vitro culture o f *Elaeocarpus sylvestris* (Lour.) *Poir. Acta Horticul. Sin.* **31**, 245 (2004).
22. Z. G. Zeng, F. M. Xiao, G. H. Bao, J. S. Ye, S. L. Wang, Y. Nie, C. H. Wang and J. S. Liu. A study on the genetic variation of seedling growth and wood traits in provenances of *Elaeocarpus sylvestris*. Acta Agri. Univ. Jiangxiensis **25**, 815 (2003).
23. H. S. Li. *Experimental Principle and Technology of Plant Physiology and Biochemistry*. (Higher Education Press, Beijing, 2000).
24. J. C. Chen and X. F. Wang. *Plant Physiology Experiment*. (South China University of Technology press, Guangzhou, 2002).
25. Y. F. Qi, F. Z. Xu, Z. H. Zhou, Y. J, Xing, L. H. Xu and D. L. Qiu. Effect of plant density on photosynthesis on photosynthesis capability of new maize hybrid luyuandan 22. *Acta Agri. Nucl. Sin.* **18**, 14 (2004).
26. A. G. Wang, C. H. Shao and G. H. Luo. Inquiry into malondialdehyde as index of peroxidation of plant lipids. *Plant Physio. Comm.* **30**, 55 (1986).
27. S. Y. Chen. Injury of membrane lipid peroxidation to plant cell. *Plant Physio. Comm.* **27**, 84 (1991).

Effect of salinity-alkalinity stress on seed germination of *Cichorium intybus* L.

Jin-feng Mao, Xue-jun Yang, Jiang-li Nie*, Guo-chao Shi, Wei Zhang, Yi Pei

College of Horticulture and Landscape, Tianjin Agricultural University, Tianjin, China

E-mail: njlnie@126.com

Taking seeds of *Cichorium intybus* L. as materials, using the method of germination in petri dish, effect on *C. intybus* L. seeds germination under different concentrations of NaCl, $NaHCO_3$ soulution and both 1:1 mixed solution stress and after the stress being removed was studied, in order to investigate salinity tolerance ability of *C. intybus* L. seeds. The results showed that seed germination rate, germination potential and relative germination rate was decreased with NaCl, $NaHCO_3$ and both 1:1 mixed concentration was increased. At high concentration, the inhibitory effect of seed was the most significant. The optimal, critical and limit value of NaCl that affected relative germination rate was 3.509‰, 5.644‰, 7.779‰; these of $NaHCO_3$ concentration was 3.448‰, 5.736‰, 8.023‰ and these of both 1:1 mixed concentration was 4.076‰, 6.654‰, 9.232‰. The results of relief stress suggested that seeds stressed of higher concentration of NaCl, $NaHCO_3$ and both 1:1 mixed still had greater germination capacity especially, it showed the seeds of *C. intybus* L. had stronger resistance to salt and alkali.

Keywords: *C. Intybus* L.; Seed Germination; NaCl; $NaHCO_3$; Stress; Recovery Germination.

1. Introduction

Salinity is an important environmental factor that it will affect the growth and output of plant, but high salt will cause production reduction or death to plant. Salt stress will affect all important life process of plant nearly, such as seed germination, growth, photosynthesis, under the salt stress all growth of plant will be inhibited but different plant have different levels on salt tolerance to death and growth rate lower[1]. Therefore, it's significant that studying the effect of salt stress on plant, improving the ability of salt tolerance on plant and improving output of salt crops drastically for reasonable exploit the resources of salt-alkali soil and water.

C. intybus L. is perennial herb of the composite family. Root is fleshy, short and thick. Stem erect, ribbed, hollow, much branch. The root and aerial part of plant available for medicinal, its Chinese traditional medicine name is endive or chicory root that they have efficacy of heat-cleaning and detoxifying, dieresis detumescence, invigorating the stomach[2]. In addition, *C. intybus* L. also can as roughage or pasture.

The experiment taken *C. intybus* L. seed as material and dealt with NaCl solution and $NaHCO_3$ solution to observe the effect of salt stress on seed germination, though determined the seed germination rate, germination potential, radical length, seeding fresh weight under different concentration and degree of salt stress and the germination after relieving salt stress, thereby that study the salt tolerance on *C. intybus* L. seed to providing a reference on planting widely and applying in landscape.

2. Materials and Methods

2.1 Materials

For experimental *C. intybus* L. seed came from AnGuo Chinese herbal medicinal planting bases of Hebei province and now is saved at landscape plant teaching and research section of college horticulture and landscape of Tianjin Agricultural University.

2.2 Experimental Methods

Using potted simulation method[3][4] that diameter of the culture dish was 9 cm and putted 2 layers filter paper into its bottom. NaCl solution, $NaHCO_3$ solution and both 1:1 mixed solution set up 5 concentrations respectively, in order they were 2‰, 4‰, 6‰, 8‰ and 10‰, the control group dealt with distilled water. Observing the condition of seed germination and recording the number of germination every 24 hours (the standard of germination was the length of radical broke through the coat of seed). Every standard of calculation formula was under the follows[4].

Germination rate (%) = the number of seed germination at the end of germination / the number of seeds for the test *100%. Germination potential (%) = the number of seed germination within the allotted time / the number of seeds for the test *100%. Relative germination rate (%) = the germination rate in the 9th day after seed treatment/ Germination rate of control *100%. Relative salt injury rate (%) = (germination rate of control – salt treatment germination rate)/ Salt treatment germination rate *100%. Germination index = ∑ the number of germination in the t day/the number of days is corresponding.Vigor index = germination index/seeding fresh weight (g).

Seed germination resistance to salt concentration is when relative germination rate reach above 75% the corresponding salt and alkali concentration (optimum value) (%).

Median inhibitory concentration is relative germination rate reach to 50% corresponding salt and alkali concentration (critical value) (‰).

Salt tolerance limit value of seed germination is relative germination rate reach to 25% corresponding salt and alkali concentration (‰).

2.3 Data Analysis

Experimental data used LSD and Duncanny(D) of SPSS17.0 analysis software to pairwise comparison and variance analysis. Using excel 2007 software to deal and draw. Using Probit module of SPSS software (Probit Regression) to analyze the relation of "concentration-relative germination rate" and the ability of C. intybus L. to salt stress[4].

3. Result and analysis

3.1 The effect of different concentration of NaCl solution for seed germination of C. intybus L.

The seed of *C. intybus* L. dealt with NaCl solution that the germination was downtrend gradually ,but the germination rate and germination potential of control group all more higher than other groups was showed in Table 1, that was to say, the germination rate and germination potential of control group were 62.67% and 37.33%, the germination rate, germination potential and relative germination rate of other treatment groups were downtrend gradually, meanwhile because of the increase of salt concentration that the inhibition for seed gradually appeared, the relative salt injury rate also present uptrend. That was to say, when the concentration of NaCl solution was 2‰, the germination rate was 59.33%, the germination potential was 34.67%, relative germination rate was 94.67%, relative salt injury rate was 5.33%; when the concentration of NaCl solution was 4‰, the germination rate was 33.33%, the germination potential was 14.67%, relative germination rate was 53.18%, relative salt injury rate was 46.82%; when the concentration of NaCl solution was 6‰, the germination rate was 29.33%, the germination potential was 14.00%, relative germination rate was 46.80%, relative salt injury rate was 53.20%; when the concentration of NaCl solution was 8‰, the germination rate was 23.33%, the germination potential was 10.67%, relative germination rate was 37.23%, relative salt injury rate was 62.77%; when the concentration of NaCl solution was 10‰, the germination rate was 1.33%, the germination potential was 0, relative germination rate was 2.12%, relative salt injury rate was 97.88%. Form the foregoing, when the salt concentration more than 10‰ that seed was inhibited mainly and the relative salt injury rate was near to 100%.

Tab. 1. Effect of NaCl stress on Seed Germination of *C. intybus* L.

NaCl concentration (‰)	Germination Rate (%)	Germination Potential (%)	Relative Germination rate (%)	Relative Salt injury rate (%)
0	62.67 aA	37.33 aA	-	-
2	59.33 aA	34.67 aA	94.67	5.33
4	33.33 bB	14.67 bB	53.18	46.82
6	29.33 bcBC	14.00 bB	46.80	53.20
8	23.33 cC	10.67 bB	37.23	62.77
10	1.33 dD	0.00 cC	2.12	97.88

Note: The differences in the same column in different small letters are significantly different ($P<0.05$); the difference in capital letters indicates that the difference is very significant ($P<0.05$). The same as in the following tables.

The germination index and vigor index of control group and treatment group with 2‰ NaCl solution were higher than other treatment groups observably was showed in Table 2, that was to say the low concentration of NaCl solution could promote the seed germination of *C. intybus* L. When concentration of NaCl solution was more than 4‰, with the increase of concentration of solution, the germination index and vigor index of seed were reduced significantly; When concentration of NaCl solution reached or more than 10‰, the vigor of seed was disappeared and seed did not germinated.

Tab. 2. Effect of NaCl stress on Germination index, vigor index and seedling fresh weight of *C. intybus* L.

NaCl Concentration (‰)	Germination index	Vigor index	Seedling fresh weight (g)
0	3.48 aA	0.16 aA	0.0454
2	3.29 aA	0.13 bAB	0.0384
4	1.85 bB	0.10 cB	0.0547
6	1.63 bcBC	0.05 dC	0.0293
8	1.29 cC	0.03 dCD	0.0233
10	0.07 dD	0.00 eD	0.0113

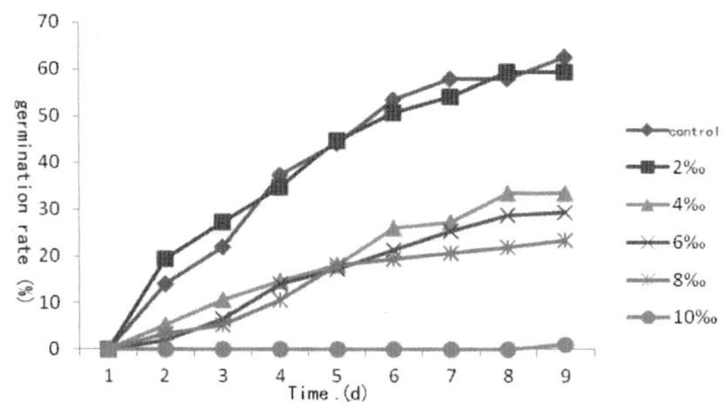

Fig. 1. The Cumulative germination curve of *C. intybus* L. seed under different concentration of NaCl solution.

The germination mainly focused on 5 days ago that C. *intybus* L. seed dealt with different concentration of NaCl solution was showed in Figure 1. The germination rate of control group was 44% in the 5th day and was 62.67% in the 9th day; the germination rate of 2‰ concentration of NaCl solution was 44.67% in the 5th day and was 59.33% in the 9th day; with the increase of concentration of NaCl solution that the germination rate was decreased; the germination rate of 4‰ concentration of NaCl solution was 18% in the 5th day and was 33.33% in the 9th day; the germination rate of 6‰ concentration of NaCl solution was 17.33% in the 5th day and was 29.33% in the 9th day; the germination rate of 8‰ concentration of NaCl solution was 18.00% in the 5th day and was 23.33% in the 9th day; the germination rate of 10‰ concentration of NaCl solution was 0 in the 5th day and was 1.33% in the 9th day. As could be see the concentration was more higher the germination rate was more lower, indicating the salt solution was inhibited on seed germination more strong.

3.2 The effect of different concentration of NaHCO3 solution for seed germination of C. intybus L.

The germination rate and germination potential of control group were 73.33% and 48.67% was showed in Table 3, with the increase of concentration of solution that the indication of every treatment groups were declined gradually but relative salt injury index was rose gradually. when the concentration of $NaHCO_3$ solution was 2‰, the germination rate was 58.67%, the germination potential was 40.00%, relative germination rate was 80.01%, relative salt injury rate was 19.99%; when the concentration of solution was 4‰, the germination rate was 52.00%, the germination potential was 36.00%, relative germination rate was 70.91%, relative salt injury rate was 29.09%; when the concentration of solution was 6‰, the germination rate was 42.00%, the germination potential was 32.00%, relative germination rate was 57.28%, relative salt injury rate was 47.72%; when the concentration of solution was 8‰, the germination rate was 22.00%, the germination potential was 20.00%, relative germination rate was 30.00%, relative salt injury rate was 70.00%; when the concentration of solution was 10‰, the germination rate was 2.00%, the germination potential was 0, relative germination rate was 2.73%, relative salt injury rate was 97.27%. Form the foregoing, with the increase the salt concentration that seed was inhibited mainly and the relative salt injury rate was near to 100%.

Tab. 3. Effect of $NaHCO_3$ stress on Seed Germination of *C. intybus* L.

$NaHCO_3$ concentration (‰)	Germination rate (%)	Germination Potential (%)	Relative Germination rate (%)	Relative salt Injury rate (%)
0	73.33 aA	48.67 aA	-	-
2	58.67 bB	40.00 abAB	80.01	19.99
4	52.00 cC	36.00 bAB	70.91	29.09
6	42.00 dD	32.00 bBC	57.28	42.72
8	22.00 eE	20.00 cC	30.00	70.00
10	2.00 fF	0.00 dD	2.73	97.27

Tab. 4. Effect of $NaHCO_3$ stress on Germination index, vigor index and seedling fresh weight of *C. intybus* L.

NaCl Concentration (‰)	Germination index	Vigor Index	Seedling fresh weight (g)
0	4.07 aA	0.19 aA	0.0468
2	3.26 bB	0.17 aAB	0.0510
4	2.89 cC	0.12 bBC	0.0421
6	2.33 dD	0.08 cCD	0.0339
8	1.22 eE	0.03 dDE	0.0251
10	0.22 fF	0.00 dE	0.0114

The germination index of control group was higher than other treatment groups observably was showed in Table 4, with the increase of salt concentration the germination index and vigor index of *C. intybus* L. seed all present downward trend, when concentration of $NaHCO_3$ solution more than 4‰, the germination index and vigor index of seed were reduced gradually; when concentration of $NaHCO_3$ solution reached or more than 10‰ that the germination index near to 0 and the vigor of seed was disappeared, illustrating the inhibition for seed closed to its limit.

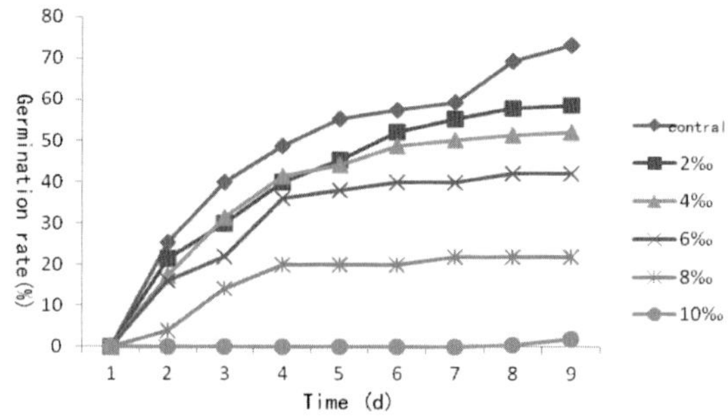

Fig. 2. The Cumulative germination curve of *C. intybus* L. seed under different concentration of $NaHCO_3$ solution.

The germination mainly focused on 5 days ago that seed dealt with different concentration of $NaHCO_3$ solution was showed in Figure 2. The germination rate of control group was 55.33% in the 5th day and was 73.33% in the 9th day; with the increase of concentration of $NaHCO_3$ solution that the germination rate was decreased; the germination rate of 2‰ concentration of $NaHCO_3$ solution was 45.33% in the 5th day and was 58.67% in the 9th day; the germination rate of 4‰ concentration of $NaHCO_3$ solution was 44.00% in the 5 th day and was 52.00% in the 9th day; the germination rate of 6‰ concentration of $NaHCO_3$ solution was 38.00% in the 5th day and was 42.00% in the 9th day; the germination rate of 8‰ concentration of $NaHCO_3$ solution was 20.00% in the 5th day and was 22.00% in the 9th day; the germination rate of 10‰ concentration of $NaHCO_3$ solution was 0 in the 5th day and was 2.00% in the 9th day. As could be see that with the increase of salt concentration the germination rate was decreased gradually, indicating the salt solution was inhibited on seed germination more strong.

3.3 The effect of different concentration of both 1:1 mixed solution for seed germination of C. intybus L.

The germination rate and germination potential of control group was 75.33% and 51.33%, with the increase of concentration of solution that the germination rate, germination potential and relative germination rate all decreased obviously was showed in Table 5, because of toxicity ions that affected seed germination in concentration were rose gradually the relative salt injury index was also rose gradually. When the concentration of both 1:1 mixed solution was 2‰, the germination rate was 72.67%, the germination potential was 40.00%, relative germination rate was 96.47%, relative salt injury rate was 3.53%; when the concentration of solution was 4‰, the germination rate was 46.67%, the germination potential was 25.33%, relative germination rate was 61.95%, relative salt injury rate was 38.05%; when the concentration of solution was 6‰, the germination rate was 40.67%, the germination potential was 24.67%, relative germination rate was 53.99%, relative salt injury rate was 46.01%; when the concentration of solution was 8‰, the germination rate was 36.67%, the germination potential was 21.33%, relative germination rate was 48.68%, relative salt injury rate was 51.32%; when the concentration of solution was 10‰, the germination rate was 11.33%, the germination potential was 6.67%, relative germination rate was 15.04%, relative salt injury rate was 84.96%. Form the foregoing, the both 1:1 mixed solution had some inhibitory action for seed germination but the inhibitory was weaker than single salt stress. When the concentration of both 1:1 mixed solution was 10‰, the seed still had the ability of germination and the relative sale injury was lower than single salt, indicating the seed had the ability of salt tolerance.

Tab. 5. Effect of both 1:1 mixed stress on Seed Germination of *C. intybus* L.

Concentration of both 1:1 mixed solution(‰)	Germination Rate (%)	Germination Potential (%)	Relative Germination rate (%)	Relative salt Injury rate (%)
0	75.33 aA	51.33 aA	-	-
2	72.67 aA	40.00 bB	96.47	3.53
4	46.67 bB	25.33 cC	61.95	38.05
6	40.67 bcBC	24.67 cC	53.99	46.01
8	36.67 cC	21.33 cC	48.68	51.32
10	11.33 dD	6.67 dD	15.04	84.96

Tab. 6. Effect of both 1:1 mixed stress on Germination index, vigor index and seedling fresh weight of *C. intybus* L.

Concentration of both 1:1 mixed solution (‰)	Germination index	Vigor index	Seedling fresh weight (g)
0	4.18 aA	0.17 aA	0.0405
2	4.03 aA	0.15 aA	0.0367
4	2.59 bB	0.08 bB	0.0318
6	2.26 bcBC	0.07 bB	0.0311
8	2.03 cC	0.06 bBC	0.0301
10	0.62 dD	0.00 cC	0.0030

The germination index and vigor index of control group and treatment group with 2‰ concentration of both 1:1 mixed solution were higher than other treatment groups observably was showed in Table 6, that was to say the low concentration of both 1:1 mixed solution could promote the seed germination of *C. intybus* L. When concentration of both 1:1 mixed solution more than 4‰, with the increase of concentration of solution, the germination index and vigor index of seed were reduced significantly; when concentration of both 1:1 mixed solution reached or more than 10‰, the vigor of seed was disappeared.

The germination mainly focused on 5 days ago that *C. intybus* L. seed dealt with different concentration of 1:1 mixed solution was showed in Figure 3. The germination rate of control group was 57.33% in the 5th day and was 75.33% in the 9th day; with the increase of concentration of 1:1 mixed solution that the germination rate was decreased; the germination rate of 2‰ concentration of both 1:1 mixed solution was 50.06% in the 5th day and was 72.67% in the 9th day; the germination rate of 4‰ concentration of 1:1 mixed solution was 37.33% in the 5th day and was 46.67% in the 9th day; the germination rate of 6‰ concentration of 1:1 mixed solution was 32.00% in the 5th day and was 40.67% in the 9th day; the germination rate of 8‰ concentration of 1:1 mixed solution was 30.67% in the 5th day and was 36.67% in the 9th day; the germination rate of 10‰ concentration of 1:1 mixed solution was 10.67% in the 5th day and was 11.33%

in the 9th day. As could be see that with the increase of salt concentration the germination rate was decreased gradually, indicating the salt solution was inhibited on seed germination more strong.

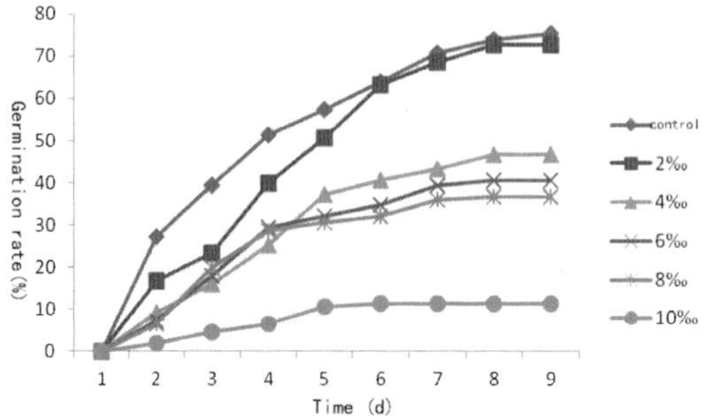

Fig. 3. The Cumulative germination curve of *C. intybus* L. seed under different concentration of both 1:1 mixed solution.

3.4 The condition of seed germination of C. intybus L. after relieving stress under different salt concentration.

Putting was not germinated seed came from different concentration of NaCl solution, $NaHCO_3$ solution and both 1:1 mixed solution into environment of distilled water and cultured 5 days to observe the condition of germination. We could learn from Table 7 that the seed could be continue to germinate after relieving stress and with the increase of concentration the rate of germination again also present upward trend, meaning the seed still had the potential of germination after relieving stress under high concentration of NaCl solution, $NaHCO_3$ solution and both 1:1 mixed solution. So, when the circumstances changed that decreased concentration of salt-alkali could recover germination, it's significant to vegetation restoration of saline soil, soil improvement and land virescence.

Tab. 7. The rate of germination again after relieving stress of *C. intybus* L. seed

NaCl Concentration (‰)	Rate of germination again (%)	$NaHCO_3$ Concentration (‰)	Rate of germination again (%)	both1:1 mixd concentration (‰)	Rate of germination again (%)
2	4.54 dD	2	4.70 dD	2	6.70 dD
4	26.60 cC	4	9.10 cC	4	31.80 cC
6	26.57 cC	6	17.20 bB	6	28.67 cC
8	47.83 bB	8	23.50 aA	8	48.40 bB
10	66.00 aA	10	24.60 aA	10	62.00 aA

3.5 Analysis the ability of salt tolerance

Using the Probit module (Probit Regression) of SPSS software to analyze the relation of "concentration-relative germination rate" and got the formula as Table 8, the results showed the appropriate value of concentration of NaCl solution to seed germination was 3.509‰ and 95% confidence limit up to 5.709‰ and down to17.710‰; critical value is 5.644‰ and 95% confidence limit up to 9.911‰ and down to 0.0155‰; limit value is

7.779‰ and 95% confidence limit up to 25.943‰ and down to 5.569‰; the appropriate value of concentration of $NaHCO_3$ solution to seed germination is 3.448‰ and 95% confidence limit up to 5.059‰ and down to 0.907‰; critical value is 5.736‰ and 95% confidence limit up to 7.661‰ and down to 3.754‰; limit value is 8.023‰ and 95% confidence limit up to12.253‰ and down to 6.425‰; the appropriate value of concentration of 1:1 mixed solution to seed germination is 4.076‰ and 95% confidence limit up to 6.205‰ and down to7.689‰; critical value is 6.654‰ and 95% confidence limit up to 11.503‰ and down to 3.578‰; limit value is 9.232‰ and 95% confidence limit up to 24.633‰ and down to7.012‰.

Tab. 8. The relativity of relative germination rate and salt concentration

Treatment	Regression equation
NaCl	y=1.783-0.316x
$NaHCO_3$	y=1.691-0.295x
Both 1:1 mixed solution	y=1.741-0.262x

Note: y is relative germination rate, x is salt concentration.

4. Discussion

4.1 The inhibition of salt concentration for seed germination

The experimental studies indicated that with the increase of salt concentration the germination rate, germination index and vigor index of *C. intybus* L. seed present the down trend. The experimental results showed that with the increase of salt concentration the inhibition action of seed germination under high salt also gradually strengthened, that was to say the stress of salt concentration to seed germination was very conspicuous. The inhibition action on seed germination was signal salt greater than mixed salt, but they all inhibited on seed germination and inhibition with the increase of salt concentration was also great.

4.2 The seed after relieving stress still had potential of germination

The experimental results of relieving stress showed that the capacity of seed germination was momentarily inhibited in salt solution, when the inhibition was relieved that the seed still had the ability of germination and inhibition active more stronger, the potential was more greater. Because of the seed in high concentration of solution that the inhibition action was the most strongest and the potential also was the most strongest, showed the seed was inhibited really and could germinated normally after relieving stress, showed the seed in good condition.

5. Conclusion

The experimental results showed different concentration of NaCl solution, $NaHCO_3$ solution and both 1:1 mixed solution all had inhibition action for *C. intybus* L. seed. Majority reaches though [5][6] NaCl solution, $NaHCO_3$ solution and both 1:1 mixed solution had inhibition action for *C. intybus* L. seed, but also others though low concentration of salt stress could promote seed germination. In this experiment with increase of concentration of NaCl solution, $NaHCO_3$ solution and both 1:1 mixed solution, the germination rate, germination potential and relative germination rate of *C. intybus* L. seed present downtrend but relative salt injury present uptrend, it showed the inhibition action of solution for seed improve gradually and there was a negative between germination rate, germination potential and relative germination rate and NaCl solution, $NaHCO_3$ solution and both 1:1 mixed solution. The experiment of relieving stress showed the seed still had potential of germination that dealt with high concentration of NaCl solution, $NaHCO_3$ solution and both 1:1 mixed solution, explaining the *C. intybus* L. seed had some ability on salt-alkali tolerance.

C. intybus L. Seed able to well grow under salt stress is the base of grow in saline soil and present good character of salt tolerance.

In conclusion, *C. intybus* L. as traditional Chinese herbal medicine if exploit it into ground covers will have a great market prospect, meanwhile also as economic crop to produce prominent economic efficiency.

References

1. Cheng Peng, The effect of salt stress on plant and the research progress of plant salt tolerance, *Shandong Commercial Vocational And Technical College*, 14(2): 123-128(2014).
2. Yan Luo, Shiqie Bai, Yan Peng and Zhang Yu, The research progress of *Cichorium intybus* L. plant resources, *Prataculturual Science*, 27(7):123-132(2010).
3. Ronghua Ji, Lei Yu, Weihua Lu and Ernie doll·Ahmatjan, The effect of salt stress on seed germination of Achnatherum splendens, *Prataculturual Science*, 28(2): 245-250(2011).
4. Yi Pei, Jiangli Nie, Huijia Liu and Junmei Chen, The effect of NaCl and $NaHCO_3$ stress on seed germination of *Celosia cristata* L., *Northern Horticulture*, 22(4): 65-68(2014).
5. Xiujuan Zhao, Yanan Han and Lu Cai, The effects of salt stress on plant physiological and biochemical characteristics, *Hubei Agricultural Sciences*, 50(19): 3897-3899(2011).
6. Minyan Lu, The effect of different salt stress on seed germination of *Festuca arundinacea*, *Prataculturual Science*, 29(7): 1088-1093(2012).

Effects of Chlorcholinchlorid on the ornamental and physiological characteristics of blueberry

Mao-lan Yue, Xia Qiu, Bo-lei Jiao and Xun Wang
No. 211, HuiMin Road, Wenjiang, Chengdu, Sichuan, P.R.C.
E-mail: wangxun0104@hotmail.com

In order to control plant height and inhibit over growth, potted blueberry were treated with Chlorcholinchlorid (CCC) by spraying on the leaves, effects of CCC on the plant height, stem diameter, canopy width, pitch spread and leaf thickness were investigated. The results showed that optimum dwarfing treatment for blueberry was application CCC solutions of 300 mg/L or 600 mg/L, the effect was most accord with dwarfing objective of pot culture.

Keywords: Blueberry; Chlorcholinchlorid (CCC); Pot culture; Dwarfing.

1. Introduction

Blueberry (*Vaccinium* SPP.) belongs to family Ericaceae. It is perennial deciduous or evergreen tree species. Blueberry is an important small fruit crop native to North America [1]. There are 91 species of 28 varieties in China, mainly distributed in Great Khingan Mountains and the southwest of China [2]. Consumption of blueberries has increased rapidly in recent years. Currently, there are more than 77,000 ha of cultivated blueberry grown worldwide, and production is expected to increase by another 46 % over the next 5 years [3]. Blueberry plantings are expanding speedily into less traditional growing regions. By 2015, total global production is predicted to reach 635,000 metric tons of blueberries per year. In China, the planting are expanding into many regions around big cities [4].

Due to their health benefits, blueberry fruits are drawing increasing attentions. As a bush plant, with pretty white flowers and attractive fruits, blueberry plants are tried growing in personal gardens or house balconies [5]. However, blueberry is in over rapid growth for a garden plant. Studies on blueberries mostly focused cultivation techniques in farmer fields [6]. Comparatively, less study was on the garden blueberry plants. As increasing interests in garden blueberry planting, it is necessary to control blueberries growth into appropriate height and scale [7].

Chlorcholinchlorid (CCC) is a plant growth regulator, which control biosynthesis of plant endogenous GA3 [8]. As the result, cell elongation is delayed, and meanwhile cell division is maintained. By using CCC, plants were shorter, stems were thicker, and leaves were darker green [9]. In order to investigate blueberry dwarfing by manual control, we analyzed different concentrations of CCC on ornamental and physiological characteristics of blueberry plants. The purpose of this study is to create garden blueberry plants, which is in perfect size and, at the same time, produce grade fruits [10].

2. Materials and methods

2.1 Materials

Three cultivars, Misty (southern highbush blueberry), Bluecrop (northern highbush blueberry) and Brigitta (northern highbush blueberry), were using.

2.2 Methods

The experiment was started in March 2015, when the blueberry plants were in the beginning of blooming. Chlorcholinchlorid were prepared in three different concentrations, 300 mg/L, 600 mg/L and 900 mg/L, and water (0 mg/L) was using as control treatment. Chlorcholinchlorid solutions were spraying on the leaves for 3 times, at an interval of 7 days, by using a hand-held compression type sprayer. Each treatment has 3 replicates.

3. Results

3.1 Ornamental characteristics

3.1.1 Effect of different concentrations of CCC on three varieties of plant height

Comparing with the control (0 mg/L), CCC solutions in 300 mg/L and 600 mg/L showed better controlled plant growth of all three blueberry cultivars (Table 1), as the plant heights were lower than the controls. It was indicated CCC solutions in 300 mg/L and 600 mg/L is enough to conduct on blueberry [11].

Tab. 1 Effect of different concentrations of CCC on three varieties of plant height (cm)

Blueberry cultivars	CCC concentrations (mg/L)			
	0	300	600	900
Bluecrop	12.2Aa	12.4Aab	12.1Ab	10.9Aa
Brigitta	10.1Aa	9.7ABa	8.8Cb	10.1Ba
Misty	22.8Aa	21.7Aab	17.3Bc	20.1ABb

Multi compared using Duncan's method, different lowercase letters said significantly different at 0.05 level, different capital letters represent 0.01 level differences significant.

3.1.2 Effect of different concentrations of CCC on three varieties of stem diameter

In Table 2, we can see that between 0-600mg/L, compared with the control, all varieties of blueberry stem diameter increased, and were higher than the control, indicating that in this concentration range, CCC can significantly increase the stem diameter of the blueberry.

Tab. 2 Effect of different concentrations of CCC on three varieties of stem diameter (cm)

Blueberry cultivars	CCC concentrations (mg/L)			
	0	300	600	900
Bluecrop	0.23Cc	0.35Bb	0.41Ba	0.51Ab
Brigitta	0.12Dd	0.18Cc	0.39Ba	0.22Ab
Misty	0.13Dd	0.17Cc	0.24Bb	0.29Aa

Multi compared using Duncan's method, different lowercase letters said significantly different at 0.05 level, different capital letters represent 0.01 level differences significant.

3.1.3 Effect of different concentrations of CCC on three varieties of canopy width

Comparing with the control, treated with CCC solutions between 0-600mg/L (Table 3), all varieties of blueberry canopy width increased, and were lower than the control, in this concentration range, CCC can significantly reduce the canopy width of blueberry.

Tab. 3 Effect of different concentrations of CCC on three varieties of canopy width (cm)

Blueberry cultivars	CCC concentrations (mg/L)			
	0	300	600	900
Bluecrop	14.7Aa	10.0Ba	7.9BCb	6.3Cb
Brigitta	20Aa	19.3ABa	15Bb	11.3Ac
Misty	22.2Aa	20.7ABab	17.9Bb	22.4Aa

Multi compared using Duncan's method, different lowercase letters said significantly different at 0.05 level, different capital letters represent 0.01 level differences significant.

3.1.4 Effect of different concentrations of CCC on three varieties of pitch spread

In Table 4, we can see that between 300-600mg/L, the growth of pitch spread of all varieties of blueberry is in a downward trend, indicating that in this concentration range, CCC can shorten the pitch spread of blueberry.

Tab. 4 Effect of different concentrations of CCC on three varieties of pitch spread (cm)

Blueberry cultivars	CCC concentrations (mg/L)			
	0	300	600	900
Bluecrop	7.2Aa	7.5Aa	5.3Bb	2.2Cc
Brigitta	7.1Ba	7.2Aa	5.7Bb	3.2Bc
Misty	6.8Bb	8.2Aa	5.9Bb	6.3Bb

Multi compared using Duncan's method, different lowercase letters said significantly different at 0.05 level, different capital letters represent 0.01 level differences significant.

3.1.5 Effect of different concentrations of CCC on three varieties of leaf thickness

Tab. 5 Effect of different concentrations of CCC on three varieties of leaf thickness (cm)

Blueberry cultivars	CCC concentrations (mg/L)			
	0	300	600	900
Bluecrop	0.032Bbc	0.035Bb	0.045Aa	0.047Ac
Brigitta	0.040Bb	0.044Bb	0.059Aa	0.035Bc
Misty	0.021Bb	0.022Bb	0.032Aa	0.020Bb

Multi compared using Duncan's method, different lowercase letters said significantly different at 0.05 level, different capital letters represent 0.01 level differences significant.

Using 300mg/L and 600mg/L of CCC solutions, the growth of blueberry leaves of all varieties increased (Table 5), and it was higher than the control, indicating that in this concentration range, CCC can significantly increase the thickness of blueberry leaves.

3.2 Physiological characteristics

3.2.1 Effect of different concentrations of CCC on three varieties of the content of malondialdehyde (MDA)

In Table 6, we can see that with the concentration increasing, the content of MDA in 3 varieties decreasing firstly and then increasing, indicating that in this concentration range, CCC can effectively reduce the damage caused by too much MDA.

Tab. 6 Effect of different concentrations of CCC on three varieties of the content of MDA (μmol/g)

Blueberry cultivars	CCC concentrations (mg/L)			
	0	300	600	900
Bluecrop	0.054Aa	0.027Aa	0.057Bb	0.088Bb
Brigitta	0.026ABb	0.024Aa	0.021BCc	0.087Bc
Misty	0.025Cc	0.023Aa	0.026Cc	0.083Cc

Multi compared using Duncan's method, different lowercase letters said significantly different at 0.05 level, different capital letters represent 0.01 level differences significant.

3.2.2 Effect of different concentrations of CCC on three varieties of the content of Chlorophyll

Increasing the concentration of CCC, the content of chlorophyll in leaves was rising and was higher than that of the control (Table 7), indicating that in 0-900mg/L, CCC can improve the content of chlorophyll in leaves of blueberry, such that leaf green discoloration.

Tab. 7 Effect of different concentrations of CCC on three varieties of the content of Chlorophyll (mg/g)

Blueberry cultivars	CCC concentrations (mg/L)			
	0	300	600	900
Bluecrop	1.438Bb	1.924Ab	2.433ABa	2.586Ac
Brigitta	1.564BCc	1.566ABb	1.893Bb	2.362Bc
Misty	2.433Cc	2.488Bc	2.589BCc	2.657Cb

Multi compared using Duncan's method, different lowercase letters said significantly different at 0.05 level, different capital letters represent 0.01 level differences significant.

4. Discussion

Blueberry has few branches but strong vigor, and it is difficulty to be cultivated as small garden plants. So it is necessary to dwarf the plant [12]. Many methods can dwarf plants, such as pruning, pinching, control of water and fertilizer, and hybrid breeding. The most simple and convenient, quick and better maintain the ornamental characteristic of dwarf method is to use plant growth regulators. Using plant growth regulators can control the tree scale, and moreover, satisfy industrialized production need.

In this study, different concentrations of CCC, Chlorcholinchlorid, has different influence on three cultivars [13]. In summary, the commendatory concentration range was 300 mg/L or 600mg/L, and the best controlled blueberry cultivar was Bluecrop.

References

1. Qing-he Wang, Ya-dong Li, Zhi-dong Zhang, et al. Ecological condition the influence of Vietnam orange flower bud differentiation. *Journal of Jilin Agriculture* 2009, 6, 705-710.
2. Rui-chi Pan. Plant physiology. Beijing, higher education press, 2003.
3. Jun-long Li. Industrialization in our country present situation and development countermeasure analysis. *Journal of Anhui Agricultural Sciences* 2006.
4. Yun-zhai Dong, Si-qing Wang. Growth retardant in ornamental plants in the research on application. *North Garden* 2004, 6, 14-16.
5. Yi-lin Wang, Zhi-bin Mao, Hong-wu Zhu. Potted flower production in our country's present situation and development prospect of. *Journal of Jiangsu Forestry Division* 1998, 25.
6. Guo-wen Li. Flower cultivation of six dwarf technology. *Modern seed industry* 2002.
7. Sheng-fa Zhou. Flower dwarf cultivation technique [J]. *Modern Agricultural Science and Technology* 2008, 8, 49.
8. Zhi-zhong Sun. Grape plants. *Chinese Flowers Miniascape* 2011, 12, 26-27.
9. San-gen Wang. The application of plant growth regulator in flower production. Beijing, shield press, 2003.
10. Jia Yin, Qi-xiang Zhang, Hui-tang Pan. PP333 and CCC, B9 on small potted harbinger dwarf effect research. *Journal of Beijing forestry university* 2010.
11. Ming-zhu Gui, Bao-zhong Hu. Small berries cultivation biology. Beijing, China agriculture press, 2002.
12. Oksana Sytar, Asel Borankulova, Irene Hemmerich, Cornelia Rauh, Iryna Smetanska. Effect of chlorocholine chlorid on phenolic acids accumulation and polyphenols formation of buckwheat plants. *Biological Research* 2014, 47, 19.
13. Shi-yong Tang, Yong-jie Wang, Xing-ying Li. Introduce five suitable varieties of potted ornamental fruit trees. *Journal of Northern Fruit Trees* 2004.

Density effect on physiological characteristics in *Michelia chapensis* seedlings

W. L. Huang[1], T. T. Zhou[2], L. Xue[3,*], J. Li[4], Z. Y. Lie[5]

College of Forestry and Landscape Architecture, South China Agricultural University, Guangzhou 510642, P. R. China

[1]278870610@qq.com, [2]1152110287@qq.com, [3]forxue@scau.edu.cn, [4]houis_56028923@qq.com, [5]876672171@qq.com

Density effect on physiological characteristics in *Michelia chapensis* seedlings was studied using potted method. The seedlings were planted with four different densities: 1, 2, 4 and 8 seedlings of each plant bag (density I, II, III and density IV). The physiological characteristics of the seedlings under different densities were determined. Results showed that density effect on contents of chlorophyll and proline was not obvious. Soluble sugar content of 1 seedling bag^{-1} was significantly greater than that of other density seedlings. With increasing density, soluble protein content increased firstly and then decreased, whereas the superoxide dismutase (SOD) activity and the malondialdehyde (MDA) content increased.

Keywords: Michelia Chapensis; Density; Physiological Characteristics; Seedlings.

1. Introduction

As one of the important selected pressures of plants in the nature [1], density is closely related with the availability of environmental resources, such as light, water and nutrients, affecting growth, morphology and survival of plants [2-4]. With increasing density, the competition among plants become violent, which changes plant morphology and growth characteristics [2-3], and then redistributes individual resources and affects biomass accumulation and allocation [5]. Moreover, seedling density also affects the physiological and biochemical characteristics of plant leaves [6-8]. At present, most researches about plant competition focus on plant growth with different densities [9] and stress-resistant physiology of plants, such as cold-resistance physiology [10-14] and drought-resistance physiology [15-19], with a few physiological characteristics examples. *M. chapensis* is an important afforestation tree species in tropical and subtropical regions. Although there are some studies about anti-inverse physiology, mixed forest, hereditary property, soil, and density effect on growth for this species [20-25], density effect on its physiological characteristics has not been reported. In this study, we planted M. chapensis seedlings with different densities, and then studied the physiological characteristics of the seedlings, which can enrich physiology knowledge of this species.

The study was partially supported by Foundation of Guangdong Forestry Department, China (No. 4400-F15050).

*Corresponding author, E-mail: forxue@scau.edu.cn

2. Materials and methods

2.1 Experimental site

The experimental field site was located in Yuejin Nursery of South China Agricultural University, Guangzhou City, South China (113°21′E, 23°09′N). It belongs to the subtropical monsoon climate. The annual average temperature and the average relative humidity are 21.9°C and 77%, respectively. The average annual rainfall is 1899.8 mm, focusing on April to October. The experimental environment is suitable for seedling growth. Determination of physiological indexes was done in College of Forestry and Landscape Architecture, South China Agricultural University.

2.2 Experimental materials

One year-old *M. chapensis* seedlings were taken as materials, which came from Guosen nursery field in Guangdong Province. These seedlings were planted in seedling bags with the size of 35 cm in diameter and 30 cm in depth. The base materials in bags were humus soil and yellow subsoil with a ratio of 1:1. Seedlings were planted according to 1, 2, 4 and 8 seedlings per bag (density I, II, III, and IV), respectively, namely 10, 20, 40 and 80 seedlings m2. Experimental period is from Mar. to Sept. 2013. When the study began, the ground diameter, height and crown width of seedlings were 0.57±0.07, 38.99±4.34 and 16.83±4.81cm, respectively.

2.3 Study methods

Functional leaves from third to eighth of each seedling were selected to determine physiological indexes. The physiological index determination was done in Sept. 2013. The contents chlorophyll, soluble protein and soluble sugar were measured by spectrophotometry, anthrone colorimetric method and coomassie brilliant blue method, respectively [26]. Proline content, superoxide dismutase (SOD) and malondialdehyde (MDA) were determined by acidic ninhydrin method, nitroblue tetrazolium (NBT) photoreduction method and thiobarbituric acid method determination, respectively [27]. All indexes were measured in triplicate.

2.4 Data analysis

All data were analyzed by using Excel software and SAS 9.3 software.

3. Results

There was not significant difference in chlorophyll content among 1, 2 and 8 seedlings bag^{-1} of *M. chapensis* leaves, but that of 4 seedlings bag^{-1} was significantly lower than other density seedlings (P<0.05) (Figure 1).

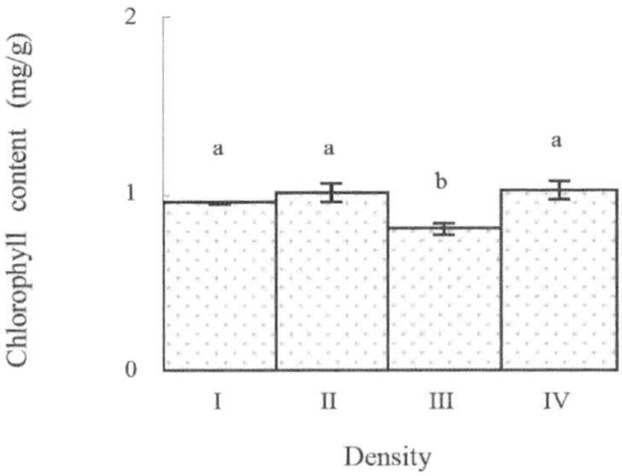

Fig. 1. Chlorophyll content I, 1 seedling bag-1; II, 2 seedlings bag-1; III, 4 seedlings bag-1; IV, 8 seedlings bag-1

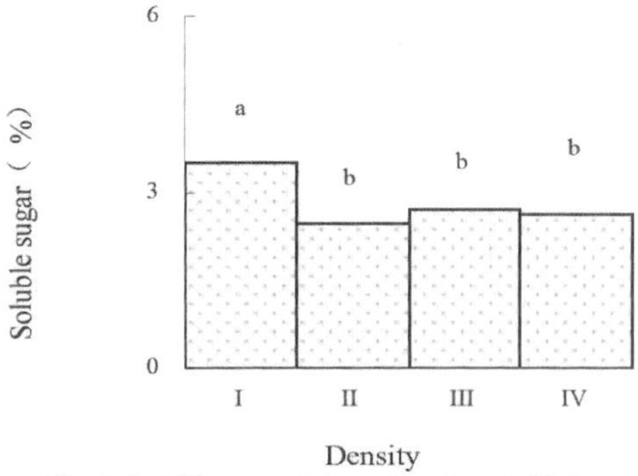

Fig. 2. Soluble sugar I, 1 seedling bag-1; II, 2 seedlings bag-1; III, 4 seedlings bag-1; IV, 8 seedlings bag-1

The content of soluble sugar decreased in the order of 1 seedling bag^{-1} >4 seedlings bag^{-1} >2 and 8 plant bag^{-1} and that in 1 seedlings bag^{-1} was significantly higher than other density seedlings (P<0.05) (Figure 2).

The content of soluble protein in seedlings decreased in the order of 4 seedlings bag^{-1} >8 seedlings bag^{-1} >1 and 2 seedlings bag^{-1}. That of 4 and 8 seedlings bag^{-1} was significantly higher than 1 and 2 seedlings bag^{-1} (P<0.05) (Figure 3).

The content of proline content in seedlings decreased in the order of 2 seedlings bag^{-1} >1 seedling bag^{-1} >4 and 8 seedlings bag^{-1}. The difference among different density seedlings was not significant (Figure 4).

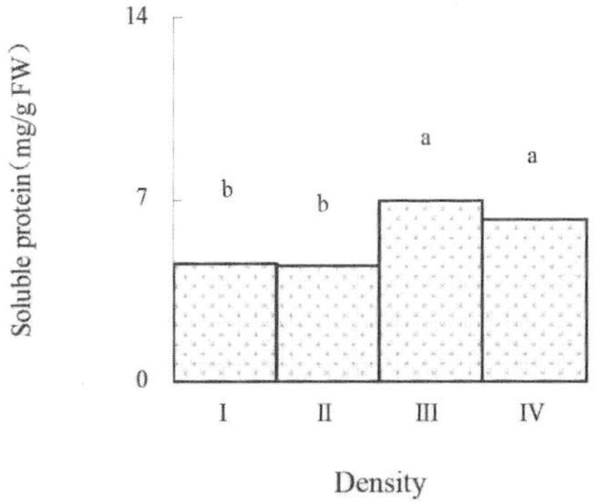

Fig. 3. Soluble protein I, 1 seedling bag-1; II, 2 seedlings bag-1; III, 4 seedlings bag-1; IV, 8 seedlings bag-1

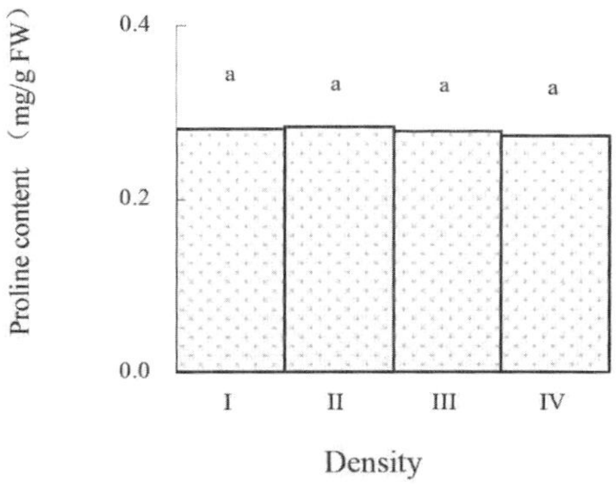

Fig. 4. Proline content I, 1 seedling bag-1; II, 2 seedlings bag-1; III, 4 seedlings bag-1; IV, 8 seedlings bag-1

The SOD activity in different density seedlings of *M. chapensis* decreased in the order of 4 and 8 seedlings bag^{-1} >2 seedlings bag^{-1} > 1 seedlings bag^{-1}. The first two were significantly lower than the latter two (P<0.05) (Figure 5).

The content of MDA increased with increasing density. The MDA content of density 4 and 8 seedlings bag^{-1} was significantly higher than density 1 and 2 seedlings bag^{-1} (P<0.05) (Figure 6).

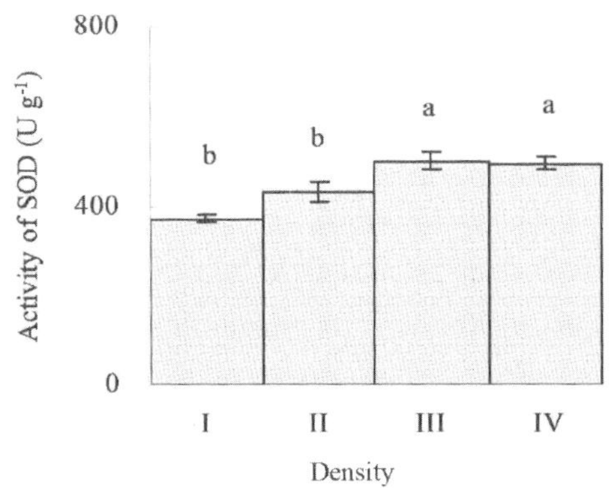

Fig. 5. SOD activity I,1 seedling bag-1; II, 2 seedlings bag-1; III, 4 seedlings bag-1; IV, 8 seedlings bag-1

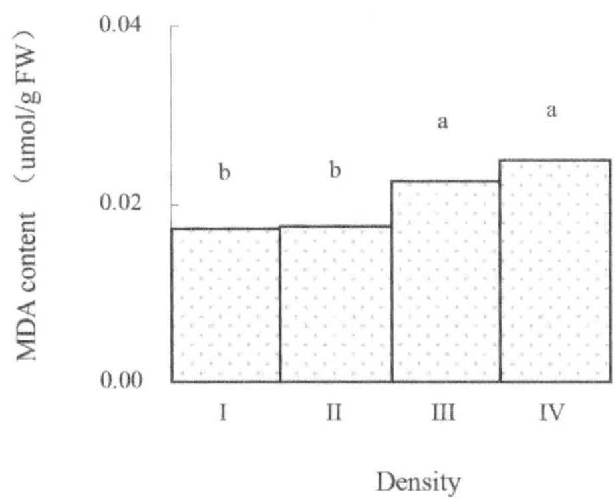

Fig. 6. MDA content I,1 seedling bag-1; II, 2 seedlings bag-1; III, 4 seedlings bag-1; IV, 8 seedlings bag-1

4. Discussion

As an important indicator of leaf quality, chlorophyll content is directly or indirectly related to the photosynthetic rate of unit leaf area, and affects plant biomass and yield. Lu et al. [28] found that leaf chlorophyll content reduced slightly with increasing planting density. In the present study, the chlorophyll content of 1 seedling bag^{-1} was significantly lower than other density seedlings, generally density effects on chlorophyll content was not obvious, the reason may be that experimental seedlings were small, shade phenomenon did not occur in high density seedlings, so that density effect on the chlorophyll content was small.

Soluble sugar and proline are the ideal osmotic adjustment substances in plant cells, which can alleviate the damage of membrane system under stress environment. Soluble sugar can be used as a signal molecule being similar to plant hormones, which play a regulation role in the process of plant growth, development, maturation and aging [29]. In the present study, the soluble sugar content of 1 seedling bag^{-1} was significantly higher than other density seedlings. In the case of low density, the individual competition effect is weakened, a high soluble sugar content is beneficial to seedling growth.

Soluble proteins can help plant cells to maintain a low osmotic potential and to resist the damage caused by stress [30]. Hu [31] found that with increasing density the content of soluble protein of spring corn decreased, the less tolerant varieties decreased greatly. The content of soluble protein in the leaves increased first and then decreased with increasing density, which indicated that 4 seedling bag^{-1} promoted increase in soluble protein to resist the environmental stresses such as light, nutrient and so on. However, 8 seedling bag^{-1} may result in increase of hydrolytic enzyme activity, which makes the protein degradation. Another reason is that density stress directly influences protein synthesis [30].

Proline is a component of plant protein and can be found in a wide range of free state. In addition to being an osmotic adjustment substance, proline plays an important role in stabilizing the biological macromolecular structure, reducing the cell acid, ammonia detoxification and regulating cell redox potential [32]. Proline content in plants increases significantly under stress condition, being a reflection of anti-resistance ability of plants. For example, the proline accumulates in many plants under drought and salt stress. In this study, density effect on proline content was not obvious, because small seedlings caused a less shade, which resulted in a weak density effect on the proline.

In the process of stress of plants, the ability to produce reactive oxygen is often greater than the ability to remove reactive oxygen. Excess intracellular reactive oxygen influences biological membrane and function or structure of other biological macromolecules. SOD is a key enzyme in the cellular defense system, having the functions of maintaining the balance of active oxygen metabolism and protecting the membrane structure [33]. It can coordinate with the leaf protective enzyme system, and the effectively remove free radicals in cells, thus delaying the leaf senescence [30]. The activity of SOD increased with increasing density, which indicated that the high density seedlings defense the stress from light, soil nutrient and water by increasing SOD activity.

When plants suffer from stress, reactive oxygen become disequilibrium, directly leading to the accumulation of reactive oxygen in plants, consequently membrane lipid peroxidation aggravates and cell membrane system is damaged. MDA is one of the most important products of membrane lipid peroxidation. MDA can be cross-linked to protein molecules, and its content denotes the degree of lipid peroxidation [20]. MDA content increased with increasing density in this study, it indicated that the degree of membrane lipid peroxidation increased with increasing density stress.

References

1. W. Japhet, D. W. Zhou, H. X. Zhang and T. Yu. Evidence of phenotypic plasticity in the response of *Fagopyrum esculentum* to population density and sowing date. *J. Plant Biol.* **52**, 303 (2009).
2. I. Müller, B. Schmid and J. Weiner. The effect of nutrient availability on biomass allocation patterns in 27 species of herbaceous plants. *Perspec. Plant Eco, Evol. Syst.* **3**, 115 (2000).
3. H. Maherali and E. H. Delucia. Influence of climate-driven shifts in biomass allocation on water transport and storage in ponderosa pine. *Oecologia.* **129**, 481 (2001).
4. J. G. Hamilton, A. R. Zangerl, E. H. DeLucia and M. R. Berenbaum. The carbon nutrient balance hypothesis: its rise and fall. *Ecol. Let.* **4**, 86 (2001).
5. Y. W. Yang, G. X. Wang, X. L. Li and H. K. Zhou. Research advances in plant density-dependent regulation and its responses to environmental change. *Chin. J. Ecol.* **30**, 1813 (2011).
6. L. S. Guo and H. Y. Sheng. On influences of population density of *Plantago virginica* on its morphological characters. *Bull. Bot. Res.* **22**, 236 (2002).
7. B. Li, C. Yang and P. Lin. *Ecology* (High Education Press, Beijing, 2000).
8. Y. X. Huang, X. Y. Zhao, H. X. Zhang, Y. Y. Luo and W. Mao. Responses of

Agriophyllum squarrosum phenoty picplasticity to the changes of soil nutrient and moisture contents and population density. *J. Appl. Ecol.* **19**, 2598 (2008).

9. L. L. Liang, L. Xue, J. D. Fu, W. G. Zheng, X. L Shi and H. F. Feng. Density effects on interactions between above and below ground competition in young *Acacia auriculaeformis* stands. *J. South China Agri. Univ.* **30**, 59 (2009).
10. Y. Xu, L. Xue and M. Qu. Physiological and ecological mechanisms of plant adaptation to low temperature. *Sci. Sil. Sin.* **43**, 88 (2007).
11. Y. J. He, L. Xue, X. R. Ren, H. Cao, L. L. Liang and Y. Xu. Effects of low temperature stress on physiological characteristics of six tree species seedlings. *Chin. J. Ecol.* **27**, 524 (2008).
12. X. R. Ren, L. Xue, X. E. Wang, T. F. Xie, H. Cao and Y. J. He. Effects of low temperature stress on the physiological processes of six afforestation seedlings. *J. Cent. South Univ. For. Techno.* **28**, 56 (2008).
13. R. Zhang, J. X. Xu, L. Xue, Z. Y. Yang, C. Q. Wu and S. H. Guo. Effects of low temperature stress and release by chilling on physiological characteristics of four broadleaf seedling types. *Ecol. Sci.* **33**, 419 (2014).
14. Y. R. Shao, J. X. Xu, L. Xue, R. Zhang, C. Q. Wu and G. C. Lu. Effects of low temperature stress on physiological-biochemical indexes and photosynthetic characteristics of seedlings of four plant species. *Acta Ecol. Sin.* 33, **4247** (2013).
15. Y.Y. An, Z. S. Liang and W. F. Hao. Growth and physiological responses of different intensity of drought stress in the seedlings of the bar. *Acta Ecol. Sin.* **13**, 716 (2011).
16. X. L. Shi, L. Xue, X. R. Ren, H. F. Feng, W. G. Zheng and J. D. Fu. Preliminary study on drought resistance of four broadleaved seedlings under water stress in South China. *For. Res.* **13**, 760 (2011).
17. H. F. Feng, L. Xue, X. R. Ren, J. D. Fu, W. G. Zheng and X. L. Shi. Physiological responses of 4 kinds of broad leaved seedlings to PEG simulated drought. *Acta Ecol. Sin.* **31**, 3829 (2011).
18. C. J. Shan, R. L. Han and Z. S. Liang. 4 native grasses in Loess Plateau under drought stress antioxidant properties. *Acta Ecol. Sin.* **24**, 170 (2012).
19. X. Y. Xie, Z. L. Ma, P. Bai and X. J. Liu. The response of pepper blossom fruit morphological and physiological drought stress. *Acta Ecol. Sin.* **34**, 3797(2014).
20. J. M. Jiang, H. J. Teng, J. L.Yuan, Q. F. Luan and Z. F. Tang. Genetic Diversity of *Michelia chapensis* Dandy populations. *For. Res.* **18**, 9 (2005).
21. D. L. Tian, G. Q. Yin, X. Fang, Q. F. Luan and Z. F. Tang. Carbon density, storage and spatial distribution under different 'Grain for Green'patterns in Huitong, Hunan Province. *J. Act. Ecol. Sin.* **30**, 6297 (2010).
22. S. K. Xu and L. Xue. The soil characteristics of 6 species of broad-leaved tree species seedlings. *J. South China Agri. Univ.* **31**, 76 (2010).
23. S. H. Guo, L. Xue, R. Zhang and Z. Y. Yang. Photosynthetic physiological responses of 4 seedlings under stress. *J. South China Agri. Univ.* **33**, 373 (2012).

24. L. H. Ye, Z. Y. Yang, L. Xue, G. W. Lie and X. L Huang. Physiological response of three tree species seedlings to submergence stress. *J. South China Agri. Univ.* **32**, 368 (2012).
25. S. M. Zhu, L.L. Xiao, L. Xue, X. He, X. D. Wang and T. C. Hu. Effects of planting density on growth and biomass of *Michelia chapensis* seedlings. *J. Cent. South Univ. For. Techno.* **35**, 77 (2015).
26. H. S. Li. *Experimental Principle and Technology of Plant Physiology and Biochemistry* (High Education Press, Beijing, 2000).
27. J. X. Chen and X. F. Wang. *Experimental Principle and Technology of Plant Physiology and Biochemistry* (South China University of Technology Press, Guangzhou, 2002).
28. H. W. Lu, C. L. Wang, R. Q. Lu, J. J. Li, B. F. Li, Y. Wu, H. Y. Zhang and Y. J. Su. Effects of planting density on leaf chlorophyll content and yield of maize. *J. Anhui Agri. Sci.* **41,** 13505 (2013).
29. K. E. Koch, Z. Ying, Y. Wu and W. T. Aviqne. Multiple paths of sugar sensing and a sugar/oxygen overlap for genes of sucrose and ethanol metabolism. *J. Exper. Bot.* **51**, 427 (2000).
30. L. D. Gong, Y. H. Tian, Y. F. Long and L. P. Wei. Effect of continuous drought stress and rewatering on osmotic regulation in leaves of *Hevea brasiliensis*. *Chin. Agri. Sci. Bull.* **28**, 35 (2012).
31. M. Hu. Effects of density on physiological and yield of photosynthesis and senescence of spring maize, Master's thesis, Northeast Agricultural University (2009).
32. Y. Li, M. G. Sun, Y. J. Kong and L. Xue. Physiological and biochemical responses of *Gleditsia sinensis* seedlings to drought stress. *J. South China Agri. Univ.* **27**, 66 (2006).
33. Z. H.Wu, F. H. Zeng, S. J. Ma, Y. J. Xie and X. Y. Lu. A review of advances in active oxygen metabolism in plants under water stress. *Subtrop. Plant Sci.* **33**, 77 (2004).

The primary cultivation test report of *Phyllostachys nidularia* Munro in Wolong nature reserve during rainy season

Xiao Hong Wang

College of Tourism and Culture Industry, Chengdu University, Shiling Town, Chengdu 610106, China

E-mail: wxh1702@163.com

Phyllostachys nidularia Munro is the favorite food of captive giant panda in China, it is necessary to build bamboo produce base for captive panda, especial after the 5.12 Wench Earthquake. In order to get effective culture technique for bamboo base building, we did cultivation test of *Ph. Nid* (*Phyllostachys nidularia* Munro) in Gengda town in Wolong nature reserve in rainy season of this year, up to now we already finished transplanting and did six times survey about the bamboo growing and got some useful data. The paper recorded and analysis the data, and the result showed that recovering, shooting and survival rate of our test are all better than normal planting pattern, especially leave 3 branches and use specific compound bamboo fertilizer contributed to higher survival rate and shooting, but the number of leaving branches has no obvious effect on shorten recovering length.

Keywords: Phyllostachys Nidularia Munro; Cultivation Test; Rainy Season.

1. Introduction

Phyllostachys nidularia es for handwork, so it is not a bamboo species with high economic value. Even so, it still is an important bamboo species in Sichuan province. Because it is one of the favorite bamboo for giant panda, especial for captive panda[2].

In order to build bamboo produce base for captive panda in Wolong nature reserve after earth quake, we did cultivation test in Shenshuping captive panda breeding base that was built in Gengda Town, Wolong nature reserve after earth quake. Up to now, the test already have achieved initial success, and we collected and collated information and data of bamboo planting and growing, in order to get useful experiment and basic direction for upper works.

2. Materials and methods

2.1 Experimental sites

Study area is located in Gengda town, Wolong national nature reserve, about 120 km away from Chengdu. Its altitude is about 1760 m, annual average temperature 13.3 °C to 14.8 °C, the annual sunshine is 900 hours, annual precipitation is 1100-1300 mm, annual average humidity is 80%; the test sites is mountain yellow brown soil with 15-30 degrees slop.

2.2 Experimental design

2.2.1 Planting time

Wolong national nature reserve lies in the eastern boundary of Shanghai-Tibet plateau climate fate, because its geographical position and topography, formed a typical subtropical inland mountain climate, dry and wet seasons is very obvious, and seasonal temperature change is obvious too. So it can be divided into the winter half year (November to April of following year) and summer half year (from may to October) two seasons, and winter half year is dry and cold, the average temperature is 8.3 °C, and the lowest temperature can reach -8.7 °C; summer half year is rainy and worm, the average temperature can reach 19.5 °C, rainfall reached more than 68% of total year. Therefore, in order to make full use of local heat and water resources, we planted *Ph. Nid* in rainy season of June and July in stead of traditional bamboo planting method in spring of April and May, try to get the best cultivation effect.

The exactly transplanting time of our test is on May 27, 2015, the highest temperature of that day is 23 °C, the lowest temperature is 16 °C, the weather is cloudy, it is fit for bamboo transplanting.

2.2.2 Planting pattern

In order to get the best planting pattern by cultivation test, we designed the different treatments including the number of kept branches, fertilization mode and planting density as control. Kept branch number just mean the number of branch that left on the bamboo trunk when we transplanting, and the number are 5, 4 and 3 in this test. Fertilization mode just mean the kind of fertilizer we use during planting, including specific compound fertilizer for bamboo and common compound fertilizer. Planting density is the number of planting bamboo in certain area, we divided total sites into 12 sample plots, each plot is 10 m x 10 m, and we dug the hole for bamboo planting in 2 m x 2 m line spacing, half of plot planted 2 *Ph. Nid* and other half planted 3. In order to make easier and convenience to records and statistics, we use capital letters and number to show different treatments. We use "A+number" means the number of keeping branches, such as A5 indicates that kept living branch number is 5, and A4 is four, and so on. We use "B+number" to show fertilization pattern, B1 means used common compound fertilizer and B2 means specific compound bamboo fertilizer. "C+number" means bamboo number per hole, such as C2 means two bamboo per hole and C3 means three. For example, if the treatment of one plot is A5×B1×C2, it exactly means bamboo in this plot were planted as kept five branches, given common compound fertilizer and two bamboo per hole. The detail of the each sample plot including the number of different items, treatment pattern and total plants are shown in Table 1.

2.3 Survey items and method

Transplanting is a special process for plant, and plant will have obvious change after transplanting, usually their leaves will dry up and turn yellow even died in certain days, we call that recovering stage. When recovering stage is end, plant will recover and begin to grow again, they will sprout new leaves, buds, or new bamboo shooting again, which just mean it already recovered from transplanting. So the length of recovering stage, survival rate and new shoot rate of different treatments can illustrate its stand or fall. We surveyed the quantity of recovered, survival bamboo and new shoot of *Ph. Nid* in each treatment every 15 days to compare their advantages and disadvantages. But because the time of recovering stage ending, shooting beginning and the survival are different, so we recorded them in different time.

3. Results and analysis

We conducted six times from transplanting to now, and the results of each time and items showed below. Because the recovering stage of bamboo is pretty long, to make sure what actually happened must after a long time. So the survival rate just based on the last data, which come from the last survey we did after 90 days of transplanting.

3.1 Survival rate

From Table 2, it can be seen that the survival rate of our test is very high, average death rate is only 3.08%, which just mean the average survival rate can up to 96.92%, this result is obvious higher than normal bamboo transplanting[3]. In different treatments, the death rate of treatment 1 is the highest, which treatment left 5 branches, used normal compound bamboo fertilizer and planted 2 bamboo per hole while transplanting; and the lowest death rates is treatment 10, which left 3 branches, used specific compound bamboo fertilizer and planted 2 bamboo per hole. The result of the survival showed that leave less branches and use specific compound bamboo fertilizer contributed to higher survival rate, the planting density has no obvious effect on survival.

3.2 The length of recovering stage

We recorded and analysis the recovering stage based on the data from the second survey to sixth, that is 30 days after transplanting to 90 days. From Table 3, it can be seen that the recovering situation of our test is pretty good, 30 days after transplanting, *Ph. Nid* began to sprout new leaves, average recovered bamboo rate up to 18.15%, 45 days later, most bamboo ended recover, average recovered bamboo rate is 75.15%, till 60 days, almost all bamboo ended recovering stage, average recovered rate up to 97.08%. The result showed that the recovering situation of our test is very good, recovering stage is much shorter than average level[3], and bamboo can recover very quickly in our test.

In different treatments, 30 days after transplanting, the highest recovering rate is 4, the lowest is 11; 45 days later, the highest is 2, the lowest is 7; 60 days later, the highest is 4, the lowest is 11; 75 days and 90 days after transplanting, the change of the rate is very little and the result is nearly same, both of the highest and the lowest are all 1. Although the highest and the lowest recovering rate in different stage is not same treatment, but the result showed that used the specific compound bamboo fertilizer and planted 3 bamboo per hole contributed to higher recovering rate, and the number of leaving branches has no obvious effect on recovering length.

3.3 Shooting situation

We recorded and analysis the new shoot number based on the data from the third time survey to sixth, that is 45 days after transplanting to 90 days. Form Table 4, it can be seen that *Ph. Nid* has very good shooting ability in our test, the shooting rate is 3.5% just after 45 days of transplanting, and it can up to 17.7% 90 days after transplanting. The shooting situation much better than normal planting pattern[3], not only shooting number but also shooting period are all better.

In different treatments, 45 days after transplanting, the highest shooting rate is 5, the lowest is 1; 60 days later, the highest is 8, the lowest is 1; 75 days later, the highest is 8, the lowest is 10; 90 days after transplanting, the highest is 8, the lowest is 1. Although the highest and the lowest shooting rate in different stage is not same treatment, but the result showed that used the specific compound bamboo fertilizer and leaving 4 branches contributed to higher shooting rate, and the number of bamboo per hole has no obvious effect on recovering length.

Acknowledgments

This paper is the phased objectives of the project "Schuan-Hongkong-Wolong sustained cooperation plan - HK07". We are thankful to OPCF Fund (Ocean Park Conservation Foundation, Hong Kong) and Hong Kong SAR Government for their support, and also thank Vanessa Hull for his language correcting.

References

1. Institution of Kunminggense Academiaae Sinicae, *Flora Yunnanica*. Science Press, Beijing, China, pp.105–106, 2003.
2. Liu, Q., He, H., Shen, Z.B., Study on the growth status and correlative environment factors of bamboo *Neosinocalamus affinis* from Chengdu region, Sichuan environment. **20(4)**, pp. 43–46, 2001.
3. Li, C.B. Research on giant panda's main diet bamboo species. Guiyang: Guizhou Science and T echnology Press, 2-180.1997.

Tab. 1 The treatment patterns of Ph. Nid cultivation testing

Treatment	1	2	3	4	5	6	7	8	9	10	11	12
The detail of treatment	A5×B1×C2	A5×B2×C2	A5×B1×C3	A5×B2×C3	A4×B1×C2	A4×B2×C2	A4×B1×C3	A4×B2×C3	A3×B1×C2	A3×B2×C2	A3×B1×C3	A3×B2×C3
Total plants	56	56	84	84	56	56	84	84	56	56	84	84

Tab. 2 Survival rate of each treatments of Ph. Nid

Treatment	1	2	3	4	5	6	7	8	9	10	11	12	Average
Death bamboo number	3	1	3	2	2	2	2	2	2	1	3	2	2
Death rate (%)	5.36	1.79	3.57	2.38	3.57	3.57	2.38	2.38	3.57	1.29	3.57	2.38	3.08

Tab. 3 The recovering stage of each treatments of Ph. Nid

Treatments	The second survey		The third survey		The fourth survey		The fifth survey		The sixth survey	
	Recovered bamboo	Recovered rate (%)	Recovered bamboo	Recovered rate (%)	Recovered bamboo	Recovered rate (%)	Recovered bamboo	Recovered rate (%)	Recovered bamboo	Recovered rate (%)
1	6	10.71	40	71.43	45	80.36	49	**87.50**	53	**94.64**
2	14	16.67	47	**89.93**	51	91.07	55	98.21	55	98.21
3	16	28.57	43	76.79	74	87.50	77	91.67	81	96.431
4	17	**30.36**	78	92.86	80	**97.62**	83	**98.81**	83	**98.81**
5	9	16.07	39	69.62	49	87.50	50	89.26	54	96.43
6	11	19.62	42	75.00	51	91.07	52	92.86	54	96.43
7	7	11.90	**50**	**59.52**	68	80.95	79	94.05	82	97.62
8	13	15.48	57	67.86	77	91.67	82	97.68	82	97.68
9	8	14.29	39	69.64	47	87.04	52	92.86	54	96.43
10	10	17.86	49	87.50	52	92.86	55	98.21	55	98.21
11	9	**10.71**	51	60.71	67	**79.76**	81	96.43	81	96.43
12	14	25.50	68	80.95	79	94.05	82	97.68	82	97.68
Average	11	18.15	50	75.15	61	88.45	66	94.60	68	97.08

Tab. 4 New shoots number of each treatments of Ph. Nid

Treatments	The third survey		The fourth survey		The fifth survey		The sixth survey	
	New shoot number	New shoot rate (%)	New shoot number	New shoot rate (%)	New shoot number	New shoot rate (%)	New shoot number	New shoot rate (%)
1	1	**1.8**	4	**7.1**	5	**8.9**	**5**	**8.9**
2	2	3.6	5	8.9	7	12.5	8	14.3
3	2	2.4	11	13.1	14	16.7	15	17.9
4	3	3.6	13	15.5	16	19.0	16	19.0
5	2	3.6	5	8.9	9	16.1	11	16.1
6	2	3.6	7	12.3	10	17.9	11	19.6
7	3	3.6	13	15.5	16	19.0	17	20.1
8	2	2.4	14	**16.7**	17	**20.2**	18	**21.4**
9	3	**5.4**	5	8.9	8	14.3	9	16.1
10	2	3.6	6	10.7	11	19.6	11	19.6
11	3	3.6	11	13.1	15	17.6	15	17.9
12	4	4.8	13	15.5	16	19.0	17	20.1
Average	2	3.5	9	12.2	12	16.7	13	17.7

The Similarities and Differences of Flue-cured Tobacco Aroma-note between Yunnan and Fujian

Zhengfeng Li[1,b], Yuzhen Xia[2], Yi Wang[1], Dingrong Mou[2] and Shaokun Lu[3,a,*]

[1]*China Tobacco. Yunnan Industrial Co. Ltd. R&D Center, Kunming, 650224, China,*
[2]*Hongta tobacco (Group) Co. Ltd.,*
Yuxi, 653100, China
[3]*College of Foundation and Information Engineering, Yunnan Agricultural University*
E-mail: [a]lsk999@126.com, [b]1208913744@qq.com

To distinguish the similarities and differences of flue-cured tobacco aroma-note between Yunnan and Fujian, the student-t test, correlation analysis and regression method were adopted to study 157 flue-cured tobacco samples collected from Yunnan and Fujian. The results indicate that fresh-sweetness, straw, and burnt are the top three aroma-notes in Yunnan and Fujian flue-cured tobacco. There is a negative correlation between fresh-sweetness and green in Yunnan and Fujian flue-cured tobacco. However, the aroma-note difference between them is also significant. The fresh-sweetness (green) in Yunnan flue-cured tobacco is stronger (weaker) than Fujian flue-cured tobacco. The differences of corresponding routine chemical components are the content of kalium and chlorine. The content of kalium (chlorine) in Yunnan flue-cured tobacco is significantly lower (higher) than Fujian flue-cured tobacco. The fresh-sweetness and green in Yunnan flue-cured tobacco are closely related to reduced sugar, kalium and the ration of kalium to chlorine, but such relationship is not obvious in Fujian flue-cured tobacco. The green in all aroma-notes could be used as a key index to distinguish the Yunnan flue-cured tobacco and the Fujian flue-cured tobacco.

Keywords: Aroma-note, Flue-cured Tobacco, Fresh-sweetness, Green, Chemical Component.

1. Introduction

Since the aroma-type and the aroma-note of flue-cured tobacco significantly impact on the flue-cured tobacco production and the industrial uses of flue-cured tobacco, many researchers extensively studied the aroma-type and the aroma-note of flue-cured tobacco. They divided the aroma-type of flue-cured tobacco into three types: full, clear and medium aroma [1-13]. In recent years, some researchers studied the aroma-note of flue-cured tobacco. For examples, Wang et al. [14] found that the baking and the spicy aroma are the most two important aroma-note in Jiangxi Province. The fresh-sweetness, burnt-sweetness, woody, resin and fruity are the main aroma-type of flue-cured tobacco which lead the spatial difference of flue-cured tobacco aroma style over Jiangxi Province. Qiao et al. [15] pointed out that full aroma-type is mainly characteristic with burnt-sweetness, woody and burnt aroma-note, the clear aroma-type is mainly characteristic by fresh-sweetness, green and woody, and the medium aroma-type is mainly characteristic with pure-sweetness, woody and spicy. Deng et al. [16] found that the aroma-note of flue-cured tobacco over Hunan Province mainly features with hay, burnt-sweetness and burnt. Wang et al. [17] indicated that the two aroma-note: burnt-sweetness and hay contribute most to the full aroma-type. Xia et al. [18] pointed that the flue-cured tobacco of Yunnan Province significantly featured with fresh-sweetness and green, but the more green will depress the fresh-sweetness. Although Yunnan and Fujian are well acknowledged as the major production region of clear flue-cured tobacco [2, 9, 15], Li et al. [19] found that there is obvious aroma-note differences between them, and the associated studies are still very few up to now. This motives us to explore the similarities and differences of flue-cured tobacco aroma-note between Yunnan and Fujian from the way of quantitative analysis. We hope that the corresponding results could be used in the development of Chinese - style cigarette.

2. Data And Methods

Data: The 127 flue-cured tobacco samples from Yunnan Province and 30 flue-cured tobacco samples from Fujian Province are provided by the technology center of Yunnan tobacco industry limited liability company. There are thirteen aroma-notes: hay, green, woody, burnt, acid, spicy, nutty, fruity, floral, herb, fresh-sweetness, pure-sweetness, and burnt-sweetness. Any a aroma-note is scaled by 5. The higher the value is evaluated, the more significant the certain aroma-note is highlighted, and vice verse.
Methods: The analysis methods include correlation analysis, regression analysis, and hypothesis testing.

3. Results

Aroma-note features: The 5 statistical indices: mean value, standard deviation, minimum, maximum and variation coefficient are calculated for the flue-cured tobacco aroma-note of Yunnan and Fujian, respectively (Table 1and Table 2).

Tab. 1 The statistical primary feature of Yunnan flue-cured tobacco aroma-note

Aroma-note	Mean Value	Standard Deviation	Minimum	Maximum	Variation Coefficient (%)
Fresh-sweetness	4.26	0.48	2.42	5.00	11.16
Pure-sweetness	0.27	0.29	0.00	1.50	106.47
Burnt-sweetness	1.03	0.35	0.33	2.58	34.15
Hay	2.35	0.66	0.75	3.42	27.89
Green	0.35	0.32	0.00	1.17	91.98
Woody	0.47	0.24	0.00	1.50	50.72
Burnt	1.19	0.34	0.50	2.60	28.85
Acid	0.65	0.24	0.00	1.08	37.08
Spicy	0.96	0.26	0.00	1.50	27.27
Nutty	0.92	0.28	0.17	2.00	30.86
Fruity	0.61	0.24	0.00	1.33	40.26
Floral	0.15	0.15	0.00	1.00	96.56
Herb	0	0	0	0	/

Tab. 2 The statistical primary feature of Fujian flue-cured tobacco aroma-note

Aroma-note	Mean Value	Standard Deviation	Minimum	Maximum	Variation Coefficient (%)
Fresh-sweetness	3.83	0.40	3.08	4.50	10.35
Pure-sweetness	0.23	0.26	0.00	1.00	114.83
Burnt-sweetness	0.72	0.25	0.42	1.42	34.75
Hay	2.10	0.77	0.75	3.00	36.54
Green	0.68	0.37	0.00	1.25	53.75
Woody	0.27	0.24	0.00	0.75	90.35
Burnt	1.09	0.31	0.58	1.88	28.84
Acid	0.44	0.29	0.00	0.92	65.61
Spicy	0.78	0.37	0.00	1.42	47.22
Nutty	0.81	0.39	0.08	1.67	48.15
Fruity	0.73	0.44	0.33	2.00	60.87
Floral	0.06	0.07	0.00	0.25	105.97
Herb	0	0	0	0	/

Tables 1 and 2 show that the sequence of mean value of Yunnan flue-cured tobacco aroma-note from high to low is fresh-sweetness >hay>burnt>burnt-sweetness>spicy>nutty>acid>fruity>woody>green>pure-sweetness>floral; and the sequence of mean value of Fujian flue-cured tobacco aroma-note is fresh-sweetness>hay>burnt>nutty>spicy>fruity>burnt-sweetness>green>acid>woody>pure-sweetness>floral. Thereinto, the maximum of mean value of aroma-note appears at Fresh-sweetness in both Yunnan and Fujian, and corresponding variation coefficients reach their minimum (lower than 20%). The results suggest that the flue-cured tobacco aroma note of Yunnan and Fujian truly belongs to fresh-sweetness, but there are some differences between their aroma-notes. the fresh-sweetness in the flue-cured tobacco aroma-note of Fujian is located between intermediately obvious and more obvious, but that in the flue-cured tobacco aroma-note of Yunnan is laid between more obvious and evident. Because the rest aroma-note indices have higher variation coefficients, the student-t test method was adopted to distinguish the difference between Yunnan and Fujian (Table 3).

Tab. 3 The significance test of the aroma-note difference between Yunnan and Fujian

Aroma-note	Aroma-note difference (Yunnan minus Fujian)	Student-t Value	Confidence Level (%)
Fresh-sweetness	0.43	4.64	99.9993
Pure-sweetness	0.04	0.76	55.2309
Burnt-sweetness	0.30	4.46	99.9984
Hay	0.25	1.84	93.2140
Green	-0.33	-4.96	99.9998
Woody	0.21	4.29	99.9968
Burnt	0.10	1.46	85.4272
Acid	0.22	4.26	99.9965
Spicy	0.18	3.16	99.8111
Nutty	0.11	1.81	92.7148
Fruity	-0.12	-2.03	95.6214
Floral	0.09	3.30	99.8781
Herb	0	0	/

It can be seen from Table 3 that the top two flue-cured tobacco aroma-note differences between Yunnan and Fujian appear at green and fresh-sweetness. In comparison with Fujian flue-cured tobacco, the fresh-sweetness of Yunnan flue-cured tobacco is mostly higher. In fact, the corresponding difference value is 0.43. However, the green of Fujian flue-cured tobacco is significantly higher than that of Fujian flue-cured tobacco: the difference value between them is -0.33. The difference value not passing significance test mainly appears at burnt-sweetness, Hay, burnt and nutty.

Given the green as the independent variable x and fresh-sweetness as the dependent variable y, the linear regression model can be established with the least square method as follows.

$$y_{Yunnan} = -0.53x_{Yunnan} + 4.45 \tag{1}$$

$$y_{Fujian} = -0.69x_{Fujian} + 4.30 \tag{2}$$

where (1) and (2) pass the F significance test at 99.99% confidence level. We can analyze the relationship between green and fresh-sweetness over Yunnan and Fujian. Equations (1) and (2) show that there is negative correlation between green and fresh-sweetness. This result means that, if the green is evaluated with higher value, the fresh-sweetness will be get lower value, and vice versa. However, the regression coefficient of Fujian is lager than Yunnan in an absolute condition, suggesting that the fresh-sweetness rapidly become less with green increasing in Fujian flue-cured tobacco in comparison with Yunnan flue-cured tobacco.

Routine chemical components: To reveal the possible chemical causes why the flue-cured tobacco aroma-notes have these differences obtained above. The same analysis method was adopted to study the difference of routine chemical components between Yunnan flue-cured tobacco and Fujian flue-cured tobacco (Table 4).

Tab. 4 The significance test of routine chemical component differences between Yunnan and Fujian

Chemical component	Mean Yunnan	Mean Fujian	Difference	Confidence Level (%)
Nicotine	2.55	2.58	-0.03	12.74
Total sugar	29.38	29.28	0.09	7.66
Reducing sugars	25.49	27.30	-1.81	97.15
Total nitrogen	2.02	1.87	0.16	98.22
Kalium	1.69	2.54	-0.84	99.99
Chlorine	0.39	0.15	0.24	99.58
Sugar/ Nicotine	13.84	13.33	0.51	28.07
Nitrogen/ Nicotine	0.88	0.81	0.07	78.96
Kalium/ Chlorine	20.02	24.89	-4.87	53.39

Table 4 shows that the differences of routine chemical components between Yunnan flue-cured tobacco and Fujian flue-cured tobacco passing significance test at 95% confidence level mainly manifest at Kalium, Chlorine, Total nitrogen, and Reducing sugars. Hereinto, the contents of Kalium and Reducing sugars in Fujian flue-cured tobacco are significantly higher than those in Yunnan flue-cured tobacco, whereas the contents of Chlorine and Total nitrogen are obviously lower than those in Yunnan flue-cured tobacco. Since the linear relationship between green and fresh-sweetness shown in

2.1 is considered, here, we further the relationship between fresh-sweetness, green, and routine chemical components over Yunnan and Fujian, respectively (Figs. 1 and 2).

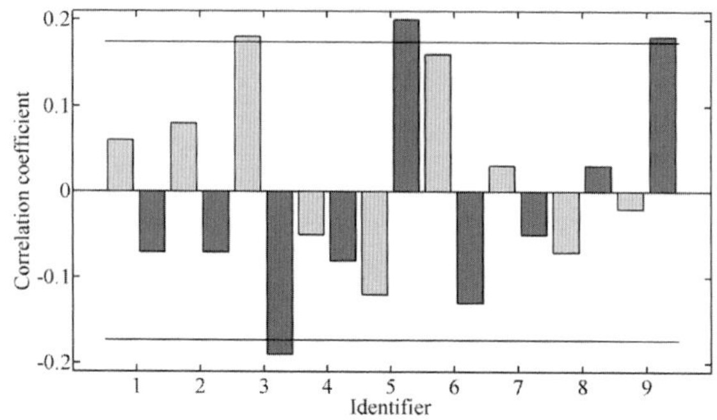

Fig. 1. The correlation coefficients between fresh-sweetness, green, and routine chemical components over Yunnan.

The light grey denotes fresh-sweetness, and dark grey denotes green. Identifier 1 denotes Nicotine, 2 Total sugar, 3 Reducing sugars, 4 Total nitrogen, 5 Kalium, 6 Chlorine, 7 Ratio of Sugar to Nicotine, 8 Ratio of Nitrogen to Nicotine, and 9 Ratio of Kalium to Chlorine. The two thick lines denote the critical value passing significance test at 95% confidence level.

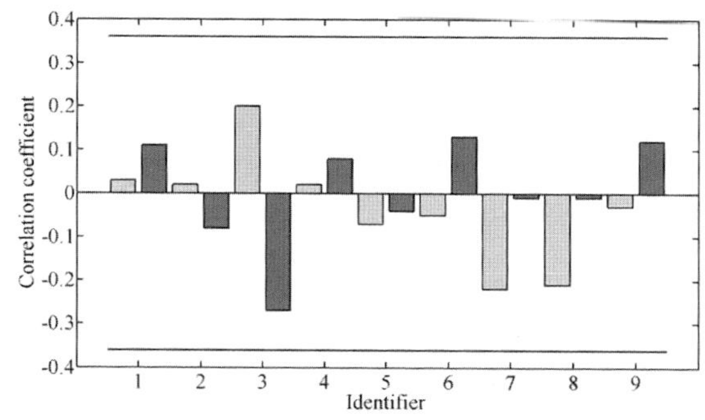

Fig. 2. The same as Fig. 1 but for Fujian.

Fig. 1 displays that the correlation coefficient between fresh-sweetness and Reducing sugars only passes the significance test at 95% confidence level. This result indicates that there is significant positive correlation between them. The content of Reducing sugars is higher, the fresh-sweetness in Yunnan flue-cured tobacco become more typical. In addition, the correlation coefficient between green, Reducing sugars, Kalium and Ratio of Kalium to Chlorine also pass the significance test at 95% confidence level, suggesting that there are significant negative correlation between green and Reducing sugars, significant positive correlation between green, Kalium and Ratio of Kalium to Chlorine. Although the correlation coefficients between aroma-note of Fujian flue-cured tobacco and routine chemical components cannot pass the significant test, the relative higher correlation coefficients are similar to those associated with Yunnan flue-cured tobacco

(Fig. 2), implying that the Reducing sugars may connect with the fresh-sweetness, and the ratio of Kalium to Chlorine may associate with the green over the two region.

4. Summary

Yunnan and Fujian belong to growing district of fresh-sweetness flue-cured tobacco, but there are larger differences of flue-cured tobacco aroma-note in the two region. The quantitative studies in this paper indicate that the pronounced differences mainly occur at fresh-sweetness and green. The grade of fresh-sweetness in the Yunnan flue-cured tobacco is above 10% higher than the Fujian flue-cured tobacco, suggesting that Yunnan flue-cured tobacco is featured with more typical fresh-sweetness than Fujian flue-cured tobacco. However, the grade of green in Yunnan flue-cured tobacco approximate 80% lower than Fujian flue-cured tobacco, that is, the green in Fujian flue-cured tobacco is more obvious than Yunnan flue-cured tobacco. The differences of routine chemical component between the two regions mainly occur at Kalium, Chlorine, Total nitrogen and Reducing sugars, in which the differences associated with Kalium and Chlorine are most significant. There are significant positive correlation between fresh-sweetness and reducing sugars, significant negative correlation between green and reducing sugars, significant positive correlation between green and Kalium, between green and Ratio of Kalium to Chlorine in Yunnan flue-cured tobacco. But there is not any significant correlation between fresh-sweetness, green and routine chemical component in Fujian flue-cured tobacco. In addition, there are some similarities between Yunnan flue-cured tobacco and Fujian flue-cured tobacco. The similarities include that the top three grades of aroma-note are fresh-sweetness, hay, and burnt, and the correlations between fresh-sweetness and green are significantly negative. In summary, green can be used as an important index to distinguish the Yunnan flue-cured tobacco from Fujian flue-cured tobacco.

Acknowledgement

The present work was supported by scientific and technological projects of China National Tobacco Corporation (CNTC) [Ts-03-20110020].

References

1. Ding R.K., Wang C.H., Zhu Z.Q. and et al., Cigarette technology. *Food Industry Press*, 49–71(1958).
2. Zhou J.H., Yang H.Q., Lin G.H., and et al., Studies on the main volatile aroma components in tobacco from different flue-cured tobacco production regions, *J. Hunan Agric. Univ. Nat. Sci.*, 30: 20–23(2004).
3. Tang Y.J., Orientation of style and characteristics of tobacco Leaves, *Chin. Tob. Sci.*, 29: 1–5(2008).
4. Zhao M. Q., Cigarette blending learning. *Science Press,* 416–417(2008).
5. Li Z.H., Wang N.R. Wang D.S., and et al., Preliminary study on important factors on aroma type of flue-cured tobacco and construction of aromas index model. *J. Anhui Agri. Sci.*, 37: 2055–2057(2009).

6. Xie J.P., The principle and application of tobacco flavor technology. *Chemical Industry Press*, 9–10(2009).
7. Liu C.Y., Liu H.X., Chang Z.L., and et al., Advances in tobacco aroma characteristics. *Chin. Tob. Sci.*, 31: 751–78(2010).
8. Ma J.X., Zhang Q., Ou. Y.W., and et al., Analysis on flavor notes of flue-cured tobacco varieties introduced from USA. *Acta Tabacaria Sinica*, 16(suppl.): 22–27(2010).
9. Shi H.Z., Liu G.S., Yang H.J., Ju X.M., Science of tobacco aroma. *China Agricultural Press*. 1–311(2011).
10. Tang Y.J., On aroma type of flue-cured tobacco. Chin. *Tob. Sci.*, 32(3): 1–7(2011).
11. Wang C.W., Research development of aroma components in flue-cured tobacco. *J. Anhui Agri. Sci.*, 39: 8582–8584(2011).
12. Wu Y., Zeng X.Y., Zhu B.K., and et al., Applicability of flavor style sensory evaluation method for Chinese style Cigarette in Different Areas. *Tob. Sci. & Tech.*, 9: 5–9(2012).
13. Chang A.X., Qu Y.S., Ji Y., and et al., Quality characteristics of flue-cured tobacco leaves for different types of aroma in Fujian production region. *Chin. Tob. Sci.*, 32(4): 1–5(2011)
14. Wang N.R., He X.K., Hui J.Q., and et al., Aromatic charms and its spatial characteristics of the flue-cured tobacco leaves in Jiangxi Province. *Chin. Tob. Sci., 33(4): 7–12(2012).*
15. Qiao X.Y., Shen Y.J., Ma Y.P., and et al., Study on characteristic aroma notes of flue-cured tobacco leaves of different flavor styles. *Tob. Sci. & Tech.*, 2: 5–14(2014).
16. Deng X.H., Deng J.Q., Xiao C.S., and et al., The distribution of flavor notes in strong flavor type tobacco leaves from Hunan province. *Acta Tabacaria Sinica,* 20: 39–46(2014).
17. Wang X.L., Zhao M.Q., Ren W., and et al., Relationships among note, aroma quality and aroma type in full-aroma-style flue-cured tobacco production areas. *Chin. Tob. Sci.*, 35(3): 95–98(2014).
18. Xia Y.Z., Wang Y., Mou D.R., and et al., Canonical correlation between aroma-note of clear aroma-type flue-cured tobacco over Yunnan and its routine chemical components. *J. Anhui Agri. Sci.*, 48: 9923–9925(2014).
19. Li Z.H., Wang N.R., Wang D.S., and et al., Preliminary study of aroma type styles of flue-cured tobacco in different ecological scale regions. *Chin. Tob. Sci., 30(5): 67–46,76(2009).*

Session 3: Animal Science

Land covers and their changes in the Amur tiger distribution regions in China and Russia

Lingjun Meng[1,2], Limei Zhang[1,3], Yiqiu Li[4,5] and Zhongke Feng[1,6,*]

[1]*College of Geographical Science, Harbin Normal University, Harbin 150025, China*

[2]*School of Continuing Education, Heilongjiang University, Harbin 150080, China*

[3]*School of History, Cultural Tourism,Heilongjiang University, Harbin 150080, China*

[4]*Ecological Security and Protection Key Laboratory of Sichuan Province, Mianyang Normal University, Mianyang 621000, China*

[5]*Institute of Geographical Sciences and Natural Resources Research, CAS, Beijing 100101, China*

[6]*Beijing Forestry University, Beijing Municipal Key Laboratory of Precision Forestry, Beijing 100083, China*

**Corresponding author: Zhongke Feng*

Email: fengzhongke@126.com

The Amur tiger (Panthera tigris altaica) is one of the most endangered animals. Reasonable protective measures have been carried out by Russia to ensure a basically stable growth of the Amur tiger population since 1930s, and the population has grown to more than 500. There are only 20 Amur tigers in China for now. Global Land Cover 2000 and GlobCover 2009 were used as the data sources to analyze and investigate the land cover status of the Amur tiger distribution regions in China and Russia. The results showed: The human activity area was decreasing in the entire Amur tiger distribution region, and this situation was more remarkable in China. On the whole, Russia had higher forest coverage and better forest quality, leading to a certain difference between the habitats for the Amur tiger in China and Russia. A comparison of land covers of the Amur tiger distribution regions in China and Russia would provide scientific reference and basis for the conservation of the Amur tiger in China.

Keywords: China and Russia; Amur Tiger; Land Cover; Global Land Cover 2000; GlobCover 2009.

1. Introduction

Today, most Amur tigers live in the Russian Far East such as Khabarovsk and Primorsky Krai, and a few distribute in the east of Heilongjiang Province and Jilin Province in China (ZHANG Ming-Hai et al., 2010). As one of the most endangered animals, the Amur tiger has been included in the Red List of the Top 10 Most Endangered Animals in the world by the International Union for Conservation of Nature (IUCN) (Sanderson et al., 2006). Reasonable protective measures have been carried out by Russia since 1930s, and the Amur tiger population has increased to more than 500, further leading to habitat saturation (Carroll C & Miquelle DG, 2006).

It was estimated that there were 200-250 Amur tigers in northeast China at the end of

1950s (WU Jian-guo, et al., 2009). However, as a result of deforestation and development of wood-processing industry, the habitat for the Amur tiger was severely threatened. Moreover, illegal poaching for their furs and bones worsened their situation. Consequently, the population of the Amur tiger reduced drastically during the 1970s and 1980s. In the early 1970s, there were only dozens of Amur tigers in northeast China and North Korea. The population was less than 20 till the early 1990s (Sun HY, et al, 2005, 2006). Currently, the Amur tiger could be only found in several separated island-like areas in the east of Heilongjiang Province and Jilin Province. They have vanished in those areas where they used to be widespread, including the Great and Lesser Khingan Mountains and the Changbai Mountains (Zhou XY, 2008). There is a distinct difference in the habitat for the Amur tiger between China and Russia. In the present study, a comparison of land cover data of the Amur tiger distribution regions in China and Russia would provide scientific reference and basis for the conservation of the Amur tiger in China.

2. Materials and Methods

2.1 The Materials

The data sources used in this study included Global Land Cover 2000 (GLC2000), GlobCover 2009, the data of the administrativc boundary between China and Russia and the vector data of the Amur tiger distribution.

Global Land Cover 2000 (GLC2000) data set divides the globe into 19 regions, and by using the "bottom-to-up" approach in each area according to the FAO land cover classification system, uses different classification methods to assess the regional land cover data, and then creates a global scale land cover data set by means of stitching operation [31-32].

The classification module of the Globcover processing chain consists in transforming the MERIS FR multispectral mosaics produced by the pre-processing modules into a meaningful global land cover map. As explicitly requested by ESA, the challenge is to produce, in an automatic, repeatable and global way, a global land cover map at 300m resolution with a legend defined and documented using the UN Land Cover Classification System (LCCS). The typology has been defined using UN LCCS with the view to be as much as possible compatible with GLC2000 (Fritz et al., 2003). The classification module has been designed by UCL-Geomatics to combine both the spectral and temporal richness of the MERIS FR time series and to be globally consistent while regionally-tuned.

The GlobCover 2009 land cover product created by the European space agency (ESA). The data source originates from the ENVISAT satellite platform MERIS data. The data set then uses the hierarchical partition classification method to perform the layered extraction of different ecological geographic regions types, and the LCCS (Land Cover Classification System) classification system to divide the global surface into 22 Land Cover types [34].

The literature and maps of the Amur tiger distribution were collected, preprocessed through scanning, image registration and projection transformation, and then digitized to capture the vector data of the Amur tiger distribution.

2.2 The Methodological Approach

Through Geographic Information System (GIS) spatial analysis, the data of different types of land covers of the Amur tiger distribution regions in China and Russia in 2000 and 2009 were extracted respectively for comparative analysis.

3. Results and discussion

3.1 Comparative analysis of the land covers in 2000

Land covers distribution of the Amur tiger distribution regions in 2000 are shown in Fig. 1.

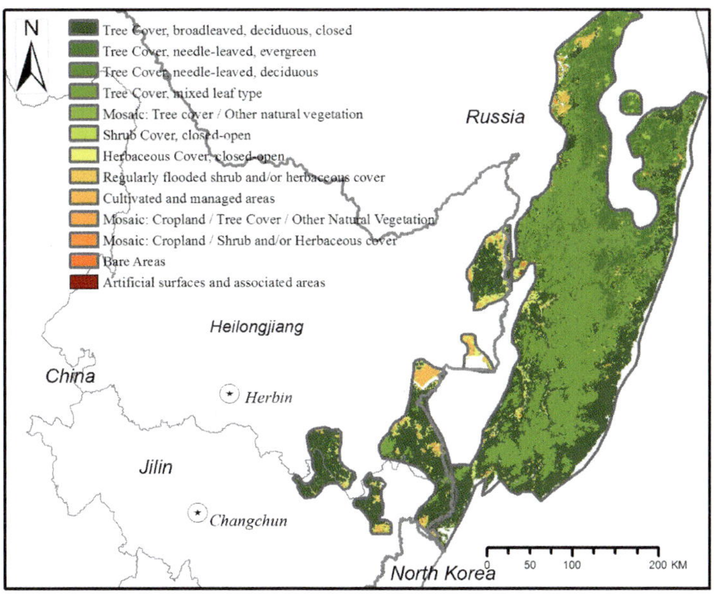

Fig. 1 Land cover map of the east-northern tiger distribution area 2000

According to Fig. 1, the land cover data of the entire Amur tiger distribution region as well as the distribution regions in Russia and China in 2000 were extracted respectively. The area percentages of different land covers are shown in Table 1.

Table1 indicated that: the entire Amur tiger distribution regions were primary Tree Cover, mixed leaf type areas, accounting for 38.02% of the entire distribution regions; it was followed by Tree Cover, broadleaved, deciduous and closed areas which account for 34.96%; Tree Cover, needle-leaved, evergreen areas and Tree Cover, needle-leaved, deciduous areas which account for 12.04% and 6.21% respectively; Cultivated and managed areas which account for 2.05% and Mosaic areas which account for 0.98%.

Tree Cover, mixed leaf type areas which account for 45.47% were dominant in Russia, while Tree Cover, broadleaved, deciduous and closed areas which account for 68.57% were dominant in China; Tree Cover areas in Russia and China were respectively 94.73% and 75.35%; Herbaceous Cover, closed-open areas in China were 11.18%, which was significantly higher than 2.47% in Russia; Cultivated and managed areas in China were 9.85%, which was also significantly higher than 0.32% in Russia.

Tab. 1 Land covers area percentage of the east-northern tiger distribution area 2000

Label type	world	Russia	China
Tree Cover, broadleaved, deciduous, closed	34.96%	27.49%	68.57%
Tree Cover, needle-leaved, evergreen	12.04%	14.69%	0.12%
Tree Cover, needle-leaved, deciduous	6.21%	7.08%	2.31%
Tree Cover, mixed leaf type	38.02%	45.47%	4.53%
Mosaic: Tree cover / Other natural vegetation	0.07%	0.07%	0.05%
Shrub Cover, closed-open	0.33%	0.36%	0.21%
Herbaceous Cover, closed-open	4.06%	2.47%	11.18%
Regularly flooded shrub and/or herbaceous cover	1.24%	1.08%	1.98%
Cultivated and managed areas	2.05%	0.32%	9.85%
Mosaic: Cropland / Tree Cover / Other Natural Vegetation	0.32%	0.33%	0.28%
Mosaic: Cropland / Shrub and/or Herbaceous cover	0.66%	0.61%	0.88%
Bare Areas	0.01%	0.01%	0.02%
Artificial surfaces and associated areas	0.03%	0.03%	0.02%

3.2 Comparative analysis of the land covers in 2009

Land covers distribution of the Amur tiger distribution regions in 2009 are shown in Fig. 2.

According to Fig. 2, the land cover data of the entire Amur tiger distribution regions as well as the distribution regions in Russia and China in 2009 were extracted respectively. The area percentages of different land covers are shown in Table 2.

Table 2 indicated that: the entire Amur tiger distribution regions were primary Open (15-40%) needleleaved deciduous or evergreen forest (>5m), accounting for 34.19% of the entire distribution regions; it was followed by Closed (>40%) broadleaved deciduous forest (>5m) areas which account for 22.30%; Closed to open (>15%) mixed broadleaved and needleleaved forest (>5m) areas which account for 18.43%; Mosaic areas which account for 19.29%; Artificial surfaces and associated areas (Urban areas >50%) which account for 0.08%.

Open (15-40%) needleleaved deciduous or evergreen forest (>5m) areas which account for 36.29% were dominant in Russia, while Mosaic forest or shrub land (50-70%) / grassland (20-50%) areas which account for 25.28% were dominant in China. Broadleaved and needleleaved forest areas in Russia and China were 80.38% and 50.59%, respectively; Sparse (<15%) vegetation areas in China were 15.82%, which was significantly higher than 3.10% in Russia; Artificial surfaces and associated areas (Urban areas >50%) in China were 0.31% which was also significantly higher than 0.02% in Russia.

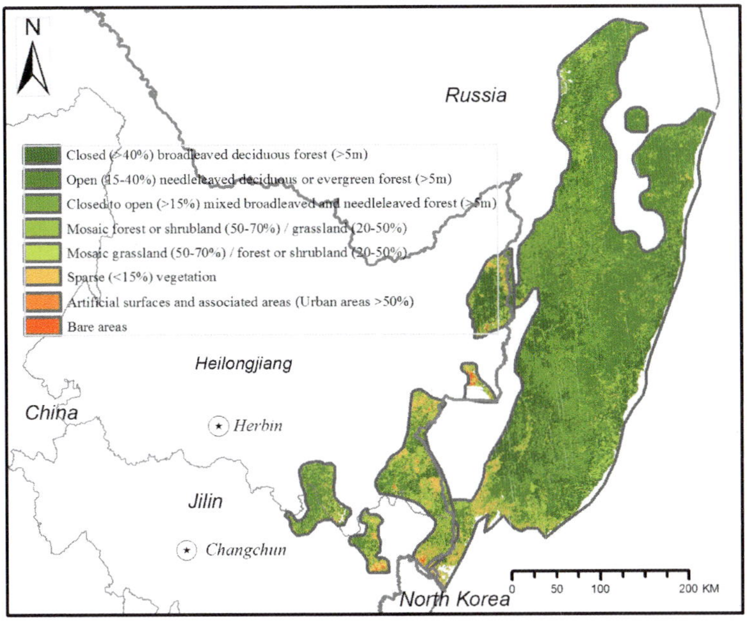

Fig. 2 Land cover map of the east-northern tiger distribution area 2009

Tab. 2 Land covers area percentage of the east-northern tiger distribution area 2009

Label type	world	Russia	China
Closed (>40%) broadleaved deciduous forest (>5m)	22.30%	22.08%	23.29%
Open (15-40%) needleleaved deciduous or evergreen forest (>5m)	34.19%	36.29%	24.79%
Closed to open (>15%) mixed broadleaved and needleleaved forest (>5m)	18.43%	22.01%	2.51%
Mosaic forest or shrubland (50-70%) / grassland (20-50%)	15.31%	13.07%	25.28%
Mosaic grassland (50-70%) / forest or shrubland (20-50%)	3.98%	3.36%	6.70%
Sparse (<15%) vegetation	5.43%	3.10%	15.82%
Artificial surfaces and associated areas (Urban areas >50%)	0.08%	0.02%	0.31%
Bare areas	0.28%	0.06%	1.29%

In conclusion, according to the land cover status of the Amur tiger distribution regions in 2000 and 2009, human activity area was decreasing in the entire Amur tiger distribution regions, and this situation was more remarkable in China. Russia had higher forest coverage and better forest quality.

4. Conclusions

The Amur tiger is a key species in this region, playing a critical role in the whole natural ecosystems. Preservation of habitat for the Amur tiger is not only important for the survival of the Amur tiger, but also important for the stability and growth of the other species, and even the stability of the ecosystem.

The living environment for the Amur tiger in Russia has been well protected since 1930s. Along with the issue of the forest law and the wildlife conservation law, the habitat for the Amur tiger in China has also been well protected. A recent survey showed that the population of the Amur tiger in China has increased from 10-14 in the early 1990s to approximately 25.

Larger territories for the Amur tiger are required due to its increasing population. The Amur tiger distribution regions in China are close to the Russian Far East with incomparable geographical advantage. As long as we continue to carry out reasonable and effective measures to improve the habitat for the Amur tiger, the northeast of our country will definitely become a beautiful home for the Amur tiger.

Acknowledgments

This research was jointly funded by the project of National Natural Science Foundation of China (grant no. 41371486), and the open project of the ecological security and protection key laboratory of Sichuan province (grant no. ESP201304, QD2014A001). We are grateful to the anonymous reviewers who made valuable and constructive comments on the manuscript.

References

1. Zhang Ming-Hai, MA Jian-Zhang. Current Status and Protection Vision of Wild Northeast Tiger in China. Chinese Journal of Zoology, 2010, 45 (1) :165-168.
2. Sanderson E, Forrest J, Loucks C, et al. Setting Priorities for the Conservation and Recovery of Wild Tigers: 2005-2015. The Technical Assessment, WCS, WWF, Smithsonian and NFWF-STF, 2006, New York, Washington DC.
3. Pikunov DG, Bragin AI, Vakhreev GI (1984) The future of tigers of Sikhote-Alin. In: Proceedings of the Applied Science Conference. Man and Nature in the Far East, pp. 124-126. Vladivostok.
4. Carroll C, Miquelle DG. Spatial viability analysis of Amur tiger (Panthera tigris altaica) in the Russian Far East: the role of protected areas and landscape matrix in population persistence. Journal of Applied Ecology, 2006 43, 1056-1068.
5. Wu Jian-guo, KOU Xiao-jun, TI AN Yu, et al. Land use pattern and its dynamic changes in Amur tiger distribution region[J]. Chinese Journal of App lied Ecology, Mar. 2009, 20 (3) : 713- 724.
6. Sun HY. The ecological corridor of the Amur tiger. China Nature, 2006, (5), 18–20.
7. Sun HY, Lu XD, Tian JL, et al. The wild population monitor of Amur tiger in Heilongjiang Province. Forestry Science & Technology, 2005, 30(6), 33–35.
8. Protection Status of Wild Amur Tiger (Panthera tigris altaica) and their Conservation Strategy. CHINESE JOURNAL OF WILDLIFE, 2008, 29(1), 40-43.
9. Bartholomé, E., and A. S. Belward. GLC2000: A new approach to global land cover mapping from Earth observation data. Remote Sens, 2005, 26, 1959-1977.

10. Friedl, M. A., Sulla-Menashe, D., Tan, B., Schneider, A, et al. MODIS Collection 5 global land cover: Algorithm refinements and characterization of new datasets. Remote Sensing of Environment, 2010, 114,168-182.

Experimental study on immune system of schisandra oral liquid in mice

W. Guo[1,2], X. Liu[1], C. M. Wang[1], H. Li[1], H. X. Sun[1], C. Y. Zhang[1], J. G. Chen[1]and J. H. Sun[1,*]
[1]*College of Pharmacy, Beihua University, Jilin 132013, China*
[2]*Jilin Central General Hospital, Jilin 132011, China*
[*]*E-mail: sunjinghui2008@126.com*

Immune function effects of Schisandra Yangzheng oral solution (SYOS) were investigated in this study. 48 Kunming female mice were randomly divided into four groups: control group, low-dose SYOS group (L- SYOS, 10mL/kg • d), medium-dose SYOS group (M- SYOS, 20mL/kg • d) and high-dose SYOS group (H- SYOS, 40mL/kg • d), and there were 12 mice in each group. All the mice in the different groups were given the corresponding agents orally and those in the control group was given the same volume of distilled water once a day continuously for 20 days. After the last administration, phagocytic index of mice in all the groups were measured by carbon clearance experiment was conducted, to examine the phagocytosis of mononuclear macrophages of Schisandra Yangzheng oral solution. 40 Female C57 mice were randomly divided into four groups: control group, low-dose SYOS group (L- SYOS, 10mL/kg • d), medium-dose SYOS group (M- SYOS, 20mL/kg • d) and high-dose SYOS group (H- SYOS, 40mL/kg • d), and there were 12 mice in each group. All the mice in the different groups were given the corresponding agents orally and those in the control group was given the same volume of distilled water once a day continuously for 20 days. After the last administration, lymphocyte transformation rates induced by ConA were measured to examine the appreciation ability of T lymphocytes. The results showed that after the administration, the phagocytosis of mononuclear macrophages and appreciation ability of T lymphocytes were enhanced in all the SYOS-treated groups, suggesting that Schisandra Yangzheng oral solution has a significant immune enhancement effect, and can be developed into a safe health functional food with a immunostimulating function, but its mechanism should be explored by further experiments.

Keywords: Schisandra Yangzheng oral solution; Immunostimulating; Mice.

1. Introduction

Nowadays, with the changing of rapid lifestyle, people's stress gradually increases and physical exercise decreases, the disorder of diet and daily work, all of these lead to the decline of immune function, and the body is often in fatigue, sub-health and non resistance. Regulation of immune function to improve the body's resistance to fatigue has become the focus of current academic and attention. Currently the sale of the majority of immune function of the product is not only expensive, but also limited to regulating short-term physical condition, the clinical effect is not ideal. Schisandra Yangzheng oral solution, based on the concept of traditional Chinese medicine health care, are a compound preparation composed of Schisandra, Raspberry, Astragalus and several other Chinese medicines, to developed to enhance the immune function of adjuvant drugs. In this study, Schisandra Yangzheng oral solution were observed and investigated by ConA induced spleen lymphocyte transformation test and mouse carbon clearance test.

2. Materials

2.1 Animals

Clean-grade female Kunming and C57 mice, weighing 22 ± 2g, were provided by Experimental Animal Research Center, Jilin University (Conformity certificate number: SCXK (Ji) 2015-0003.

2.2 Drugs and Instruments

Schisandra Yangzheng oral solution (SYOS, Jilin Province Schisandra Development and Industrialization Engineering Research Center, prepared into a suspension with distilled water just before the experiment); India ink (Shanghai Ruiyong Science and Technology Co., Ltd. Lot number: 150413, 100mL); Trypan blue (Beijing Dingguo Changsheng Biotechnology Co., Ltd. Lot number: 150303, 100mL); Concanavalin-A (ConA, Sigma, USA, Lot number: 150607); Hanks' solution (GENVIEW, USA, Lot number: 150512); RPMI1640 (Thermo Fisher Biochemical Products (Beijing) Co., Ltd. Lot number: 150303); Calf serum (Zhejiang Tianhang Biological Technology Co., Ltd. Lot number: 150211); 2-mercaptoethanol (Beijing Dingguo Changsheng Biotechnology Co., Ltd. Lot number: 141103); Isopropanol (Tianjin Damao Reagents Factory, Lot number: 141216).

Microplate spectrophotometer (TECAN, Switzerland, Type Infite M200); Thermostatic biochemical incubator (Shanghai Boyuan Industrial Co., Ltd. Type SPX-250B-Z).

3. Experimental Methods

3.1 Mouse carbon clearance tests

48 Female Kunming mice were randomly divided into four groups: control group, low-dose SYOS group (L- SYOS, 10mL /kg • d), medium-dose SYOS group (M- SYOS, 20mL / kg • d) and high-dose SYOS group (H- SYOS, 40mL/kg • d), and there were 12 mice in each group.

All the mice in the different groups were given the corresponding agents orally and those in the control group was given 0.2mL/10g of distilled water once a day continuously for 20 days.

At 60 mine after the last administration, mice in all the groups were injected of diluted India ink 0.1mL/10g via tail vein. Blood samples were collected by punctuating the retro-orbital sinus at 2, 10 min after the India ink injection, and add it into 0.1%Na2CO3 2mL solution immediately. Na2CO3 solution as the blank control, optical density value A2 and A10 were measured at 600nm wavelength using microplate spectrophotometer. After calculation of carbon clearance index K, the mice were sacrificed and the liver and spleen were weighed to calculate the phagocytic index α. The related calculation process and equations showed in Eqs. (1) and (2).

$$K = (logA2 - logA10)/(t10 - t2) \tag{1}$$

$$\alpha = \frac{body\ weight}{(liver\ weight+spleen\ weght)} \times K^{\frac{1}{3}} \tag{2}$$

3.2 ConA induced mouse spleen lymphocyte transformation test

40 Female C57 mice were randomly divided into four groups: control group, low-dose SYOS group (L- SYOS, 10mL /kg • d), medium-dose SYOS group (M- SYOS, 20mL / kg • d) and high-dose SYOS group (H- SYOS, 40mL/kg • d), and there were 12 mice in each group.

Take the spleen and then placed in a 3-5mL of sterile Hanks' liquid plate, and covered a piece of gauze. Grind spleen gently with a large syringe core to make into single cell suspension. After through 200mesh sieve, suspension washed 2 times with Hanks' liquid and centrifugal for 10min at 1000r/min each time to acquire cell suspension, then the cells were suspended into the culture and the number of viable cells was counted by using the phenol blue staining.

The cell suspension was divided into 24holes and two holes were added into the culture plate and each of which was 1mL. A hole with 75uL ConA solution (equivalent to 7.5ug/mL) and another hole as a control, then put the plate into 5% CO2, 37°C CO2 incubator for 72 h. 4 h before the end of the culture, each hole supernatant is aspirated 0.7mL and then added 0.7mL calf serum-free RPMI1640 medium, while adding MTT (5mg/mL) 50uL into each hole. After incubation, each hole was added 1mL acidic isopropanol, ultrasonic shock 2 sec to dissolve purple crystal, then packaging to 96 hole culture plate, each hole packing 3 hole as parallel control sample, optical density value were measured at 570nm wavelength using microplate spectrophotometer.

4. Statistical methods

The data were analyzed with SPSS16.0 statistical software. The measurement data were expressed as mean ± standard deviation (mean ± S). Differences among multiple groups were compared with analysis of variance, heterogeneity of variance was tested with Knlskal-Wallis nonparametric test. P <0.05 indicated a statistically significant difference.

5. Results

5.1 Effect of SYOS on mouse carbon clearance tests

Tab. 1 Effect of SYOS on mouse carbon clearance tests

Groups	n	Phagocytic index	P
Control	12	4.98±0.53	
L- SYOS	12	5.88±0.73*	0.041
M- SYOS	12	6.09±0.84*	0.024
H- SYOS	12	6.47±1.21**	0.008

Notes: Compared with those in the control group,
**: P<0.05.*
***: P<0.01.*

Phagocytic index of mice in L- SYOS, M- SYOS and H- SYOS groups were significantly higher than those in the control group ($P<0.05$, $P<0.05$, $P<0.01$), suggesting that SYOS has an obvious improve the phagocytic ability of mice. The results are shown in Table 1.

5.2 Effect of SYOS on ConA induced mouse spleen lymphocyte transformation test

Lymphoblast optical density of mice in L- SYOS, M- SYOS and H- SYOS groups were significantly higher than those in the control group ($P<0.05$, $P<0.01$), suggesting that SYOS has an obvious improve the lymphocyte transformation rates of mice. The results are shown in Table 2.

Tab. 2 Effect of SYOS on ConA induced mouse spleen lymphocyte transformation test

Groups	n	optical density	P
Control	12	0.91±0.15	
L- SYOS	12	1.27±0.23*	0.048
M- SYOS	12	1.50±0.62*	0.033
H- SYOS	12	2.04±0.83**	0.006

Notes: Compared with those in the control group,
**: P<0.05.*
***: P<0.01.*

6. Discussion

Schisandra Yangzheng oral solution is composed of traditional Chinese medicines, such as Schisandra, Astragalus,Raspberry and other compositions, which are in combination with each other in accordance with the assistant and guide principles of composing prescription from the angle of supplement of "Qi" and "Yang".

The related pharmacological studies suggest that Schisandra can enhance non-specific immune function and humoral immue function[1-3]. Astragalus can significantly increase the total number of white blood cells in the blood, promote the phagocytic function and bactericidal capacity of neutrophils and macrophages. Astragalus can significantly enhance cellular immunity, promote PHA, ConA, PWM induced lymphocyte transformation[4-6]. Raspberry has obvious lymphocyte proliferation and increased intracellular cAMP action[7].

The immune system can quickly remove a variety of pathogenic substances, is an important system to maintain the stability of the environment in vivo. When the T lymphocytes cultured in vitro, it will transformation to the lymphatic cells, such as cell volume increased, high level of metabolism, protein and nucleic acid synthesis increased after stimulated by non-specific mitogen (eg PHA, ConA)[8-10]. The level of lymphocyte transformation tare may reflect the level of immune cells, it can be used as an index measuring the body's immune function[11-12]. The results of this study showed that after the administration, all the SYOS-treated groups can stimulate the transformation of T lymphocytes, improve the appreciation ability of T lymphocytes by compared with those in the control group.

In normal mice, the phagocytic cells of the liver can be swallowed up to remove the 90% carbon particles, and the spleen macrophages are about 10% carbon. Quantitative intravenously to mice India ink (carbon particle suspensions), a certain time interval repeatedly take blood, measuring the concentration of carbon particles in the blood, according to the blood stream clearance of carbon particles is speed, to determine the function of macrophages. The results of this study showed that after the administration, the phagocytic ability of mononuclear macrophages was enhanced in all the SYOS-treated groups, in which compared with those in the control group, suggesting that Schisandra Yangzheng oral solution has a significant immune enhancement effect, and can be developed into a safe health functional food with an immunostimulating function, but its mechanism should be explored by further experiments.

Acknowledgments

This research work was supported by project of Jilin Provincial Science & Technology Department (YYZX201124-3) and Science and Technology Bureau of Jilin City (201464062).

References

1. Y. Li, S. C. Qu, W. J. Sun, Protectine effect of CPSC on immunosuppressive mice. J. N. BETHUNE UNIV. MED. SCI. Vol. 16 (Changchun. China, 1995), pp. 583–585.
2. Y. X. Cao, J. P. Zheng, J. Liu, H. P. Yao, Shenqi Schisandra capsules sedative, anti-stress and immunomodulatory effects. China Journal of Chinese Materia Medica. Vol. 30 (Beijing. China, 2005), pp. 1631–1633.
3. M. S. Miao, X. Y. Fang, Effect of Polysaccharide immune function in normal mice. Chinese Journal of Traditional Medical Science and Technology. Vol. 10 (Chongqing. China, 2003), pp. 100, 87.
4. S. Fu, J. Zhang, F. Menniti-Ippolito, X. Gao. Huangqi Injection (a Traditional Chinese Patent Medicine) of Chronic Heart Failure: A Systematic Review. PLoS ONE. Vol. 6 (SanFrancisco. USA, 2011), pp. 708-711.
5. Q. Y. Liu, Y. M. Yao, Y. Yu, N. Dong, Z. Y. Sheng. Astragalus Polysaccharides Attenuate Postburn Sepsis via Inhibiting Negative Immunoregulation of CD4+ CD25(high) T Cells. PLoS ONE. Vol. 6 (SanFrancisco. USA, 2011), pp. e19811.
6. G. H. Niu, X. Sun, C. M. Zhang, J. S. Zhang. Effect of compound Astragalus recipe on lymphocyte subs immunoglobulin and complernents in patients with myasthenia gravis. Chinese Journal of Integrative Medicine. Vol. 29 (Beijing. China, 2009), pp. 305-308.
7. K. H. Chen, J. Fang, B. Gong, Q. Z. Mo, H. P. Quan, W. P. Sun. The Enhancing Effect of Rubus chingii Hu on the Lymphocyte Proliferation and its Relationship with Cyclic Nucleotides. Vol. 15 (Shanghai. China, 1995), pp. 302-304.
8. R.M. Loria, Immune up-regulation and tumor apoptosis by androstene steroids, Steroids, Vol. 67 (New York. USA, 2002), pp. 953–966.

9. P. O. Livingston, G. Ragupathi, C. Musselli. Autoimmune and antitumor consequences of antibodies against antigens shared by normal and malignant tissues. Journal of clinical immunology, Vol. 20 (Amsterdam. Netherlands, 2000), pp. 85–93.
10. S. Tsuyako, N. Isao, M. Eidi, M. Kaori, O. Saburo, T. Hirokuni. Autoimmune pancreatitis as an initial manifestation of systemic lupus erythematosus. Modern Rheumatology, Vol. 14 (Tokyo. Japan, 2004), pp. 309-313.
11. C. A. Scanga, V. P. Mohan, K. Yu, H. Joseph, K. Tanaka, J. Chan, J. L. Flynn. Depletion of CD4 T cells causes reactivation of murine persistent tuberculosis despite continued expression of interferon gamma and nitric oxide synthase 2. Journal of Experimental Medicine, Vol. 192(New York. USA, 2000), pp. 347-358.
12. U. E. Schaible, F. Wright, P. A. Sieling, K. Fischer, H. L. Collins, K. Hagens, R. L. Modlin, V. Brinkmann, S. H. Kaufmann. Apotosia facilitates antigen presentation to T lymphocytes through MHC-I and CD8 in tuberculosis. Nature Medicine, Vol. 9 (New York. USA, 2003), pp. 1039-1046.